Neurogenesis of Central Respiratory Rhythm

CONFERENCE DATA

This volume contains the Proceedings of the International Symposium of the Centre National de la Recherche Scientifique (CNRS) on Neurogenesis of Central Respiratory Rhythm : Electrophysiological, Pharmacological and Pathological Aspects, held in Ile de Bendor, BANDOL, FRANCE, September 18-22, 1984.

Organizing Committee

Armand L. BIANCHI	Marseille
Monique DENAVIT-SAUBIE	Gif-sur-Yvette

Local Organizing Committee

Jean-Claude BARILLOT	Marseille
Jean CHAMPAGNAT	Gif-sur-Yvette
Michel DUSSARDIER	Marseille
Patrick GAUTHIER	Marseille
Gérard HILAIRE	Marseille
Roger MONTEAU	Marseille
Marie-Pierre MORIN-SURUN	Gif-sur-Yvette

Secretariat and Material Organization

Eliane BOUDINOT	Gif-sur-Yvette
Daniel CATALIN	Marseille
Elisabeth DUSSAUZE	Marseille
Anne-Marie LAJARD	Marseille
Jocelyne ROMAN	Marseille

NEUROGENESIS OF CENTRAL RESPIRATORY RHYTHM

EDITED BY

Armand L. BIANCHI, Dr. Sci.
Professeur de Physiologie
Département de Physiologie et
Neurophysiologie
LA CNRS 205
Faculté des Sciences St. Jérôme
Univ. AIX-MARSEILLE III

13397 MARSEILLE CEDEX 13
France

Monique DENAVIT-SAUBIE, Dr. Sci.
Maître de Recherche au CNRS
Département de Neurophysiologie
Appliquée
Laboratoire de Physiologie
Nerveuse
Centre National de la Recherche
Scientifique
91190 GIF-SUR-YVETTE
France

Proceedings of the CNRS International Symposium
held at
Ile de Bendor, BANDOL, FRANCE
September 18–22, 1984

MTP PRESS LIMITED
a member of the KLUWER ACADEMIC PUBLISHERS GROUP
LANCASTER / BOSTON / THE HAGUE / DORDRECHT

Contents

SECTION 2   Neuronal organization of respiratory centers.

SECTION 5   Neuropharmacology of respiratory centers.

List of Participants

D. BALLANTYNE
I. Physiol. Dept.
University of Heidelberg
Im Neuenheimer Feld 326
D-6900 HEIDELBERG
F.R. GERMANY

J.C. BARILLOT
Dept. de Physiologie &
Neurophysiologie
Fac. Sci. et Tech. St Jérôme
Rue Henri Poincaré
13397 MARSEILLE CEDEX 13
FRANCE

P. BARTHELEMY
Faculté de Médecine
Lab. de Médecine Expérimentale
27, bd. Jean Moulin
13385 MARSEILLE CEDEX 5
FRANCE

A. BERGER
School of Medicine
Dept. of Physiology & Biophysics
University of Washington
SEATTLE, Wa 98195
U.S.A.

A.L. BIANCHI
Dept. de Physiologie &
Neurophysiologie
Fac. Sci. et Tech. St Jérôme
Rue Henri Poincaré
13397 MARSEILLE CEDEX 13
FRANCE

G. BOHMER
Physiologisches Institütet
Universität Mainz
Saarstraße. 21
D-6500 MAINZ
F.R. GERMANY

M.BONORA
Lab. Physiologie
Faculté de Médecine St Antoine
27, rue Chaligny
75571 PARIS CEDEX 12
FRANCE

R.T. BROUILLETTE
The Childrens Memorial Hospital
Division of Neonatology
2300 Children's plaza
CHICAGO, Illinois 60614
U.S.A.

D. CAILLE
CNRS
Lab. de Physiologie nerveuse
91190 GIF SUR YVETTE
FRANCE

W.E. CAMERON
Dept. of Anatomy and cell Biol.
University of Pittsburg
School of Medicine
Pittsburg, Pennsylvania 15261
U.S.A.

J. CHAMPAGNAT
CNRS
Lab. de Physiologie nerveuse
Dept. de Neurophysio. appliquée
91190 GIF SUR YVETTE
FRANCE

M.I. COHEN
Dept. of Physiology, U 303
Albert Einstein College of
Medicine
Bronx, N.Y. 10461
U.S.A.

R.O. DAVIES
972, Maloney Bldg.
Univ. of Pennsylvania
3400 Spruce Street
PHILADELPHIA, Penn. 19104
U.S.A.

M. DUSSARDIER
Dept. de Physiol. & Neurophy.
Fac. des Sci. et Tech. St Jérôme
Rue Henri Poincaré
13397 MARSEILLE CEDEX 13
FRANCE

P. DEJOURS
Lab. de Physiologie Respiratoire
C.N.R.S.
23, rue Becquerel
67087 STRASBOURG
FRANCE

C. Von EULER
Nobel Institute for
Neurophysiology
Karolinska Institutet
Box 60400
S-104 01 STOCKHOLM
SWEDEN

M. DENAVIT-SAUBIE
CNRS
Lab. de Physiologie nerveuse
Dept. de Neurophysiologie
91190 GIF SUR YVETTE
FRANCE

L. FEDORKO
Institute of Physiology
Medical Academy
Krakowskie Przedmiescie 26/28
00-927 WARSAW
POLAND

T.E. DICK
University of Washington
School of Medicine
Dept. of Physiology & Biophysics
SEATTLE, Washington 98195
U.S.A.

J. FELDMAN
Department of Physiology
Northwestern University
303 East Chicago Avenue
CHICAGO, Illinois 60611
U.S.A.

J.DROUET
Direction des Recherches Etudes et
Techniques
26 Bd Victor
75996 Paris Armées
FRANCE

A. FELTZ
Lab. de Physiologie Respiratoire
C.N.R.S.
23, rue Becquerel
67087 STRASBOURG
FRANCE

B. DURON
Lab. de Neurophysiologie
Université de Picardie à Amiens
12 rue Frederic Petit
80036 AMIENS CEDEX
FRANCE

J. FLOREZ
Dept. of Pharmacol. & Therap.
Faculty of Medicine
University of Santander
Nat. Med Center "Valdecilla"
SANTANDER
SPAIN

D.T. FRAZIER
Dept. of Physiology and Biophysics
Un. of Kentucky, Coll. of Med.
Chandler Medical Center
LEXINGTON KY 40536
U.S.A.

Ch. GRIMAUD
Faculté de Médecine
Lab. de Médecine expérimentale
27, bd. Jean Moulin
13385 MARSEILLE CEDEX 5
FRANCE

C. GAULTIER
Lab. Physiologie
Hop. Antoine Bèclère

92141 CLAMART
FRANCE

A. HARFSTRAND
Karolinska Institutet
Department of Histology
P.O. Box 60400
S-104 01 STOCKHOLM
SWEDEN

P. GAUTHIER
Dept. de Physiologie &
Neurophysiologie
Fac. Sci. et Tech. St Jérôme
Rue Henri Poincaré
13397 MARSEILLE CEDEX 13
FRANCE

R.M. HARPER
Depts. of anatomy & Physiology
& the Brain Research
Institute U.C.L.A.
LOS ANGELES, Ca 90024
U.S.A.

P.A. GETTING
Dept. of Physiology and
Biophysics
University of IOWA
IOWA CITY, IA 52242
U.S.A.

D. HENDERSON-SMART
Dept. of Perinatal Medicine
King George V Memorial Hospital
CAMPERDOWN, NSW 2050
AUSTRALIA

M. GILBEY
Department of Physiology
Royal Free Hosp. Sch. of Med.
Rowlorel Hill Street
LONDON NW3 2PF
ENGLAND, U.K.

G. HILAIRE
Dept. de Physiologie &
Neurophysiologie
Fac. Sci. et Tech. St Jérôme
Rue Henri Poincaré
13397 MARSEILLE CEDEX 13
FRANCE

P. GOGAN
Unité de Rech. Neurobiologiques
INSERM U6
280, Bd. Ste. Marguerite
13009 MARSEILLE
FRANCE

A. HUGELIN
Lab. Physiol. Nerveuse
C.N.R.S.
Lab. de Physiologie Nerveuse
91190 GIF SUR YVETTE
FRANCE

M. HURLE
L.P.N. C.N.R.S.
Neuropphysiologie appliquée
91190 GIF SUR YVETTE
FRANCE

S. ISCOE
Dept. of Physiology
Queen's University
KINGSTON, Ontario K7L 3N6
CANADA

Y. JAMMES
Faculté de Médecine
Lab. de médecine expérimentale
27, bd. Jean Moulin
13385 MARSEILLE CEDEX 5
FRANCE

P. JOHNSON
Nuffield Depart. of Obstetrics
John Radcliffe Hospital
HEADINGTON
OXFORD OX3 9DU
ENGLAND

D. JORDAN
Dept. of Physiology
Royal Free Hospital Medical School
Pond Street
LONDON NW3
ENGLAND

W.A. KARCZEWSKI
Lab. of Neurophysiology
Polish Academy of Sciences
Medical Research Center
Dworkowa 3
00-784 WARSAW
POLAND

M. KHATIB
Dept. Physiologie et
Neurophysiologie
Faculté St Jérôme
Rue Henri Poincaré
13397 MARSEILLE CEDEX 13
FRANCE

H.P. KOEPCHEN
Physiologisches Institut der
Freien
Universität Berlin - Arnimallee 22
1 Berlin 33 (West)
F.R. GERMANY

D. KURTZ
Serv. d'Exploration Fonctionnelle
du Sytème Nerveux E.E.G.
Hôpital Civil
1 Place de l'Hopital
67091 STRASBOURG
FRANCE

H. LAGERCRANTZ
Nobel Institute for
Neurophysiology
Karolinska Institute, Box 60400
S-104 01 STOCKHOLM
SWEDEN

P. LALLEY
I.Physilogisches Institut
Universität Heidelberg
Im Neuenheimer Feld 326
D-6900 HEIDELBERG
F.R. GERMANY

E. LAWSON
University of NORTH CAROLINA
Clinical Sciences Building 229
CHAPEL HILL NC 27514
USA

J.G. LEATHERMAN
Department of pediatrics
Clinical Sciences Building
Chapel Hill, North Carolina 27584
U.S.A.

D.MARLOT
Faculte de Medecine
12, rue F. Petit
80036 AMIENS CEDEX
FRANCE

L. LEGER
Dept. de Médecine expérimentale
Université Claude Bernard
2, avecnue Rockfeller
69373 LYON CEDEX 2
FRANCE

R. MARTIN-BODY
Dept. of Physiology
School of Medicine
University of Auckland
AUCKLAND
NEW-ZEALAND

B.G. LINDSEY
Dept. of Physiology
University of South Florida
Medical Center Box 8
12901 North 30th Street
TAMPA, Florida 33612
U.S.A.

S. MIFFLIN
I. Physiol. Dept.
University of Heidelberg
Im Neuenheimer Feld 326
D-6900 HEIDELBERG
F.R. GERMANY

J. LIPSKI
The John Curtin School of
Medical Research
Experimental Neurology unit
P.O. Box 334
CANBERRA CITY, ACT 2601
AUSTRALIA

J. MILIC-EMILI
McGill University
Meakins Christie Labs.
3775 University Street
MONTREAL, PQ, H3A 2B4
CANADA

D. LUNDBERG
Dept. of Anesthesiology
University Hospital
S-221 85 LUND
SWEDEN

R. MILES
Dept. of Physiology & Biophysics
The University of Texas Med.
Branch
GALVESTON, Texas 77550
U.S.A.

D.R. McCRIMMON
Department of Physiology
Northwestern University
303 E. Chicago Avenue
CHICAGO, Illinois 606 11
U.S.A.

D.E. MILLHORN
University of NORTH CAROLINA
Medical Research Wing 206 H
CHAPEL HILL NC 27514
USA

R. MONTEAU
Dept. de Physiol. & Neuroph.
Fac. Sci. et Tech. St Jérôme
Rue Henri Poincaré
13397 MARSEILLE CEDEX 13
FRANCE

M.P. MORIN-SURUN
Lab. de Physiologie Nerveuse CNRS
Dept. de Neurophysiologie
appliquée
91190 GIF SUR YVETTE
FRANCE

I.R. MOSS
Pulmonary Div., Dept. of Pediat.
Room 808, Kennedy Center
Albert Einstein Coll. of Med.
1300 Morris Park Avenue
BRONX, N.Y. 10461
U.S.A.

M. MOULINS
Lab. de Neurobiologie Comparée
Place du Dr. B. Peyneau
33120 ARCACHON
FRANCE

R.A. MUELLER
Dept. of Anesthesiology
University of North Carolina
Chapel Hill, N.C. 27514
U.S.A.

R. NAQUET
Lab. Physiologie Nerveuse
CNRS
91190 GIF SUR YVETTE
FRANCE

J.J. QUATTROCHI
Dept. of neuroscience
Harvard Medical School
Children's Hospital Medical Center
300 Longwood Avenue
BOSTON, Mass. 02115
U.S.A.

M.F. RADVANYI-BOUVET
INSERM U.29
Cent. de Rech. de Biol. du Dev.
Hôpital Port-Royal
123, bd. de Port-Royal
75014 PARIS
FRANCE

J.E. REMMERS
Dept. of Medicine
FOOTHILL Hospital
1403, 29th Street N-W
CALGARY, ALBERTA T2N2T9
CANADA

D. RICHE
Lab. de Physiologie nerveuse
C.N.R.S. - L.P.N. 1
91190 GIF SUR YVETTE
FRANCE

C.A. RICHARDSON
Room 1386 HSE
Department of Anesthesia
University of California
613 Parnassus Avenue
SAN FRANCISCO, CA 94143
U.S.A.

D.W. RICHTER
I. Physiologisches Institut
Universität Heidelberg
Im Neuenheimer Feld 326
D-6900 HEIDELBERG
F.R. GERMANY

M. RUNOLD
Nobel Institute for Neurophysilogy
Karolinska Institute,Box 60400
S-104 01 STOCKHOLM
SWEDEN

W.M. ST JOHN
Dept. of Physiology
Dartmouth Medical School
HANOVER, NH 03756
U.S.A.

N. SALES
Dept. Chimie Organique
U 266 INSERM ERA 613 CNRS
4 Av. de l'Observatoire
75270 PARIS CEDEX 06
FRANCE

P. SCHIMDT
Physiologisches Institutet
University Mainz
Saarstraße 21
D-6500 MAINZ
F.R. GERMANY

T.A. SEARS
Sobell Dept. of Neurophysiology
Institut of Neurology
Queen Square
LONDON WC1N 3BG
ENGLAND

D.C.SHANNON
Pediatric Pulmonary Unit
Massachussetts General Hospital
BOSTON, Mass. 02114
U.S.A.

R. SHANNON
Univ. of South Florida, Coll.
Med., Dept. of Physiol., Box 8
12901 North 30th Street
TAMPA, Florida 33612
U.S.A.

G.C. SIECK
Dep. of Respiratory Diseases
City of Hope National Medical
Cent.
1500 East Duarte Road
DUARTE, California 91010
U.S.A.

J. SIMMERS
Lab. de Neurobiologie Comparée
Place du Dr. Peyneau
33120 ARCACHON
FRANCE

D.F. SPECK
Dept. of Physiology and
Biophysics
Univ. of Kentucky, coll. of Med.
Chandler Medical Center
LEXINGTON, KY 40536
U.S.A.

K.M. SPYER
Department of Physiology
Royal Free Hospital Medical School
Pond Street
LONDON NW3
ENGLAND

T. TRIPPENBACH
Department of Physiology
McGil University
3655 Drummond Street
MONTREAL, PQ H3G 1Y6
CANADA

J.F. VIBERT
CNRS - LA 204
Avenue de la Terrasse
91190 GIF SUR YVETTE
FRANCE

M. WOOD
Dept. of Physiology
Royal Free Hospital
School of Medicine
Rowland Hill Street
LONDON NW3 2PF
ENGLAND

Acknowledgements

This was the 8th of a series of symposia on Neural Control of Breathing held every two or three years in Europe and North America since the First Ciba Symposium (London in 1969) on this subject. The other previous international symposia on Neural Control of Breathing and related topics were held in Warsaw, Poland (1971) ; Oxford, Great-Britain (1973) ; Amiens, France (1976) ; Stockholm, Sweden (1978) ; Heidelberg, FRG (1980) and Chicago, USA (1982). None of the proceedings has been published in book form since the meeting in Stockholm.

Many important responsibilities were assumed by our associates and colleagues both at the Faculty of Sciences St Jérôme of the University of Aix-Marseille III and at the laboratory of Nervous Physiology of the Centre National de la Recherche Scientifique at Gif-sur-Yvette. Mrs. E. DUSSAUZE, J. ROMAN, E. BOUDINOT and A.M. LAJARD had a great deal to do for months, were secretaries and hostesses, and participated as well in the general organization. Drs. J.C. BARILLOT, J. CHAMPAGNAT, M. DUSSARDIER, P. GAUTHIER, G. HILAIRE, R. MONTEAU and M.P. MORIN gave valuable assistance in both organization of the physical facilities and program planning, and greatly contributed to the success of the meeting. We must also mention the efficient contribution of D. CATALIN who solved many equipment problems. We are also grateful to all the authors for their cooperation in submitting their camera-ready papers in time.

This meeting would not have been possible without the generous financial support of the "Centre National de la Recherche Scientifique" (CNRS). We also received financial support from the "Direction des Recherches et Etudes Techniques" (DRET), the "Ministère des Relations Extérieures (French Foreign Office) and the "Institut National de la Santé et de la Recherche Médicale" (INSERM) which sponsored the simultaneous translation.

Substantial funding was also provided by SERVIER Laboratory and other funds were received from RHONE-POULENC Santé (Paris), ROUSSEL UCLAF (Paris), PHYMEP S.A. (Paris) and SCHERING AG (Berlin). We are also indebted to the City of Bandol and to RICARD SA for their help.

Preface

This volume reviews the most recent knowledge on the neurogenesis of central respiratory rhythm. The purpose of this meeting was to bring together people interested in this topic, and with different research backgrounds such as : electrophysiology, neuroanatomy, neuropharmacology and clinical medicine. It is to our knowledge the first time that such a multidisciplinary approach has been undertaken on this subject. Thus this book contains a great variety of papers with a common interest, that is to provide a better understanding of the neuronal substrate of respiration. Papers from the first session deal with the generation of rhythmic activity and with the methodologic problems raised by this type of study. The following papers were more specifically devoted to spinal and bulbar respiratory neurones, and their electrophysiology and circuitry. In some cases these neurons were studied in vitro in brainstem slices. Several papers deal with the afferent control of the central respiratory networks. A complete session was devoted to the neuro-chemical and neuropharmacological mechanisms involved in the neurogenesis and control of the central respiratory rhythm, while the pathological states and their potential treatments were the topics of the last session.

The meeting of so many experts leads to a synthesis of new concepts dealing with mainly the organization of respiratory neurons and their membrane properties, and with the involvement of chemical substances in respiratory rhythmogenesis. Therefore this book should serve as an up to date source of information for researchers concerned with neurophysiology, neuropharmacology and respiratory physiology, as well as for clinicians.

A.L. BIANCHI and M. DENAVIT-SAUBIE

1.P. Gauthier - 2. S. Iscoe - 3. G. Hilaire - 4. P. Halley - 5. M.F. Radvanyi-Bouvet - 6. S. Mifflin -
7. D. Ballantyne - 8. R.M. Martin-Body - 9. M. Wood - 10. W.A. Karczewski - 11. M. Hurlé - 12. J.C. Barrilot -
13.Gilbey - 14. K.M. Spyer - 15. M.P. Morin - 16. J. Champagnat - 17. J. Feldmann - 18. T.A. Sears -
19. J.F. Vibert - 20. H.P. Koepchen - 21. D. Jordan - 22. D. Caille - 23. J. Drouet - 24. J.E. Remmers -
25. J. Florez - 26. M.I. Cohen - 27. H. Lagercrantz - 28. C. Von Euler - 29. S. Delpiene - 30. A.L. Bianchi -
31. M. Bonora - 32. A. Hargstrand - 33. M. Runold - 34. M. Denavit-Saubié - 35. W.M. St-John - 36. M. Khatib -
37. A. Berger - 38. D.W. Richter - 39. M. Dussardier - 40. Not identified - 41. T. Trippenbach - 42. D. Riche -
43. R.O. Davies - 44. J.G. Leatherman - 45. R. Monteau - 46. Barthelemy - 47. E. Lawson - 48. R.T. Brouillette -
49. P. Johnson - 50. R.M. Harper - 51. Y. Jammes - 52. T.E. Dick - 53. D. Henderson-Smart - 54. P.A. Getting -
55. C.A. Richardson - 56 - G. Bohmer - 57. D.F. Speck - 58 - P. Sc hmidt - 59 - B.G. Lindsey - 60 - D.R. McCrimmon.

Section 1
Neurogenesis of Rhythmicity: Comparative Aspects and Methodology

1

An Overview of Backgrounds and Themes for Research in Neurogenesis of Respiratory Rhythm

M. DUSSARDIER

The aim of all research on respiratory neurogenesis is, in fact, to provide a model of the respiratory centre. The model can be merely mathematical or physical, in an attempt to explain the global properties of the system without reference to its components, for example by using black boxes (rhythm generator, ramp generator, switching systems, etc...). However, more appropriate would be a physiological model taking into account the involved biological mechanisms. Unfortunately, in spite of a rapid increase in our knowledge of respiratory neurones, we are still unable to propose a satisfactory model. What facts or concepts are lacking ? Here are some thoughts or suggestions.

1 - THE EQUIVALENT NEURONE

The neurones contributing to respiratory neurogenesis are so numerous that we cannot put all of them in a tentative scheme of the respiratory centre. This difficulty is overcome by postulating that many neurons have the same properties and functions, i.e., each group of similar neurones constitutes an homogeneous population which can be replaced by an equivalent neurone. If the number of populations is scarce, it becomes possible to think in terms of a paucineuronal network. This has resulted in efforts to classify respiratory neurones into different types. But there are many pitfalls.

a) A valid taxonomy must only consider truly significant features. A perfect description would be based on a knowledge of all inputs and outputs for each neurone and of the functions of both. Unfortunately such a study is unrealistic. Therefore, taxonomy is based on other criteria, having questionable value. In addition to the localization of neurones and at times determination of axon destination, the more commonly used criteria are the patterns of discharge, and the neuronal responses to different influences. But, even in a population which seems to be perfectly defined, such as the inspiratory bulbo-spinal neurones or laryngeal motoneurones, unitary records show great variability in the responses and patterns of discharge. If it were possible to examine each single neurone with all the known tests, and to consider all the nuances of the responses, the number of neurone types, and therefore neurone populations, would be so high as to become unmanageable and probably meaningless. The number has to be restricted, but the choice of appropriate populations is not so simple.

Obviously, we have to carefully distinguish true qualitative differences from false differences due to statistical distribution. The distinction is critical. Due to non linearity of fondamental neurone properties, such as the excitation threshold, excitation or inhibition in a single population can lead to discontinuous behaviour of groups of neurones giving the impression of different populations. For example, "paradoxical" inspiratory laryngeal motoneurones (ILN) excited by lung inflation contrary to "classical" ones, seem simply to constitute the most synaptically excitable fringe of ILN (1).

b) The concept of an equivalent neurone implies that the great number of neurones in vertebrates is a superfluous luxury (except for the provision of reliability) since a few neurones could play the same role. In fact, the concept of a neurone population being homogeneous and stable in time is perhaps a false one. We can imagine that a population is composed of neurones which possess the same main property in one situation. However, these neurones could differ on the basis of other properties which, although accessory in the first situation, become fundamental in another situation. If such were the case, a set of neurones could belong to the same population at one time, and to a different population at another time, this shift being due to nervous or humoral influences which rewire the system.

c) To require that the equivalent neuron possesses all the properties of the whole population (recruitment, relative synchronization between neurones inside the population, etc..) it would probably be necessary to confer on it some properties that differ from those of the individual neurones of the population. That is an additional difficulty.

2 - LOGIC OF THE SYSTEM

a - <u>Reciprocal inhibition</u>

The old idea that phase switching is due to reciprocal inhibition between inspiratory and expiratory neurones has been discarded because the overlap between their respective discharges seems to be insufficient. Nevertheless, inversion of IPSPs by hyperpolarization of the neurones confirms the existence of inhibitory phenomena amongst respiratory neurones.

Possibly, reciprocal inhibition does not play a classical role in phase switching, which has another origin. If each group of neurones functioned as an independent pattern generator, then this inhibition could represent a mechanism which improves the efficacity of the system by preventing, or at least restraining, the autonomous discharge of one kind of neurones when the neurones of the opposite function are firing. In this case, the inhibition could work as an exclusion (or contrast) mechanism. It is perhaps why iontophoretic application of strychnine or bicuculline onto inspiratory neurones (2) does not hinder the arrest of their inspiratory discharge, but nevertheless permits the recovery of some activity during expiration. It would be important to emphasize this concept if it appeared to be valid so as to avoid misdirection of research on the mechanisms of phase switching.

b - <u>Pattern generator</u>

A pattern generator (PG) could consist of a self limiting system represented by Fig. 1. It would fire under tonic influences ; have an excitatory feed back (+) which would increase the discharge, producing a ramp generator ; and also have a modulating feed back mechanism ($\sim$) to stop the discharge.

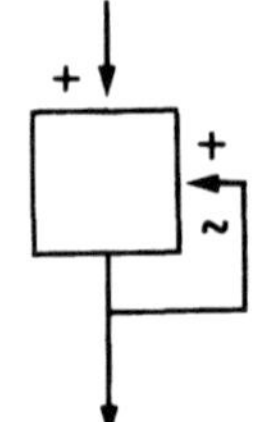

FIG. 1. <u>Model of a pattern generator</u>
    (+) : excitation
    ($\sim$) : modulation of the capacity
          to discharge rythmically

If we accept this hypothesis, a model of the respiratory system (Fig. 2) could consist of two PGs, inspiratory and expiratory respectively, coupled by a contrast generator as suggested above. In this model, each PG has its own mechanism for the arrest of its discharge, but moreover when it is active it prevents the discharge of the other PG by means of reciprocal inhibition.

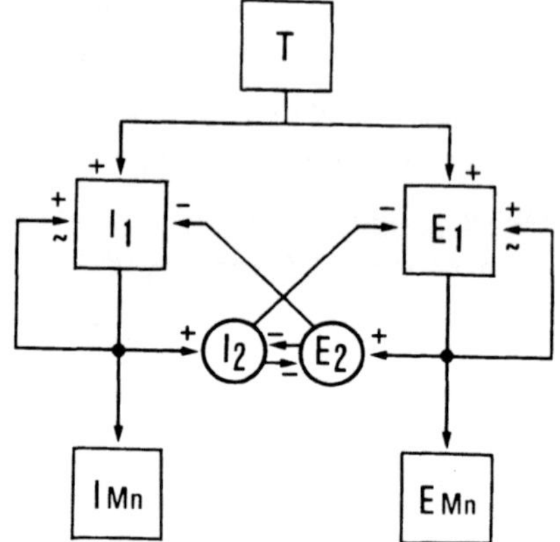

FIG. 2. <u>Tentative model of respiratory generator</u>
I_1, E_1 : inspiratory and expiratory pattern generators
I_2, E_2 : inspiratory and expiratory contrast generators
IMn, EMn : motoneurones
T : tonic generator
(-) : inhibition
(+), ($\sim$) : see FIG. 1

Figs. 1 and 2 do not prejudge the elementary mechanisms of excitation and modulation. For instance, these could be :

- intracellular mechanisms such as those giving pace maker or adaptation properties ; although they are not sufficient by themselves to explain synchronisation between neurones.

- dendrodendritic interactions by electrical or chemical synapses, or by non synaptic liberation of a chemical substance ;

such mechanisms afford a better understanding of synchronisation inside a group of neurones, but do not explain synchronisation between remote neurone groups.

        - interactions by axons or axon collaterals seem necessary to perform this last function.

        In any case, the involved mechanisms are not necessarily of a classical type, that is by EPSPs, IPSPs or PADs. In particular, modulation is not necessarily an inhibition through an inhibitory interneurone which has to be demonstrated. For instance, we can hypothesize that the system uses a neuromodulator, i.e., a substance giving a slowly developing and long lasting effect by a mechanism different from that producing postsynaptic potentials. It could modify the capacity of neurones to rythmically discharge by modification of ionic conductances responsible for depolarizing or hyperpolarizing post-potentials. It is conceivable that a single substance gives both excitation and modulation by fixation at two different membrane receptors, or that one neurone liberates two different substances, i.e., a classical excitatory neuromediator and an "off-switch" neuromodulator.

        The hypothesis that the arrest of discharges is provoked by liberation of a neuromodulator could provided a better explanation than the classical theories for the following fact : at every moment of expiration, mesencephalic stimulation triggers the production of an inspiratory patterned discharge, but this discharge is shorter for a stimulation at the beginning of expiration, than for a stimulation which occurs later (3, 4). The reason for this could be the persistence and gradual elimination during expiration of the neuromodulator responsible for the arrest of inspiration.

3 - SEVERAL MECHANISMS FOR THE CESSATION OF FIRING

        If the discharge of each neurone depends on an equilibrium between excitation and neuromodulation, the direct cause of the cessation of firing is not necessarily the same for all the neurones. If at a given time some neurones or groups of neurones have a dominant role in respiration neurogenesis, cessation of firing of these neurones by an increase in neuromodulation could probably produce a cessation of firing of the other neurones by defacilitation. Reciprocal inhibition from neurones of the opposite function could accessorily stop the lingering discharges.

4 - SEVERAL FUNCTIONS FOR ONE NEURONE

It has been proposed that Iβ neurons of the dorsal respiratory nucleus (DRN) belong to the vagal inspiratory "off-switch" system because they are excited by vagal sensitive impulses. But doubt was cast on this concept when it became clear that these neurones are bulbo-spinal neurones. In fact, Fig. 3 shows that there is a priori no reason for these neurones not to have several functions, performing as premotor (bulbo-spinal) neurones, part of the ramp generator, and part of the "off-switch" system. Possibly, the increase in excitation of phrenic motoneurones produced by Iβ neurones when they are excited by vagal impulses is only an accessory function. The dominant function of these neurones would be to increase the neuromodulation leading to the arrest of inspiratory discharges inside the DRN. The same "off-switch" effect exists perhaps also for Iα neurones, but under the influence of other hypothetical afferents (Fig. 3).

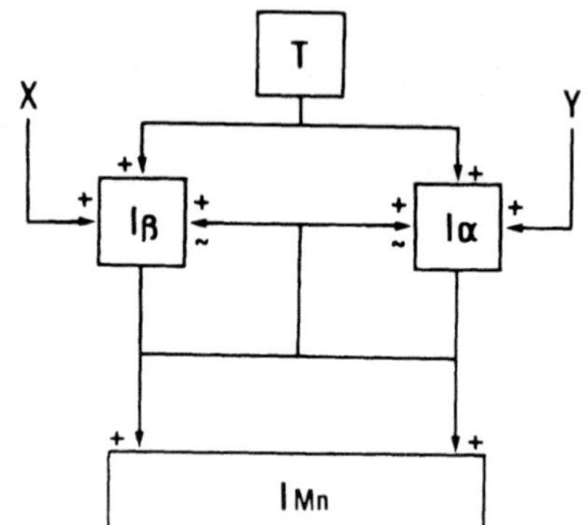

FIG. 3. **Several functions for one neurone**
Iα and Iβ : inspiratory bulbo-
        spinal neurones from dor-
        sal respiratory nucleus
X : vagal afferents
Y : other hypothetical afferents
Other symbols as in Figs 1 and 2

5 - ROLE OF HIGHER CENTRES

Some results obtained on invertebrate neurones (5, 6) are particularly suggestive : two ganglions ggl 1 and ggl 2 are interconnected, and in such conditions, in vitro, ggl 2 neurones can retain a bursting activity for a long time. If conduction in the connecting nerve is suppressed, their bursting activity can disappear at once, or after a delay varying from zero to several hours. It reappears if one of the neurotropic substances found in ggl 1 is added to the bath. Furthermore, the type of activity obtained, including the phase relations between different neurones, depends on the nature of the added substance. Therefore, ggl 2 appears to be rewired by different substances coming from ggl 1.

We can hypothesize that, in the same manner, the higher centres influence the respiratory medullary oscillator by production of neuromodulators which serve to rewire the oscillator. In this case, the respiration related activity of neurones in these higher centres could be a secondary phenomenon due to the fact that these centers are themselves reciprocally controlled by rhythmic impulses coming from the medullary oscillator. As a result the neuromodulation is short term.

6 - SIMULATION OF THE MODEL

Searchers have built complex circuits composed of interconnecting "neuromimes", i.e., analog circuits or computer programs which mimic neuronal properties. In this situation, all is known about the number of neurones composing the network, their connections and their properties in terms of cellular physiology. In spite of these perfect conditions, when the number of interconnected neurones is more than 5 or 6, it is very often impossible to predict what the properties of the system will be, owing to the complex logic of non linear systems. Could it not be imagined that, in the same manner, we have all the main necessary information about components and connections in the respiratory centers, and that we essentially lack an understanding of their logic ? It would probably be usefull to test hypotheses about this logic by using neuromimes in the role of the "equivalent neurones". But care must be taken to stay as close as possible to the physiological realities.

To be valid, a model must not necessarily possess all the

properties of the respiratory center because, probably, in different situations, such a complex system can have different configurations It is possible to imagine that each configuration has to be represented by a different model. But, of course, each model has to be compatible with the others, and we have to explain how the system can shift from one configuration to another one.

IN CONCLUSION

I admit that the above ideas are purely speculative and somewhat provocative. But every body knows that research work consists chiefly of a search for one thing in order to find another one. The most important is perhaps to have a starting hypothesis to confirm or to destroy. Skill, intelligence and luck will do the rest... especially if cats are cooperative ! Here is my hope.

REFERENCES

1 - BARILLOT, J.C. and DUSSARDIER, M. (1973). Modalités de décharge des motoneurones laryngés inspiratoires dans diverses conditions expérimentales. J. Physiol., Paris, 66, 593-629. See Fig. 9.

2 - CHAMPAGNAT, J., DENAVIT-SAUBIE, M., MOYANOVA, S. and RONDOUIN, G. (1982). Involvement of amino-acids in periodic inhibitions of bulbar respiratory neurones. Brain Res., 237, 351-365.

3 - GAUTHIER, P., MONTEAU, R. and DUSSARDIER, M. (1983). Inspiratory on-switch evoked by stimulation of mesencephalic structures : a patterned response. Exp. Brain Res., 51, 261-270. See Fig. 4 A.

4 - GAUTHIER, P. and MONTEAU, R. (1984). Inspiratory on-switch evoked by mesencephalic stimulation : activity of medullary respiratory neurones. Exp. Brain Res., 56, 475-487. See Fig. 1.

5 - MOULINS, M. and COURNIL, I. (1982). All-or-none control of the bursting properties of the pacemaker neurons of the lobster pyloric pattern generator. J. Neurobiol., 13, 447-458.

6 - MOULINS, M. and NAGY, F. (1984). Extrinsic inputs and flexibility in the motor output of the lobster pyloric neural network. In : SELVERSTON, A.I. (ed.) Model neural network and behavior. In press (Plenum press).

2

Rhythm Generation in an Invertebrate Nervous System

M. MOULINS

As known for long time for the respiratory system, it is now clear for a large number of preparations that neurogenic rhythmic motor activity is fundamentally organised by a central network of neurones, the central pattern generator (CPG) (see Delcomyn, 1980). In some cases where a CPG comprises only a few neurones, it has been possible (1) to identify all the neurones of the network, (2) to determine the synaptic relationships within the network and (3) to study the main intrinsic properties of each neurone. Such results sometimes allow us to understand how a CPG works (Selverston and Moulins, 1984) thereby providing a model to be tested with more complex CPG's.

In the present paper, results obtained for the small CPG which controls the digestive movements of the pyloric chamber of Crustacea are summarized. It is shown that the patterned rhythmic output of this network is organised fundamentally by the intrinsic regenerative properties of the component neurones and that the flexibility observed in the output of the CPG is dependent on extrinsic modulatory control of the expression of these regenerative properties.

1/ Functional organization of the crustacean pyloric CPG.

In lobsters, electromyographic recordings from the intact animal show that the rhythmic motor activity of the pyloric chamber consists of a repetitive triphasic sequence in which the dilator muscles (D), the anterior constrictor muscles (C1) and the posterior constrictor muscles (C2) are successively active (Fig. 1A, B) (Rezer and Moulins, 1983). After isolation in a Petri dish, the stomatogastric nervous system, which innervates the foregut, is still able to produce a similar triphasic rhythm (Fig. 1C, D) involving the motor neurones which innervate the pyloric muscles (PD for D, LP for C1 and PY for C2).

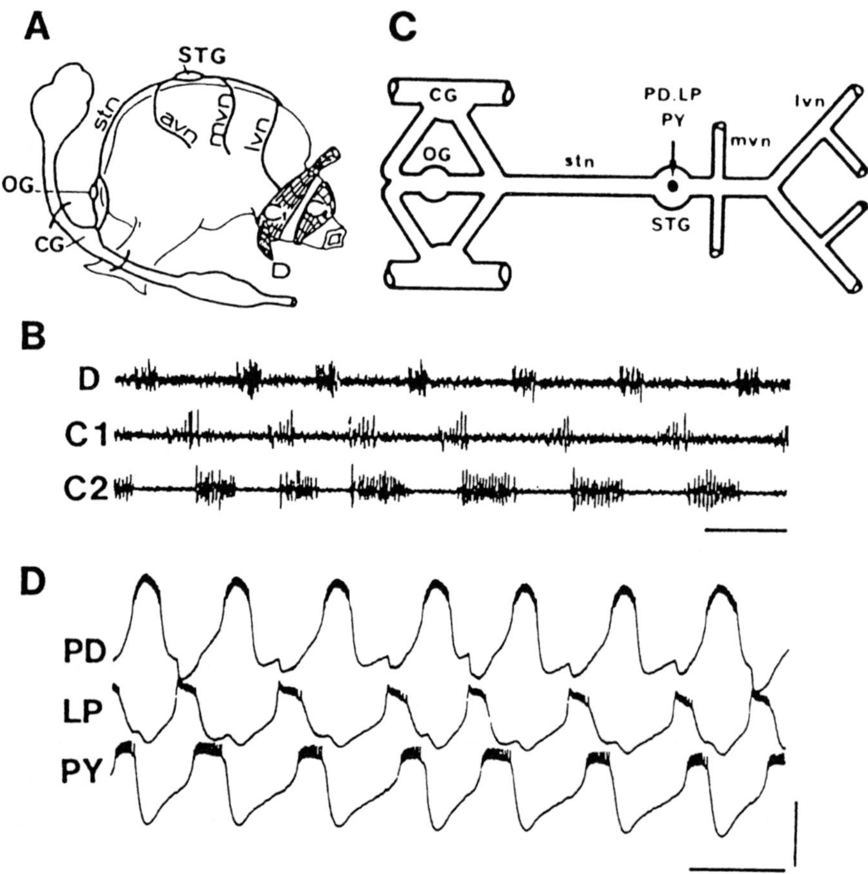

Fig. 1 The lobster pyloric neuromuscular system and the pyloric motor rhythm. **A :** In situ disposition of the pyloric chamber at the rear of the stomach. The pyloric dilator (D), anterior constrictor (C1) and posterior constrictor (C2) muscles are innervated by the stomatogastric nervous system. **B :** Electromyographic recording of the triphasic pyloric motor pattern in the intact animal. **C :** Isolated stomatogastric nervous system as used to record in vitro. The pyloric CPG is located in the stomatogastric ganglion (STG). **D :** Simultaneous intracellular recording from PD, LP and PY motorneurones (innervating D, C1 and C2 muscles) obtained from an isolated preparation of the stomatogastric nervous system. Note the similarity with **B**. CG, commissural ganglion ; OG, œsophageal ganglion ; lvn, lateral ventricular nerve ; mvn, medial ventricular nerve ; stn, stomatogastric nerve. Horizontal bar : 1s ; vertical bar : 20mV.

With such an in vitro preparation it has been possible to show that these motor neurones are organised in the stomatogastric ganglion in a cellular network which constitutes the pyloric CPG (Maynard and Selverston, 1975).

The patterned rhythmic output of the CPG is due primarily to the intrinsic properties of the neurones themselves. The dilator neurones (2 PD cells electrically coupled to an AB interneurone) have an endogenous bursting capability ; ie. from a given level of depolarisation their membrane potential oscillates spontaneously and a burst of spikes occurs on the peak of each oscillation (Fig. 2A). The constrictor neurones (the single LP and the 8 electrically coupled PY) also have regenerative properties which promote bursting. They do not manifest spontaneous oscillations but if their membrane potential reaches a given threshold, they develop a sudden regenerative depolarisation which gives rise to a plateau that is then terminated by a spontaneous regenerative hyperpolarisation (Russell and Hartline, 1978). Firing occurs on the plateau (Fig. 2B). Thus the discharge pattern of such plateauing cells is determined by active jumps in potential between a hyperpolarised non-spiking level and a depolarised spiking level.

By considering the synaptic relationships which exist between the pyloric neurones (Fig. 2C), it is possible to understand how this network produces a triphasic rhythm (Fig. 2D). When a spontaneous burst of spikes occurs in the dilator (PD) neurones, all the constrictor neurones are strongly inhibited. This inhibition terminates the plateau of PY which repolarises suddenly and slows down the depolarisation of LP which was recovering from PY inhibition. LP eventually reaches its threshold for plateauing shortly after the end of the PD burst and produces a spike generating plateau. PY which was recovering from initial inhibition by PD is now inhibited by LP. It also finally reaches threshold for plateauing and fires its burst. This PY burst in turn cuts off the LP plateau by inhibition and is then itself terminated by a new spontaneous burst of PD (and a new cycle commences).

In summary for the pyloric CPG the actual **rhythm** itself is determined by the endogenous burster properties of the dilator neurones while the **pattern** results mainly from the plateauing properties of the constrictor neurones. The synaptic relationships are involved mainly in triggering these active jumps in potential of the constrictor neurones. It will now be shown that it is via an extrinsic control of the expression of these cellular properties that enables flexibility in the expression of the CPG.

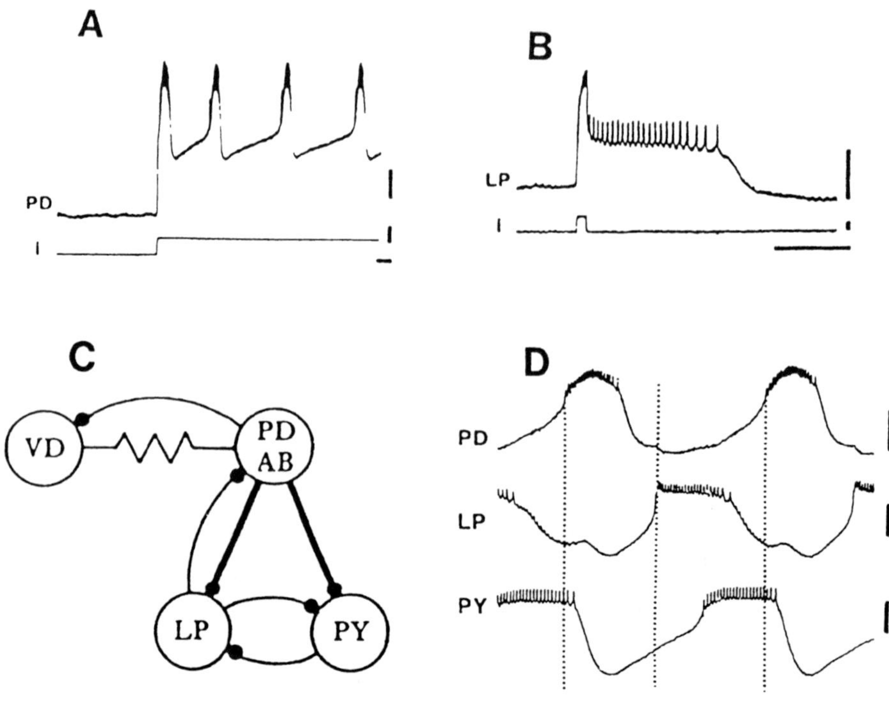

Fig. 2 Properties of the neurones of the pyloric CPG and organisation of the pyloric output. **A :** The dilator (PD, AB) neurones are endogenous bursters ; a burst of spikes occurs on the peak of each spontaneous oscillation in the membrane potential. **B :** The constrictor (LP, PY) neurones are 'plateau' cells. A brief pulse of current injected into the soma is sufficient to provoke a spiking plateau of long duration. **C :** Wiring diagram of the synaptic relationships between neurones in the pyloric CPG. The black dots represent chemical inhibitory synapses and the resistor an electrotonic coupling. **D :** Membrane potential trajectories and firing relationships of the PD, LP and PY neurones during spontaneous pyloric activity (see text). Horizontal bars ; 1s ; Vertical bars : 10mV and 1.5   A.

2/ Flexibility in the expression of the Crustacean pyloric CPG.

Long-term electromyographic recordings from the intact animal show that the pyloric output consists of several different patterns. This does not refer to cycle-by-cycle flexibility which can result from sensory inputs, but long term modifications resulting in large and sustained changes in the repetitive motor sequence. By considering only the wiring diagram of the network (Fig. 2C) it is impossible to understand how such a "fixed" circuit can produce such a large repertoire of outputs. However, it has been shown recently that extrinsic inputs coming from more rostral ganglia (see Fig. 1C) are able to modify the kinetics

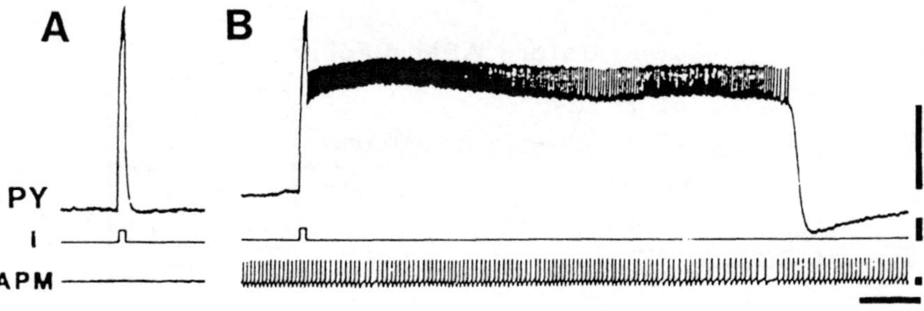

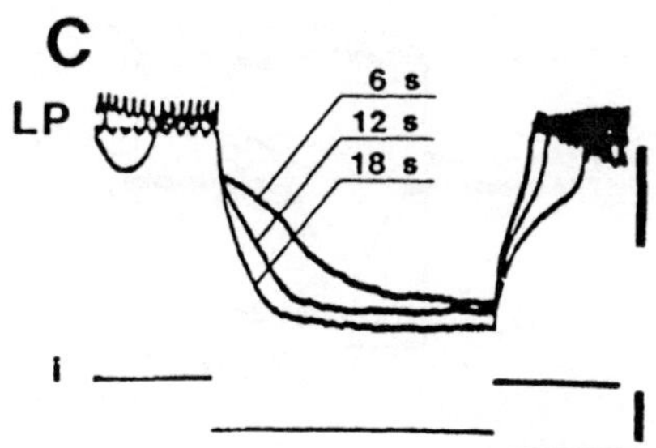

Fig. 3 APM can induce plateau properties (**A, B**) or cause long-term modification in the kinetics of the active jumps in potential of the constrictor neurones. **A, B** : A short pulse of current injected into a PY neurone results only in a passive response when APM is silent (**A**) but causes a long lasting plateau when APM is experimentally depolarized and firing (**B**). **C** : A long duration hyperpolarising pulse is injected into the cell body of LP to provoke active repolarizing changes in potential. Superimposed membrane potential trajectories obtained 6s, 12s and 18s after the end of a 6s discharge (at 5Hz) of APM. Horizontal bars : 2s in A, B ; 0,5s in C ; Vertical bars : 20mV and 2.5 A.

of the active jumps in potential of the neurones of the pyloric network (Dickinson and Nagy, 1983) and thereby control the temporal characteristics of the output sequence (Nagy and Dickinson, 1983).

Firstly that extrinsic inputs play a major role in determining the expression of the intrinsic properties of the pyloric neurones is demonstrated by experimental deafferentation of the stomatogastric ganglion. A sucrose block of impulse conduction in the single input nerve to the ganglion (the stomatogastric nerve) always results, with a variable delay, in a complete loss of both the oscillatory properties of the dilator neurones (Moulins and Cournil, 1982) and of the plateauing properties of the constrictor neurones (Russell and Hartline, 1978). Restoration of the expression of these properties (and of pyloric cycling) can be obtained by unblocking the stomatogastric nerve or by tonic stimulation of the nerve at the rear of the block.

Secondly, one of these inputs has been identified recently (Nagy et al., 1981) and has allowed demonstration that these inputs are not only **permissive**

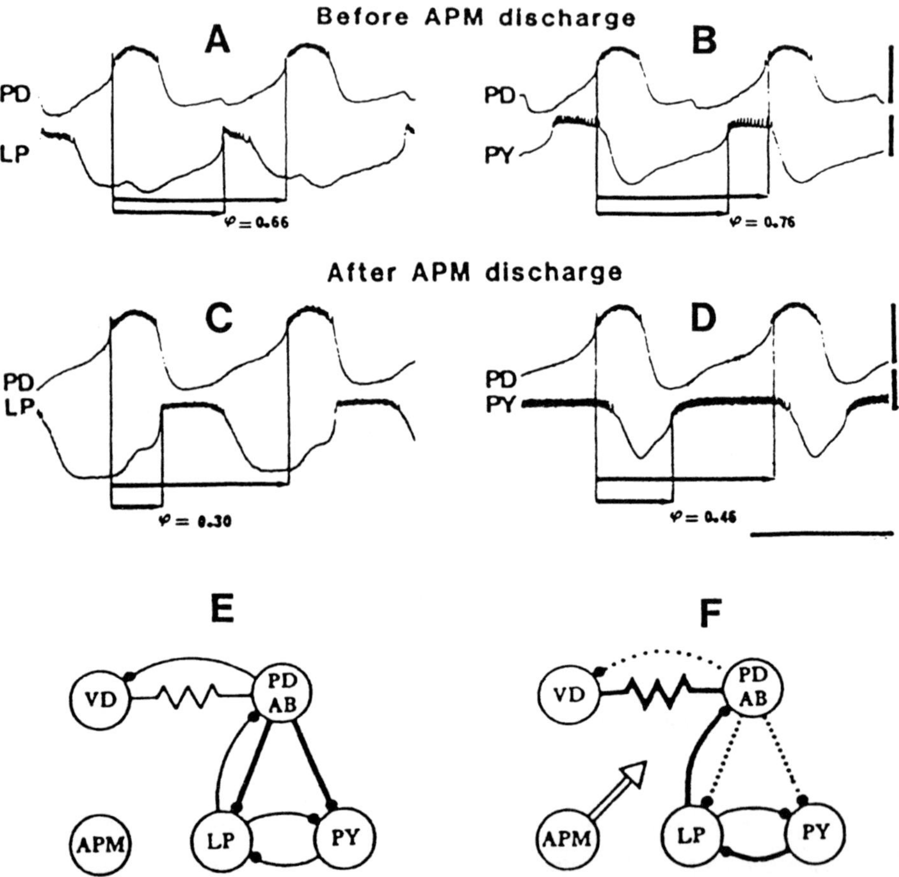

Fig. 4 APM causes long term modification in the phase relationships between the discharges of pyloric neurones. **A – D** : After a 6s discharge of APM at 5Hz the phase of LP bursting in the pyloric cycle shifts from 0.66 to 0.30 (compare **C** to **A**) while the phase of PY bursting shifts from 0.76 to 0.46 (compare **D** to **B**). **A** and **B** are recorded 10s before the onset of APM discharge, **C** and **D** are recorded 6s after the end of APM discharge. **E, F** : After APM discharge the pyloric network is functionally rewired (compare **E** to **F**); heavy lines indicate the synapses which appear to be the most effective, dotted lines the less effective. Horizontal bar : 1s ; vertical bars : 20mV.

(unmasking of cellular properties) but also **organising** (determining of the temporal characteristics of the sequence) in function. The cell body of the identified input neurone (APM, anterior pyloric modulatory neurone) is located in the œsophageal ganglion (see Fig. 1C) and its axon projects to the stomato-gastric ganglion. Tonic firing of this cell results in a restoration of the plateauing properties of the constrictor neurones (if these properties were not operating ; Fig. 3A, B) or in a modification of the time course of the active jumps in potential (if the plateauing properties were already operating ; Fig. 3C). As said previously it is via these step changes in potential that each neurone moves from the non-firing state to the firing state ; thus any modification in the kinetics of these jumps, as caused by APM, would result in a phase shift of the firing of each constrictor neurone in the pyloric cycle (Fig. 4A - D). Moreover APM acts as a modulatory input with long lasting effects (Fig. 3C) ; and a short duration discharge of APM induces a long term modification of the pyloric pattern (Fig. 4A - D).

Although we have no indication that this modulatory input produces a direct modification in the strength of the synapses within the CPG network, the phase shifts observed mimic changes in synaptic efficiency. That is, APM firing apparently results in a functional "rewiring" of the pyloric network (Fig. 4E - F). Similar apparent changes in synaptic efficiency within the network have also been obtained by application of several neuromodulatory substances (Marder, 1984). This implies that there are probably several input pathways acting in parallel (and differentially) to "shape" the pyloric pattern in such a way that the CPG can produce a large repertoire of motor patterns. These results show that a CPG, under extrinsic influences, is a dynamic structure and that can no longer be considered as a fixed circuit (Moulins and Nagy, 1984).

References

1. Delcomyn, F. (1980). Neural basis of rhythmic behavior in animals. <u>Science</u>, <u>210</u>, 492-8.

2. Dickinson, P.S. and Nagy, F. (1983). Control of a central pattern generator by an identified modulatory interneurone in Crustacea. II. Induction and modification of plateau properties in pyloric neurones. <u>J. exp. Biol.</u>, <u>105</u>, 59-82.

3. Marder, E. (1984). Mechanisms underlying neurotransmitter modulation of a neuronal circuit. <u>Trends Neurosci.</u>, <u>7</u> : 48-53.

4. Maynard, D.M. and Selverston, A.I. (1975). Organization of the stomato-gastric ganglion of the spiny lobster. IV. The pyloric system. <u>J. Comp. Physiol.</u>, <u>100</u>, 161-182.

5. Moulins, M. and Cournil, I. (1982). All-or-none control of the bursting properties of the pacemaker neurons of the lobster pyloric pattern generator. <u>J. Neurobiol.</u>, <u>13</u>, 447-458.

6. Moulins, M. and Nagy, F. (1984). Extrinsic inputs and flexibility in the motor output of the lobster pyloric neural network. In : Selverston, A.I. (eds). Model neural networks and behavior. (New-York : Plenum) (in press).

7. Nagy, F. and Dickinson, P.S. (1983). Control of a central pattern generator by an identified modulatory interneurone in Crustacea. I. Modulation of the pyloric motor output. J. exp. Biol., 105, 33-58.

8. Nagy, F., Dickinson, P.S. and Moulins, M. (1981). Modulatory effects of a single neuron on the activity of the pyloric pattern generator in Crustacea. Neuroscience Letters, 23, 167-173.

9. Rezer, E. and Moulins, M. (1983). Expression of the crustacean pyloric pattern generator in the intact animal. J. comp. Physiol., 153, 17-28.

10. Russell,D.F. and Hartline, D.K. (1978). Bursting neural networks : a re-examination. Science, 200, 453-6.

11. Selverston, A.I. and Moulins, M. (1984). Oscillatory neural networks. Ann. Rev. Physiol., 47, 29-48.

3

Neurogenesis of Respiration in Crustacea

A. J. SIMMERS and B. M. H. BUSH

Respiratory gas exchange in decapod Crustacea requires the pumping action of a bilateral pair of scaphognathites, specialized blade-like appendages whose rhythmic dorso-ventral movements cause ventilatory water currents to flow across the gill surfaces. Each scaphognathite is controlled by 10 muscles (5 depressors/5 levators) which are classified into four subgroups according to function and timing of activity **(1).**

The scaphognathite can pump water in either of two directions. In the prevalent forward mode of ventilation, water is drawn in an anterior direction through the branchial chamber. Periodically however, the scaphognathite briefly changes its pumping motion so as to force water in the opposite direction. Reversed ventilation appears to have a cleaning function and is dependent upon an inversion of the normal sequence in which the four scaphognathite muscle subgroups are recruited during each movement cycle.

The motor pattern of the ventilatory rhythm consists of alternating bursts in the two main antagonistic populations of motoneurones. The levator and depressor bursts are each composed of overlapping discharge in a number of different axons (2-3 per muscle), corresponding to the recruitment sequence of the 10 muscles and the subgroups to which they belong. The scaphognathite motor programme originates within a central pattern generator since the isolated CNS can continue to produce ventilatory output in the absence of sensory feedback. The intent here is to summarize recent intracellular data on the crustacean respiratory system, particularly that of the shore crab <u>Carcinus,</u> and consider some of the major cellular and synaptic mechanisms involved in the genesis of this rhythmic behaviour.

<u>Burst generation in ventilatory motoneurones.</u>

During spontaneous rhythmic activity, ventilatory motoneurones display large (8-30mV) oscillations in membrane potential, with each depolarizing phase giving rise to a burst of action potentials (2). For levator and depressor motoneurones alike, the membrane waves do not arise from endogenous properties of the motoneurones themselves, but are driven extrinsically by strong periodic inhibition at chemical synapses. This derives largely from the observation that the ongoing oscillations of either cell type can be phase-inverted by the injection of hyperpolarizing current. There is no evidence for periodic excitatory inputs to ventilatory motoneurones.

The source of synaptic inhibition is not the motoneurones themselves. Despite the existence of inhibitory collateral interactions between antagonistic motoneurones, these couplings appear to coordinate rather than determine rhythmicity. The basic motor output pattern for ventilation therefore originates at the interneuronal level, the motoneurones being driven by the cyclic inhibitory output of a premotor oscillator.

<u>Gating of motoneurone bursting.</u>

When the ventilatory motor output switches from the forward pattern to that underlying reversed ventilation, there is a corresponding alteration in the recruitment of motoneurones innervating two of the scaphognathite muscle subgroups. This shift to the reversal motor score depends on the central activation of a population of motoneurones which is normally silent and different from that driving the same muscle subgroups during the forward rhythm (1).It appears that both sets of "forward" and "reversal" motoneurones are driven continuously by periodic inhibition from the same central oscillator, but that bursting activity in these sets is gated by the application/removal of tonic inhibition impinging directly at the level of the motoneurones them-selves (3).

Thus during production of the normal forward pattern, the reversal moto-neurones are held relatively hyperpolarized by an independent inhibitory input which overrides that of the oscillator and suppresses bursting activity in these cells. When switching to the reversal mode occurs, this tonic inhibition is removed and the reversal motoneurones are then able to fire bursts in response to the oscillator's ongoing drive. At the same time the forward motoneurones are inhibited and remain silent during the reversal episode. The motoneurones which innervate the remaining two scaphognathite muscle subgroups do not receive gating inputs and continue to burst throughout both rhythm modes.

<u>The ventilatory oscillator : non-spiking interneurones and graded synaptic interactions.</u>

Despite evidence for strong cyclic inhibitory inputs to ventilatory moto-neurones, a feature of motoneuronal recordings is the lack of discrete spike-evoked PSP's underlying membrane oscillation. The basis for this apparent "smooth" waveform has been suggested by experiments where tetrodotoxin was used to suppress all impulses in rhythmically active preparations **(2)**. Under these conditions, the continuation of synaptically-driven oscillations in the motoneurones indicate that (i), they arise via continuously graded chemical synapses and (ii), action potentials are not required for the operation of the premotor oscillator. Accordingly, interneurones of the ventilatory oscillator have been identified which function normally without impulses **(4, 5)**. Mendelson **(4)** suggested that for lobsters and hermit crabs, a single, endogenously oscillating, non-spiking interneurone alternately drives the levator and depressor motoneurones of the scaphognathite. This scheme has been recently shown to be incorrect for the ventilatory system of the shore crab **(5)** where two distinct types of non-spiking endogenously (?) oscillating interneurone have been found. These two cell types have reciprocal phase relations with the motor output pattern ; in the first type, the depolarizing phase of oscillation is synchronized with depressor activity, while the second type depolarizes with levator bursts. Manipulation of either cell with injected current causes strong perturbation of the entire motor output pattern, indicating that they play a mutual but functionally antagonistic role in generating the ventilatory rhythm.

There exists a further type of non-spiking ventilatory neurone whose non-periodic output modulates the overall frequency of the ventilatory rhythm **(4, 5)**. Depolarizing this element increases the frequency of the rhythm, while hyperpolarization causes a decrease, or stops the rhythm altogether. The influence of this non-spiking neurone on pattern generation suggests that it serves as an interface, converting various spiking inputs into a continuously graded output to the oscillator network.

The fact that this cell's membrane potential fluctuates, albeit weakly, along with ventilatory motor output suggests it receives feedback from the network it controls. This input may originate from the motoneurones them-selves. Intracellular injection of current pulses and steady current into moto-neurones can reset the ventilatory rhythm **(6)** and increase or decrease the rhythm frequency **(2, 6)**. Thus the motoneurones have output connections with the premotor neurones controlling them and are therefore elements in the ventilatory rhythm generating system. Furthermore, this functional coupling does not depend on the production of spikes in the presynaptic motor cell.

In conclusion, the generation of rhythmic bursting in the ventilatory moto-neurones of crustaceans is determined mainly by cyclic synaptic inhibition. It appears that this extrinsic control results from the graded release of transmitter from a premotor oscillator mechanism involving non-spiking interneurones. Additional tonic inhibitory inputs serve to gate motoneurones to the oscillator's ongoing drive to produce motor output appropriate to different modes of ventilatory behavior. Further coordination and reinforcement of the motor rhythm is provided by central inhibitory connections between antagonistic motoneurones and by feedback connections from the motoneurones to the interneuronal system driving them. These ouputs also appear to be graded, in the same way as the premotor inputs to the motoneurones, and that the generation of impulses remains solely as the means of conducting motor output to the distant ventilatory muscles. Non-spiking transmission of information within and between neurones is now recognized as an important mechanism of communication within the nervous systems of both invertebrates and vertebrates alike **(7).**

References

1. Young, R.E. (1975). Neuromuscular control of ventilation in the crab Carcinus maenas. J. comp. Physiol., 101, 1-37.

2. Simmers, A.J. and Bush, B.M.H. (1983a). Central nervous mechanisms controlling rhythmic burst generation in the ventilatory system of Carcinus maenas. J. comp. Physiol., 150, 1-21.

3. Simmers, A.J. and Bush, B.M.H. (1983b) Motor programme switching in the ventilatory system of Carcinus maenas : The neuronal basis of bimodal scaphognathite beating. J. exp. Biol., 104, 163-181.

4. Mendelson, M. (1971). Oscillator neurons in crustacean ganglia. Science, 171, 1170-1173.

5. Simmers, A.J. and Bush, B.M.H. (1980). Non-spiking neurones controlling ventilation in crabs. Brain Res., 197, 247-252.

6. Di Caprio, R.A. and Fourtner, C.R. (1984). Neural control of ventilation in the shore crab, Carcinus maenas. I. Scaphognathite motor neurons and their effect on the ventilatory rhythm. J. comp. Physiol., (In press).

7. Roberts, A. and Bush, B.M.H (eds) (1981). Neurones without Impulses. Cambridge : Cambridge University Press.

4

Some Organizational Features of the Respiratory Pattern Generator and its Output as Revealed by Focal Cold Block of Different Medullary Structures

C. VON EULER, K. BUDZIŃSKA, T. PANTALEO, Y. YAMAMOTO
and F. F. KAO

INTRODUCTION

Respiration is profoundly influenced from many structures throughout
the central nervous system modulating respiratory depth and rate and
controlling the participation in, and relative contribution to the act
of breathing by the different sets of respiratory muscles.

Despite the recent rapid growth of our knowledge concerning the
bulbar respiration-related (RR) neurons, their anatomical localization,
connections and patterns of postsynaptic and discharge activities under
various conditions, we still have but little understanding of the neuronal construct of the central pattern generator (CPG) for respiration,
nor of the functional significance of the various groups of RR neurons.
Best known are perhaps the inspiration-related (IR) and expiration-
related (ER) bulbospinal neurons conveying the outputs from the bulbar
respiratory mechanisms to the different spinal motoneuron pools. Nevertheless our understanding of their functional roles in, and contributions to the respiratory motor synergy is far from clear. Whereas the
ER bulbospinal projections seem to derive, form a single group of
neurons located in the nucleus retroambigualis (nRA) in the caudal part
of the ventral respiratory group (VRG), the IR bulbospinal neurons are
found in two separate groups: one in the para-ambigual nucleus, nPA,
in the intermediate part of the VRG, and the other in the dorsal respiratory group (DRG) located in the region of the tractus solitarius nuclear
complex (nTS). [The DRG is not confined only in the ventrolateral part

of the nTS but should include also the RR neurons of the medial
aspects of nTS (1)]. The functional significance of this double representation of the output to the inspiratory muscles remains obscure.
There is no evidence showing that any of the different groups of bulbospinal 'output' neurons form parts of CPG itself.

The dense group of ER neurons found in the most rostral part of
VRG, referred to as the Bötzinger Complex (Böt.C.), has received considerable attention as a possible part of the basic CPG for respiration
(2, 3). It still remains uncertain, however, whether it plays such a
role. That structures in the rostral medulla have important functions
in control of respiratory activity have been evidenced by several
studies involving lesions and transections. Although commonly interpreted as the results of impairments of parts of the CPG itself, the
effects of such procedures might also be due to damaging structures of
importance for the tonic 'drive' inputs to the CPG from the intracranial
and peripheral chemoceptive structures.

In the studies summarized here (4, 5, 6) we have attempted to elucidate some of the above mentioned problems by means of graded focal
cold block of small restricted areas within the medulla with the aid
of needle thermodes directed stereotaxically to the various groups of
neurons identified by neurophysiological methods. The technique of
graded focal cold block allows selective blocking of synaptic transmission at thermode temperatures of around 20° C without any impairment
of impulse transmission in fibres _en passage_ (7). Conduction block of
nerve fibres can be achieved at lower thermode temperatures (10° - 0° C).

METHODS

Experiments were performed on cats under pentobarbitone anaesthesia.
Most animals were bilaterally vagotomized, with bilateral pneumothorax
and ventilated artificially. Phrenic nerve activity, external and internal intercostal activity were recorded bilaterally together with end-
tidal PCO_2, tidal volume, intratracheal pressure and blood pressure
(4).

RESULTS

<u>In the region of DRG</u> focal cooling to 20° C gave pronounced apneustic effects also when focal cold block was applied unilaterally (Fig. 1A). The apneustic effects were much stronger in vagotomized than in animals with intact vagus nerves. Usually there were no concomitant depression of the initial rate of rise in inspiratory activity. The induction of marked apneusis by focal cold block in the region of DRG seems to be in some variance with the work of Speck and Feldman (8) but is in general agreement with the results of Koepchen et al. (9). Apneustic effects were not confined to the region of the DRG but could be induced also from some other areas especially in the dorsal aspects of medulla caudal to the obex probably as the result of reducing the afferent inputs to the mechanisms subserving inspiratory off-switch functions.

<u>In the VRG</u> focal cooling to 20° C of its intermediate parts (Im VRG) often caused a depression of the slope of the inspiratory 'ramp' activity whereas, as mentioned above, cooling in the DRG did not cause depression of the initial rate of rise of the inspiratory activity. The effect on timing by focal cooling in the Im VRG were usually relatively small; both increases and decreases in respiratory rate were encountered.

All respiratory motor output effects of unilateral focal cooling in the medulla were bilaterally symmetrical. This was true both for focal cold blocks in the DRG and in the Im VRG. This suggests the existence of widespread and powerful cross connections between the respiratory neuron pools on each side in the cat. No differences were discerned between effects on the phrenic and intercostal inspiratory activity neither when cooling in the DRG nor in the Im VRG, suggesting that these two groups of IR bulbospinal neurons do not mediate the differential control of the phrenic and external intercostal motoneuron pools. The results further suggest that neural mechanisms of the DRG are relatively more involved in control of timing than those of the Im VRG which seem to be relatively more important for the intensity control. However, the neural building blocks, or part functions, constituting the respiratory CPG seem to be organized in a rather dispersed manner and with a considerably degree of "degeneracy" since neither cold block nor electrolytic lesions(8) in the DRG or the VRG cause complete arrest of rhythmic activity.

'Release' of rhythmic expiratory muscle activity was another very prominent response to focal cold block of some structures mainly in the dorsal aspects of the middle part of the medulla (Fig. 1B). The rhythmic expiratory activity could be released even in conditions when no such activity was recruited in the control situation. It was not dependent on any simultaneously occurring effects on expiratory duration and could not be related to any concomitant effects on the inspiratory activity. The release of rhythmic expiratory activity showed a marked enhancement with increasing PCO_2. In contrast, the tonic activity in abdominal and internal intercostal muscles which could be released by focal cooling in more rostroventral structures was CO_2-independent: it could be obtained also at extreme degrees of hypocapnoea and did not increase significantly with increasing levels of PCO_2. These results have suggested the existence of a dispersed system of neural mechanisms capable of controlling the recruitment threshold and intensity of rhythmic expiratory activity whereas other mechanisms seem to control the appearance and intensity of tonic activities in the abdominal and internal intercostal muscles. The latter might reflect mechanisms involved in the control of the many non-respiratory functions of these muscles (6).

In the ventrolateral parts of rostral medulla in the region of n. paragigantocellularis and preolivaris (nPG-PO) unilateral focal cooling down to 20° C caused strong depression or complete apnoea. Usually apnoea caused by focal cold block in this region was not associated with any changes in blood pressure or only with very small depressions (Fig. 1C). Our results have suggested the possibility that the structures in the region of nPG-PO are crucially involved in the transmission of the tonic chemoceptive (and possibly other) 'drive' inputs (6) without which rhythmic respiratory activity fades away. Such apnoea could sometimes, but not at all consistently, be associated with a release of tonic activity in the abdominal muscles. When present, this tonic activity was very much stronger than that which can be obtained during hypocapnic apnoea, and it proved to be CO_2-independent.

Focal cooling within the group of neurophysiologically identified Böt.C. ER neurons (located slightly dorsal to the structures of the nPG-PO) consistently caused an increase in respiratory rate, often combined with an increase in tidal phrenic amplitude and in rhythmic expiratory activity.

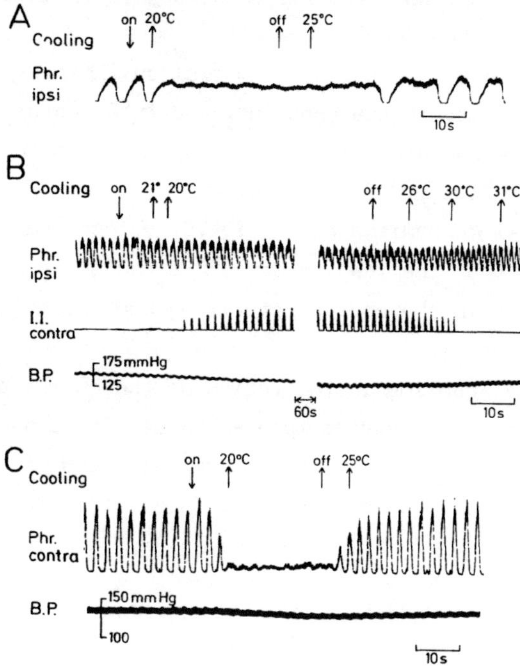

FIGURE 1   Three main types of effect on the output motor activities
(moving averages) in phrenic nerve, Phr, and internal intercostal muscle, I.I., in response to unilateral focal cooling
to 20° C at three different sites in the medulla. A: apneusis
caused by cooling in the region of the DRG (from ref. 5).
B: 'release' of expiratory motor activity caused by cooling
in the intermediate part of the dorsolateral medulla (from
ref. 6). C: apnoea caused by cooling in the region of the
paragigantocellular nucleus (from ref. 4).(Contra: contralateral, ipsi: ipsilateral to the side of the cooling. B.P.:
blood pressure). From vagotomized cats with bilateral pneumothorax and artificially ventilated.

SUMMARY

The results of these studies have failed to show any differences in
effects of focal cooling on phrenic and external intercostal activity
to focal coolings in DRG and VRG. DRG and VRG mechanisms seem to exert
somewhat different effects on the respiratory pattern: DRG mechanisms
appear to be relatively more involved in the control of the inspiratory
off-switch which, however, does not seem to be confined to the DRG.

Im VRG mechanisms, on the other hand, appear to be more concerned with gain control of the inspiratory output.

Different neural substrates control rhythmic expiratory activity and tonic activity in the abdominal and intercostal muscles. These mechanisms can operate fairly independently from those controlling inspiratory activity.

The neural mechanisms of the Böt.C. do not seem to constitute an essential part of the respiratory CPG although it appears to play an important role in the control of timing and intensity of rhythmic respiratory activity.

Unilateral cooling in the region of the nuclei paragigantocellularis lateralis could cause complete apnoea. It is suggested that these structures are involved in the transmission and integration of tonic 'drive' inputs to the respiratory CPG necessary for respiratory rhythmogenesis.

REFERENCES

1. Pantaleo, T. and Corda, M. (1984). Respiration-related neurons of the medial region of the nucleus tractus solitarius of the cat. Neuro-Science Letters, Suppl. 18, comm. S84.

2. Lipski, J. and Merrill, E.G. (1980). Electrophysioloigcal demonstration of the projection from expiratory neurones in rostral medulla to contralateral dorsal respiratory group. Brain Res., 197: 521-524.

3. Merrill, E.G., Lipski, J., Kubin, L. and Fedorko, L. (1983). Origin of the expiratory inhibition of nucleus tractus solitarius inspiratory neurones. Brain Res., 263: 43-50.

4. Budzińska, K., Euler, C. von, Kao, F.F., Pantaleo, T. and Yamamoto, Y. (1985a). Effects of graded focal cold block in the solitary and para-ambigual regions of the medulla in cat. Acta Physiol Scand, in press.

5. Budzińska, K., Euler, C. von, Kao, F.F., Pantaleo, T. and Yamamoto, Y. (1985b). Effects of graded focal cold block in rostral areas of medulla. Acta Physiol Scand, in press.

6. Budzińska, K., Euler, C. von, Kao, F.F., Pantaleo, T. and Yamamoto, Y. (1985c). Release of expiratory muscle activity by graded focal cold block in the medulla. Acta Physiol Scand, in press.

7. Bénita, M. and Condé, H. (1972). Effects of local cooling upon conduction and synaptic transmission. Brain Res., 36: 133-151.

8. Speck, D.F. and Feldman, J.L. (1982). The effects of microstimulation and microlesions in the ventral and dorsal respiratory groups in medulla of cat. J.Neuroscience, 2: 744-757.

9. Koepchen, H.P., Lazar, H., Klüssendorf, D. and Hukuhara, T. (1985). "Medullary apneusis" by lesions and cooling in the ventrolateral solitary tract region and genesis of respiratory rhythm. J.Autonomic Nervous System, in press.

ACKNOWLEDGEMENT

This work was supported by the Swedish Medical Research Council (project No. 14X-544) and Magn. Bergvalls Stiftelse. K.B. is indepted to the Swedish Institute for a fellowship within the agreement for scientific exchange between this institute and the Polish Academy of Sciences. T.P. is indepted to Consiglio Nazionale delle Ricerche, Italy, for a fellowship.

5

Multimodal Distribution of Respiratory Period Suggests a Multi-Oscillator Origin of Breathing Rhythm

A. HUGELIN

Experiments combining brain stem lesion, Hering-Breuer reflex elimination and neural depression using drugs have shown that several bulbo-pontine structures control respiratory rhythmogenesis. Furthermore suppression of all but anyone of these respiratory subsystems does not suppress periodicity (1, 2). Two kinds of hypothesis may explain these results. It could be assumed that, after elimination of the normal oscillating circuit and rearrangement of connections within the pattern generator, a new oscillating loop is constructed that takes over rhythmogenesis. In a second hypothesis, referred to as a "multi-oscillator theory" (*), it can be supposed that several pattern generators are normally working in parallel and that when one of them is eliminated rhythmicity is modified but not suppressed.

(*) It must be stressed that the term "oscillator" refers here to an abstract idea which designate an operation and not to a physical system. Accordingly,a multi-oscillator network may designate anatomically, either a set of interconnected nuclei, or several circuits of neurons organized differently within a given structure. In the latter case it should be noted that some of the operational oscillators may share a number of common neurons.

It is well established that in most biological systems
periodicity proceeds from multi-oscillator networks. Heart
beat, movement of the digestive tract, locomotion, alpha
rhythm, endocrine and nervous circadian activity constitute well documented examples of multi-oscillator organization. The present study was carried out to examine whether
the same principle may apply to the neurogenesis of breathing.

Schematically, two types of multi-oscillator have been
considered previously. 1) In the model proposed by Pavlidis (3), oscillation originates from a population of tightly coupled selfsustained oscillators. Each of these has
an individual intrinsic period but, due to a strong coupling, the ensemble oscillates at a common pace. 2) A variety of other multi-oscillator models have been proposed
which are made up of non-hierarchically organized pacema-
kers. In this type of multi-oscillator the coupling is
weak, allowing internal desynchronization in some experimental conditions. The "splitting" of circadian rhythms
has been explained in this way.

In the case of respiration, I propose the following
intermediate hypothesis. 1) Breathing movement arises from
a set of coupled oscillators having different intrinsic
frequencies. 2) In a given experimental condition, the
ensemble is entrained by one of the oscillators which ad-
vances or postpones transition of the ongoing phase in the
other oscillators. 3) If oscillation amplitude increases
in one of the entrained oscillators, the latter may become
the entraining oscillator and the period of the ensemble
shifts to a new value. Accordingly when the experimental
state changes in a non-stationary environment a multimodal
period distribution is to be expected, provided the intrinsic periods of the various entraining oscillators are different. In the present paper, multimodality of respiratory

period (RP) distribution was investigated in long term recordings carried out in man and cat.

LONG TERM RESPIRATORY PERIOD RECORDING IN MAN
Non-invasive impedance pneumography was used to analyze RP
distribution in seven subjects. With emitting, recording
and reference electrodes conveniently located around the
chest and abdomen, it is possible to record continually
over 24 hours, respiratory movements in subjects standing,
sitting, recumbent and during active behavior. In these series, during the 24 hr recordings subjects were working in
the lab during day time, reading novels or news papers
while sitting or recumbent during rest time. Travel to and
from work was by subway or by driving a car. In three subjects, RP was analyzed during several 15 mn sessions of
80-100 watts exercise performed on a Fleisch bicycle at
constant speed. In one subject, RP period was analyzed
each day for a week, before, during and after a 22 Km bicycle ride in the country. The subject was instructed to
perform the ride in about one hour at his own cycling pace by freely selecting gears.

In confirmation of previous authors, non invasive recording of respiratory movements showed an extreme variability of both frequency and tidal volume (4). We have
observed however that, in controlled behavioral conditions
RP histograms show consistent and significant differences
of distribution between the following states : a) at rest
while sitting or recumbent, b) during mental activity,
c) during exercise, d) after exercise, e) during quiet
sleep and f) during active sleep (fig. 1). In one subject
RP modes were found identical after a six months interval.
Lastly, comparison between individuals showed very similar features of RP distribution in the various behavioral
situations.

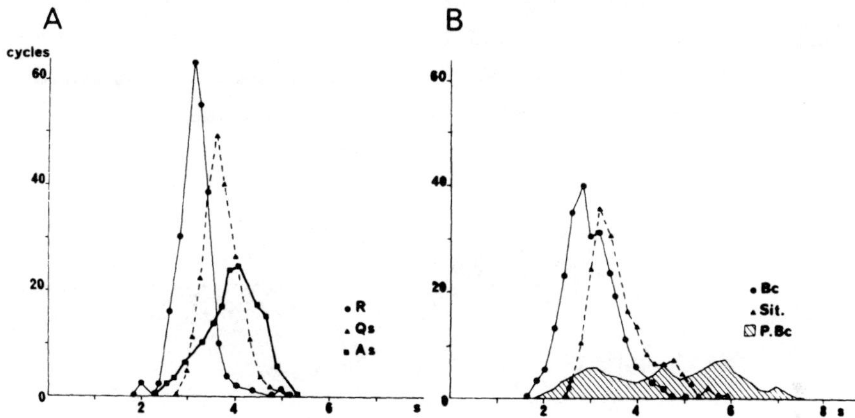

FIGURE 1 Respiratory period (RP) histograms established
from impedance pneumography recording in man.
Each histogram was constructed from five 15 mn
sequences recorded : A, during a 24 hr session
while at rest reading a novel (R), during quiet
sleep (Qs) and active sleep (As) ; B, in the same
subject during experiments of muscular exercise
carried out on five successive days, while sitting before exercise (Sit), during bicycle exercise (Bc) and post-exercise (P.Bc).
Note the multimodal distributions observed in B
with peaks falling roughly in coincidence with
peaks observed on the other curves.

In awake subjects, it was frequently difficult to obtain monomodal histograms. This was due to sudden jumps of
RP from a given range to definitely higher or lower values.
Fig. 2 shows several examples of RP shifts during post-
exercise. Interestingly, peaks of the multimodal histograms occurred roughly at the same RP values (Fig. 1 B).

Jumps were observed in awake subjects while resting and
during mental and physical tasks in uncontrolled situations. They were very frequent following exercise and during the minutes following a short series of movements.
They were not observed during quiet sleep or active sleep.
On many occasions period shift coincided with spontane-

ous change in behavior, when the subject was instructed
to modify his behavior or when he was induced to change it
unconsciously. Fig. 2 B and C shows an example of dramatic RP change from post-exercise bradypnea to sustained
polypnea induced by attracting subject's attention.

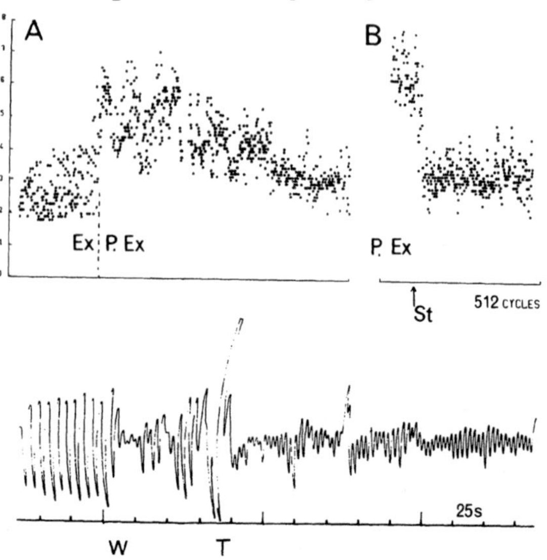

FIGURE 2 A : dot display representation of RP as a function of rank during (Ex) and post exercise (P.Ex)
Note the sudden period shifts in P.Ex condition.
B and C : after identical exercise with an additional attention test ; after a warning signal
(W), the subject was instructed to press a key
at the presentation of a tone (T). Note the RP
decrease to lower values and its persistence during at least 20 minutes.

These observations may suggest that, in accordance with
the multi-oscillator hypothesis, several oscillators having different frequency ranges, alternately determine the
respiratory rhythm and that they are selected as a function of the behavioral state. It must be recognized however that, even though the hypothesis could hardly be maintained in the absence of a multimodal distribution, observation of the latter does not constitute a proof "per se".
It is indeed possible that the same multimodality would be
observed in the case of step changes in the input to a

single oscillator, though step changes occurring repeatedly at identical levels in the same subject seem very unlikely. More convincing evidence would therefore be obtained from experiments in which discontinuity could be observed when a ramp input was applied to the system. This
condition was investigated in chronic and acute preparations of the cat.

LONG TERM RP RECORDING IN CHRONIC CAT
Pneumotachographic recording of breathing was carried out
in 16 cats chronically implanted with venous catheter and
cortical and subcortical electrodes for analysis of sleep.
During recording sessions which lasted from four to eight
hours, animals were restrained by means of a head plate
cemented to the skull, limbs moving freely in a veterinary
pouch. Recording sessions were repeated with one to several weeks interval. The observation period extended from
three to 18 months.

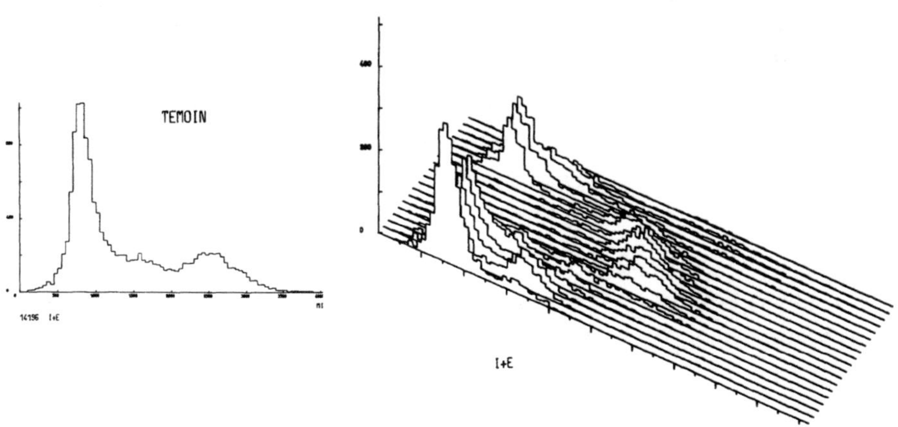

FIGURE 3 RP distribution in a chronic cat during a seven
        hours recording session. Left : histogram cons-
        tructed from the whole record. Right : 3-D repre-
        sentation of moving partial histograms.

RP histogram computation typically showed a three modal distribution (fig. 3 A). The shorter period range (0.5-1s) was observed during arousal and excitation, the intermediate range (1-1.8s) during quiet wakefulness or drowsiness and the larger range (2-4s) corresponded to sleep. By constructing partial RP histograms for 1 hour sequences and recalculating histograms each 20 mn (moving histograms) it was possible to compare RP shifts with sleepwakefulness variations. The 3-D representation mode shown in fig. 3 B clearly showed two discontinuities in RP distribution corresponding to a slow shift of state from arousal to drowsiness and from drowsiness to deep sleep.

Another argument supporting the multi-oscillator theory was furnished by observations under anesthesia. Firstly, general anesthetics act selectively on certain respiratory brain stem structures. For instance light pentobarbital anesthesia (7.5-12 mg/kg) significantly reduces the proportion and modulation of respiration related units in pontine reticular formation, whereas this dosage has no effect on other structures (6). Secondly anesthetic action is supposed to modify activity in each structure in a dose dependent manner. Accordingly we assumed that, in the case of a multi-oscillator organization, administration of anesthetics by inhalation or infusion at linearly increasing concentration, should suddenly shift RP from the actual range to one of the two others. We therefore decided to study the effects of slow induction of anesthesia using various drugs. It was observed that, whereas EEG showed a progressive change with deepening of anesthesia, RP remained unchanged until it suddenly shifted to another range. Nitrous oxide and halothan (7) produced tachypnea whereas a slow breathing rhythm was observed under pentobarbital and alfathesin. The threshold of RP switch was reached at N2O 66 % and halothan 2.5 %. Above these values RP was locked on 1 and 1.3 s respectively. When pento-

barbital was infused at the dose of 0.33 mg/kg/mn, the
threshold of EEG changes was reached after 15 mn of infusion, corresponding to a cumulative dose of 5 mg/kg. In
contrast, the threshold of RP change was not reached before 45 mn, ie. a cumulative dose of 15 mg/kg.

RAMP VARIATION OF CO2 DRIVE IN ACUTE PREPARATION
Another method for progressively changing afferent input
to the respiratory system consisted of slowly increasing
or decreasing PACO2 in immobilized cats. Animals were decorticated, bivagotomized and spinalized at the C6 level.
PACO2 was varied by electronically controlling inflation
volume and inflation and deflation duration, while the
animal was inhaling pure oxygen. RP remained quite stable
at rest provided PACO2 was kept constant in normocapnia
(fig. 4). In contrast, when PACO2 was changed within about
30 minutes from the normocapnic level to hypo- or hypercapnia, RP first slowly shifted to lower or higher values.
In contrast, above 42-46 mmHg and below 24-28 mmHg, RP
plateaued while the amplitude of the integrated phrenic
discharge still varied. This observation may be explained
by differing responses to CO2 of three oscillators. It can
be assumed that in normocapnia RP is determined by a first
oscillator whose frequency is CO2 dependant. On the other
hand, in hyper- and hypocapnia two other oscillators would
entrain the network. For each of them the response to CO2
increase would mainly consist of an increase in oscillation amplitude without period change. This explanation is
backed by the total lack of effect of PACO2 changes on RP
when the rhythm was locked on a high frequency by halothan inhalation. When RP was locked on large values by
nembutal infusion, change in PACO2 produced variations within the limits of the high range but did not succeed in
bringing it to intermediate or lower ranges.

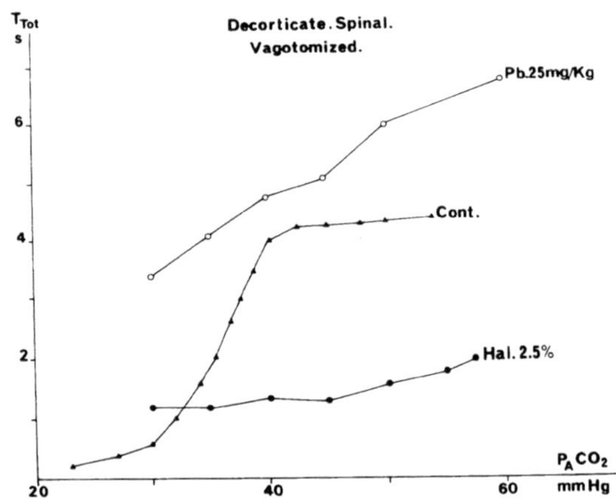

FIGURE 4 Effect of PACO2 ramp variation on RP in decorti-
cate, spinal, bivagotomized and paralyzed cat.
Cont. : without anesthesia. Hal : under halothan
anesthesia ; Pb : after injection of pentobarbital.

CONCLUSION

Multimodal distribution of RP does not constitute a deci-
sive argument in support of the multi-oscillator hypothesis. On the other hand, discontinuity occuring in case of
apparent random fluctuation or ramp variation of the input
to the respiratory control system is inconsistent with the
hypothesis of a single oscillator. One has therefore to
turn towards the multi-oscillator hypothesis which has be-
en shown valuable for every other biological oscillating
systems.

In relation to the physiological significance of this
type of organization, it can be recalled that the control
of breathing is a motor command whose behavioral function
has been reemphasized recently. According to results presented in this paper, it can be suggested that the respiratory motor program is selected among a small repertoire
as a function of the behavioral condition. Ventilation
would be secondarily adjusted to metabolic needs whatever
the period through chemical reflex control of the tidal
volume.

REFERENCES

1. St-John , W.H., Glasser, R.L. and King, R.A. (1971).
Apneustic breathing after vagotomy in cats with chronic
pneumotaxic center lesions. Respir. Physiol., 12, 239-250.
2. Gautier, H. and Bertrand, F. (1975). Respiratory
effects of pneumotaxic center lesions and subsequent vagotomy in chronic cats. Respir. Physiol., 23, 71-85.
3. Pavlidis, J. (1969). Population of interacting oscillators and circadian clocks. J. Theor. Biol., 22, 418-436.
4. Benxiden, H.H., Smith, G.M. and Mead, J. (1964). Pattern of ventilation in young adults. J. Appl. Physiol.,
19, 195-198.
5. Foutz, A.S. and al. (in preparation).
6. Caille, D., Vibert, J.F., Bertrand, F., Gromysz, H. and
Hugelin, A. (1979). Pentobarbitone effects on respiration
related units ; selective depression of bulbo-pontine reticular neurones. Respir. Physiol., 36, 201-216.
7. Caille, D., Briois, I., Poujeol, C., Foutz, A.S., Vibert, J.F. and Hugelin, A. (1985). Brainstem unit discharges during slow and fast respiratory rhythms induced by
halothan. (this symposium).

6

Respiratory Rhythmogenesis Depends Upon Three Main Subsystems

J.-F. VIBERT and D. CAILLE

Recent studies supported the hypothesis of multiple sites for ventilation neurogenesis in the central nervous system. In previous papers (1,2) we suggested that the respiratory generator was made up of at least three components: 1/ the Hering-Breuer loop and the dorsal and ventral bulbar respiratory nuclei, 2/ the pneumotaxic system, and 3/ the bulbopontine reticular formation. The contribution of the phasic vagal pulmonary afferents is unquestionable. The involvement of the pneumotaxic system and the reticular formation are still under discussion. We hypothetized that the contribution of each of these systems might differ according to physiological states. When either of these is eliminated and the others still operate, inspiratory discharge is normally switched off. In contrast, when two are eliminated, the respiratory rhythm is perturbed and may lead to an apneusis. Under barbiturate anesthesia, such apneusis was observed in spontaneously breathing, vagotomized cats with or without pneumotaxic lesions. Apneusis can also be produced when high doses of barbiturates are injected into otherwise intact animals. This paper explores the hypothesis that the respiratory rhythm could result from a weighed influence from vagal, pneumotaxic and reticular systems on an inspiratory off-switching mechanism. However, this hypthesis did not exclude the possibility that some finite level of tonic non specific afferent volleys was necessary to reach the off-switching threshold. Vagal influence was transiently suppressed by withholding a single lung inflation. Pentobarbital was used as a tool to decrease the reticular formation influence. Pneumotaxic influence was eliminated by electrocoagulation. Induced apneusis duration was related to the pentobarbital dose when injected as repeated bolus every 2 mins in several successive courses, with or without phasic vagal influence, with or without additional pneumotaxic lesion.

62

These procedures were carried out in order to shed light on the processes which changed in the pattern generator when the normal respiratory pattern turns into an apneustic pattern.

In intact cats immobilized and ventilated through a respiratory cycle controlled pump, the relative contribution of phasic vagal afferents and central respiratory drive was tested by withholding a single lung inflation (resulting in the opening of the Breuer-Hering loop plus asphyxia) one minute after injection of a bolus dose of 2.5 mg/kg of sodium pentobarbital. The phrenic burst duration in open loop (TI-Open) and in closed loop (TI-Closed) conditions were determined. In the first series, TI-Open remained the same as TI-Closed below the cumulated dose of 7.5-10 mg/kg ; above this value an apneustic plateauing was observed. TI-Open/TI-Closed increased proportionally with the cumulated dose. After injections were discontinued the ratio TI-Open/TI-Closed returned to control values within 15-30 mins. A repetition of these courses of pentobarbital administration caused new increases in the TI-Open/TI-Closed ratio : for a given pentobarbital dose, TI-Open/TI-Closed became progressively higher in the successive series. Partial lesions of the pneumotaxic complex led to higher TI-Open/TI-Closed values.

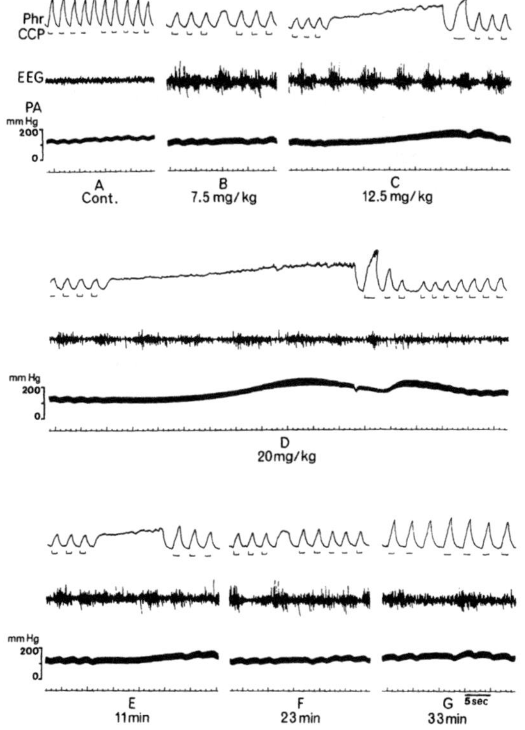

Figure 1
Effects of withholding a single pump inflation on integrated phrenic nerve discharge, EEG and blood pressure during a first series of pentobarbital bolus injection (2.5 mg/kg). A: Control; B: 1 min after the third bolus (cumulated dose 7.5 mg/kg); C: 1 min after the 5th bolus (12.5 mg/kg); D: 1 min after the 8th bolus (20 mg/kg); E, F and G: lung inflation withholding 11, 23 and 33 min after the last bolus respectively. Phr: integrated phrenic activity; CCP: cycle controlled pump with bars corresponding to inflations;

In conclusion, a short lasting dose dependent sensitivity of one or several components of the respiratory rhythm generator exists at high pentobarbital concentration. Pentobarbital delays the triggering of the otherwise unaffected inspiration off-switching mechanism, thus leading to an apneusis. The relative contribution of the vagal and the tonic and/or phasic central inputs to the respiratory rhythm generator appears to be unbalanced by anesthesia. The effect of pentobarbital on bulbar structures supporting the Breuer-Hering reflex could be less important than the effect of anesthesia on the higher brain-stem structures involved in respiratory control or genesis. A desequilibrium in favor of vagal afferents

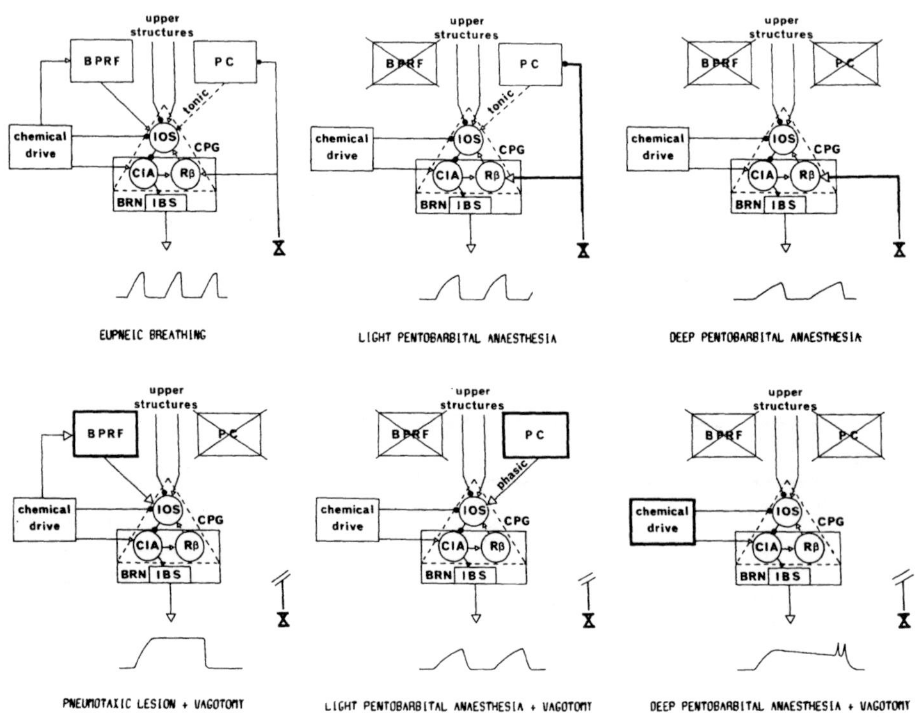

Figure 2

The main systems involved in respiratory rhythmogenesis in some experimental conditions. Anatomical formations: BPRF: Bulbo-pontine reticular formation; PC: pneumotaxic complex; BRN: bulbar respiratory nuclei; X: vagus nerve. Mechanisms and neurons: CPG: central pattern generator; IOS: inspiratory off-switch; CIA: central inspiratory activation; R b: beta neurons; IBS inspiratory bulbo-spinal neurons. Black dot: inhibition; open arrows: excitation. Thick black lines indicate the predominant sytem.

would be induced during deep anesthesia. Two evidences support these hypotheses. The nucleus of the tractus solitarius, in which are located inspiratory vagal neurons receiving inputs from pulmonary stretch receptors is less sensitive to deep anesthesia than reticular formation and pneumotaxic center (3). The greater facility to drive the respiratory generator through vagal afferents when the animal is under deep anesthesia (4) compared to light anesthesia (5) or non anesthetized animals (6) reveals a preponderant role of the vagal afferents for respiratory phase alternance. Therefore, the respiratory rhythm depends mainly upon the phasic vagal afferents in the deeply anesthetized animals while it depends upon central inputs from the pneumotaxic center and/or reticular formation in the non-anesthetized animal.

References

1 - Caille, D., Vibert, J.-F., Bertrand, F., Gromysz, H. and Hugelin, A. : (1979) Pentobarbitone effects on respiration related units;selective depression of bulbopontine reticular neurones. Respir. Physiol. , 36 201-216

2 - Caille, D., Vibert, J.-F. and Hugelin, A. : (1981) Apneusis and apnea after parabrachial or Kolliker Fuse N. lesion ; influence of wakefulness. Respir. Physiol. , 45 79-95

3 - Caille, D. : (1982) Contribution a l'etude du reseau neuronique a modulation respiratoire. Role de la formation reticulaire dans la genese et/ou le controle du rythme respiratoire.These de Doctorat es Sciences Naturelles Universite Pierre et Marie CURIE, PARIS VI. 391 p. - 63 fig. .

4 - Baconnier, P. , Benchetrit, G., Demongeot, J. and Pham-dinh, T. : (1983) Entrainment of the respiratory rhythm: I.Experiment and physiological model. In Modelling and control of breathing, B.J. Whipp and D.M. Wiberg, Editors. Elsevier ed. 126-133

5 - Vibert, J.-F., Caille, D. and Segundo, J.P. : (1981) Respiratory oscillator entrainment by periodic vagal afferents : an experimental test of a model. Biol. Cybernetics. , 41 119-130

6 - Benchetrit, G., Caille, D., Pham-dinh, T. and Vibert, J.-F. : (1977) Synchronisation et entrainement de la decharge phrenique par les afferences pulmonaires. J. Physiol., Paris , 73 139a

7

Multiple Brainstem Sites for Ventilatory Neurogenesis

W. M. ST. JOHN

The observation that the medulla is the most caudal portion of the brainstem capable of supporting rhythmic ventilatory activity, albeit gasping, has led to the hypothesis that a single medullary mechanism underlies the neurogenesis of all patterns of automatic ventilatory activity. If this hypothesis is correct, there should be basic similarities between eupnea and gasping. To examine this question, we studied eupneic and gasping activities in the same animals (1,2). Gasping was produced reversibly by cooling the brainstem of a decerebrate cat, via a fork thermode, at the pontomedullary junction; irreversible gasping was established by freezing the brainstem at this site. The rate of rise of phrenic activity was much faster in gasping than eupnea. Reflecting these different rates of rise, the manner in which phrenic motoneuronal activities were assembled to produce the phrenic burst differed markedly in eupnea and gasping. In eupnea, discharges of phrenic motoneurons recruited "early" and "late" during neural inspiration underlie the massed phrenic discharge. Both groups of motoneurons commence activities at the same time in gasping. Another difference between eupnea and gasping is that gasping was unaltered by stimuli acting upon the peripheral and central chemoreceptors. These comparisons of eupnea and gasping provide no support for the concept that a single medullary mechanism is responsible for the genesis of both automatic ventilatory patterns. Rather, our data allow for the possibility that eupnea and gasping may be generated at separate sites.

We evaluated the hypothesis that a site, critical for the neuro-genesis of gasping, could be localized in the medulla (3). By block-ade of activities with cold or lidocaine, lesions with radiofrequency power and lesions of neurons with kainic acid, we established that neither the dorsal nor ventral medullary respiratory nucleus is re-quired for gasping. Rather, a critical site was found in the lateral tegmental field. A lesion in this site eliminated gasping activities (Fig. 1), whereas electrical stimulation induced premature gasps.

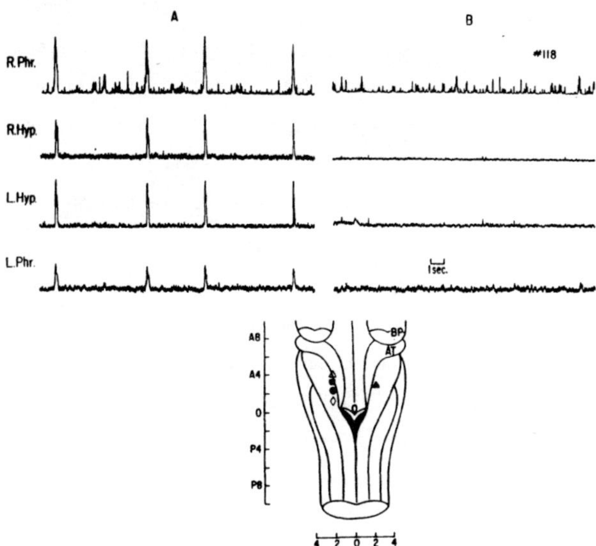

FIGURE 1   Elimination of gasping activities by a single radiofrequency lesion in the lateral tegmental field of medulla in a de-cerebrate cat. Left panel shows gasping activities of right (R) and left (L) phrenic (Phr.) and hypoglossal (Hyp.) nerves;   right panel shows elimination of activities by lesion. Filled symbols in drawing are medullary sites in which lesions, 3 mm below the surface, were effective in various cats. Open symbols are ineffective loci BP=brachium pontis, AT=auditory tubercle. From reference 3.

Success in eliciting gasps required the delivery of stimuli at an in-terval after a spontaneous gasp which was longer than a refractory period (3). The duration of this refractory period was inversely related to the frequency of spontaneous gasps (4);   this relationship is similar to that for pacemaker elements in other systems. Indeed, responses of neurons in the lateral tegmental field fulfill criteria consistent with those of a central pattern generator which contains pacemaker elements (4).

It is important to note that the critical area for gasping, described above, appears to play no role in neurogenesis of eupnea. Thus, neither lesions nor stimulation in this area caused alterations of the eupneic rhythm (3). Likewise, results of lesion and stimulation studies in other laboratories have provided no support for the possibility that the neurogenesis of eupnea depends upon the dorsal and/or ventral medullary respiratory nuclei (5,6,7). These negative findings, taken with our results, have two interrelated implications: 1) the neurogenesis of eupnea may depend upon medullary sites not yet examined or upon pontile mechanisms; 2) there are multiple sites for ventilatory neurogenesis. The latter implication has been confirmed in our studies.

Using decerebrate cats having spinal cord transections at the first cervical level, we found that the two recurrent laryngeal nerves acquired independent rhythmic discharges after a complete midsagittal division of the brainstem (8). In animals exhibiting gasping following a functional brainstem transection between pons and medulla, independent rhythmic discharges of the recurrent laryngeal nerves were recorded after a midsagittal medullary section (8). These data confirm and extend results of Gromysz and Karczewski (9,10), demonstrating that asynchronous bilateral phrenic rhythms can be established in rabbits and monkeys following midsagittal sections in the floor of the IVth ventricle. Our results for complete midsagittal brainstem sections demonstrate that there is at least one potential site for ventilatory neurogenesis in each half of the brainstem and in each half of the medulla. Likewise, Aoki and his colleagues have presented evidence that spinal mechanisms may generate rhythmic, respiratory-related discharges independent of the brainstem (11).

We have recently completed studies in which we have examined if respiratory rhythm generation can occur in pons, following its separation from medulla (12). In decerebrate cats, we recorded activities of the phrenic nerve and mylohyoid branch of the trigeminal nerve. The respiratory-related activity of the trigeminal nerve in eupnea is linked to that of the phrenic; peak mylohyoid activity typically occurs at the end of the period of phrenic activity (13). Since the

trigeminal motoneurons are within pons, efferent activity of the trigeminal nerve can serve as a probe to assess pontile respiratory-
related activity. Following a complete brainstem transection between
pons and medulla, the trigeminal and phrenic nerves discharged rhythmically, but with independent rhythms (12). In some animals, which
had no spontaneous trigeminal discharges after the transection, such
discharges appeared, with a rhythm different from phrenic, after doxa-
pram, strychnine or protriptyline had been administered (12). A
critical level of generalized tonic activity may therefore be necessary
for pontile respiratory activities to be manifested. Nevertheless, our
data conclusively establish that organized rhythmic activities can be
generated in pons independent of medulla and suggest the possibility
that pontile mechanisms may underlie the neurogenesis of eupnea.

A role for pontile mechanisms in respiratory rhythm generation, and
hence the neurogenesis of eupnea, has previously been rejected based
primarily upon three experimental observations: 1) an inability to
record respiratory-modulated neuronal activities in pons of anesthetized
animals (14); 2) in decerebrate cats, an acquisition of respiratory-
modulated discharge patterns of many pontile units only when periodic
activation of vagal pulmonary afferents was prevented (15); 3) an
elimination of respiratory-modulated pontile neuronal activities and
those of the trigeminal and facial nerves following brainstem transections between pons and medulla (16,17). Why results in previous experiments involving brainstem transections were negative is uncertain.
However, as noted above, stimulants were required for pontile respiratory modulated activities to be expressed following the transection
between pons and medulla in some animals of our study (12).

Concerning anesthetics, Caille et al. (18) found that modest doses
of pentobarbital eliminated respiratory-modulated activities from the
pontile reticular formation but not from the pneumotaxic center
located in the nucleus parabrachialis medialis and Kolliker-Fuse
nucleus. In this same context, we found that a blockade by cold of
rostral pontile activities always altered ventilatory activity in
decerebrate cats even after large doses of halothane, chloralose-
urethane or pentobarbital had almost eliminated eupneic phrenic
discharges (Fig. 2). This finding indicates an influence of the
pneumotaxic area on the breathing pattern even under deep anesthesia.

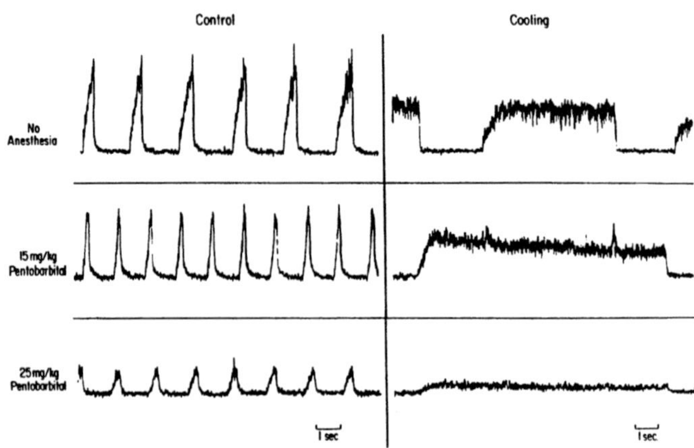

FIGURE 2   Influence of cooling of the rostral pons upon phrenic activity of decerebrate cat before and after administrations
of pentobarbital. Note that after administration of 25 mg/kg
of pentobarbital, phrenic activity was much reduced, but
cooling of pons still altered this activity.

Finally, the presence of intact vagi do not preclude a role for
pontile mechanisms in ventilatory control since lesions of the pneumotaxic center markedly alter the ventilatory pattern in either anesthetized or decerebrate cats with intact vagi (19,20). Also, phasic
respiratory-modulated neuronal activities are numerous in the pneumotaxic center of chronic, unanesthetized cats with intact vagi (21).

The above evidence demonstrates that there are multiple potential
sites for ventilatory neurogenesis in pons, medulla and spinal cord.
Moreover, there is no definitive evidence to exclude the possibility
that eupneic ventilatory activity is generated by mechanisms inherent
to pons. Within pons, we do not consider it likely that rhythmic
respiratory activity arises either within the trigeminal motor nuclei
or the reticular formation, but ventilatory neurogenesis within the
pneumotaxic center is a distinct possibility. Trigeminal systems are
considered improbable since activity of the trigeminal nerve can be
eliminated by diazepam or hypocapnia, while phrenic discharge continues (13). As discussed above, respiratory-modulated activities of
pontile reticular units, but not those of the pneumotaxic center, are
markedly reduced in anesthesia (18). A model for respiratory rhythm
generation by mechanisms inherent to the pneumotaxic center has been
proposed by Bertrand et al. (22).

Following ablation of the pneumotaxic center, neurons in another

site, possibly that of the lateral tegmental field of medulla (3),
will assume responsibility for generating the respiratory rhythm. The
existence of multiple sites for ventilatory neurogenesis would provide
an explanation for the observation that ventilatory activity very
similar to that of eupnea may be reestablished in animals having
chronic pneumotaxic center lesions and vagotomy (19,20). In this same
context, Hukuhara has claimed that eupnea, and not gasping, is obtained
following a brainstem transection at the pontomedullary junction (17).
However, examination of his experimental records reveals a pattern of
phrenic activity following the transection which differs greatly in
the amplitude and timing of bursts from the eupneic pattern recorded
before transection (17).

In summary, we propose that there are multiple sites for ventilatory neurogenesis in the mammalian brainstem. Mechanisms inherent to
the pontile pneumotaxic center may underlie the neurogenesis of eupnea.
This proposal for the neurogenesis of ventilatory activity does not
exclude a considerable role for neural circuits. Hence, we fully
agree with Cohen (23) that, at a minimum, such circuits are required
for coordinating activity between various sites for ventilatory neurogenesis and shaping the pattern of activities of respiratory motoneurons in the brainstem and spinal cord.

ACKNOWLEDGMENTS

The studies of the author were supported by grants 20574 and 26091
from the National Heart, Lung and Blood Institute, National Institutes
of Health (U.S.A.).

REFERENCES

1. St. John, W.M. and Knuth, K.V. (1981). A characterization of the
respiratory pattern of gasping. J. Appl. Physiol. 50, 984-993
2. St. John, W.M. and Bartlett, D. (1981). Comparison of phrenic
motoneuron activity in eupnea and gasping. J. Appl. Physiol. 50,
994-998
3. St. John, W.M., Bledsoe, T.A. and Sokol, H.W. (1984). Identification of medullary loci critical for the neurogenesis of gasping.
J. Appl. Physiol. 56, 1008-1019
4. St. John, W.M., Bledsoe, T.A. and Tenney, S.M. (1985). Charac-

terization by stimulation of medullary mechanisms underlying gasping neurogenesis. _J. Appl. Physiol._ (in press)

5. Speck, D.F. and Feldman, J.L. (1982). The effects of microstimulation and microlesions in the ventral and dorsal respiratory groups in medulla of cat. _J. Neurosci._ _2_, 744-757

6. Feldman, J.L., McCrimmon, D.R. and Speck, D.F. (1984). Effect of synchronous activation of medullary inspiratory bulbo-spinal neurones on phrenic discharge in cat. _J. Physiol. (London)_ _347_, 241-254

7. Berger, A.J. and Cooney, K.A. (1982). Ventilatory effects of kainic acid injection of the ventrolateral solitary nucleus. _J. Appl. Physiol._ _52_, 131-140

8. St. John, W.M. (1983). Independent bilateral sites for ventilatory neurogenesis. _J. Appl. Physiol._ _55_, 785-793

9. Gromysz, H. and Karczewski, W.A. (1981). The effects of brain stem transection on respiratory activity in the rabbit. _Acta Neurobiol. Exp._ _41_, 225-235

10. Gromysz, H. and Karczewski, W.A. (1982). Phrenic motoneurone activity in split-brain stem cats and monkeys. _Respir. Physiol._ _50_, 51-61

11. Aoki, M., Mori, S., Kawahara, K., Watanabe, H. and Ebata, N. (1980). Generation of spontaneous respiratory rhythm in high spinal cats. _Brain Res._ _202_, 51-63

12. St. John, W.M. and Bledsoe, T.A. (1985). Genesis of rhythmic respiratory activity in pons independent of medulla. _J. Appl. Physiol._ (submitted)

13. St. John, W.M., Bledsoe, T.A. and Bartlett, D. (1985). Characterization of respiratory-related activity of the trigeminal nerve. _Respir. Physiol._ (in preparation)

14. Salmoiraghi, G.C. (1963). Functional organization of brainstem respiratory neurons. _Ann. N.Y. Acad. Sci._ _109_, 571-583

15. Feldman, J.L., Cohen, M.I., Wolotsky, P. (1976). Powerful inhibition of pontine respiratory neurons by pulmonary afferents activity. _Brain Res._ _104_, 342-346.

16. Cohen, M.I. (1958). Intrinsic periodicity of pontile pneumotaxic mechanisms. _Am. J. Physiol._ _195_, 23-27

17. Hukuhara, T. (1976). Functional organization of brain stem respiratory neurons and its modulation induced by afferences. In:

Duron, B. (ed.). Respiratory Centres and Afferent Systems, pp.41-53.
(Paris: INSERM)

18. Caille, D., Vibert, J.F., Bertrand, F., Gromysz, H. and Hugelin,
A. (1979). Pentobarbitone effects on respiration related units: selective depression of bulbopontine reticular neurones. Respir Physiol.
36, 201-216

19. St. John, W.M., Glasser, R.L., and King, R.A. (1972). Rhythmic
respiration in awake, vagotomized cats with chronic pneumotaxic center
lesions. Respir. Physiol. 15, 233-244

20. Gautier, H., and Bertrand, F. (1975). Respiratory effects of
pneumotaxic center lesions and subsequent vagotomy in chronic cats.
Respir. Physiol. 23, 71-85

21. Sieck, G.C. and Harper, R.M. (1980). Pneumotaxic area neuronal
discharge during sleep-waking states in the cat. Exp. Neurol. 67, 79-
102

22. Bertrand, F., Hugelin, A. and Vibert, J.F. (1971). A stereologic
model of pneumotaxic oscillator based on spatial and temporal distribution of neuronal bursts. J. Neurophysiol. 37, 91-107

23. Cohen, M.I. (1979). Neurogenesis of respiratory rhythm in the
mammal. Physiol. Rev. 59, 1105-1173

8

Split-Brainstem and Respiratory Rhythmogenesis

W. A. KARCZEWSKI, H. GROMYSZ, W. A. JANCZEWSKI,
J. KULESZA and M. MALINOWSKA

Separation of both halves of the medulla by a midline
section in rabbits and monkeys (1, 2) shows that the Respiratory Pattern Generator (RPG) is composed of two symmetrical sub-generators. They are normally synchronized
by crossing connections, but after separation each of
them is capable of generating its "own" respiratory pattern and of responding to various reflex and chemical stimuli. The results from cats are conflicting: a midline
transection of the medulla stops all respiratory movements
in eupnoeic cats by interrupting decussating bulbo-spinal
axons (3) and, presumably, by increasing the threshold
for rhythmic firing of the RPG itself (4). Irregular,
asynchronous firing was however observed in recurrent laryngeal nerves in hypercapnic cats (5). Although the midline section reduces frequency and amplitude of respiratory volleys in both RPGs of the eupnoeic rabbit, their
relative sensitivity to changes in vagal and chemical inputs does not diminish and may even increase. It has been
shown that the vagal input can drive, excite or inhibit
only one respiratory pattern sub-generator without affec-
ting the pattern generated by the contralateral RPG (6, 7).
Also respiratory reflexes elicited from the chest wall
and limbs can be limited to one half-medulla (8). A complete functional separation of both RPGs requires, however,

a caudal extension of the midline section to $C_5 - C_6$ level of the spinal cord (6) since otherwise some traces of mutual interference between both phrenic outputs can still be observed.

Hypoxia (mean PaO_2 49.0 $\pm$ 6.6 mmHg) and hypercapnia (mean $PaCO_2$ 52.9 $\pm$ 5.7 mmHg) elicit an increase in the amplitude and/or frequency of inspiratory volleys in both RPGs. It is of interest that one half of the "split-respiratory centre" can sometimes choose a different strategy of response than the other: an increase in the electrophysiological equivalent of minute ventilation (amplitude of "integrated" phrenic nerve activity times frequency of respiratory volleys) can be achieved by increasing amplitude of discharges generated by one RPG (with little or no change in frequency), whereas the contralateral RPG simultaneously demonstrates a frequency response.

An additional transverse hemisection stops both spontaneous generation of rhythmic activity and the responses to hypoxia and hypercapnia, but rhythmic firing can still be elicited by stimulation of the vagal input in the half-medulla ipsilateral to hemisection. The contralateral RPG remains spontaneously active and responds vigorously to the hypoxic and hypercapnic stimulus.

It is concluded that both RPGs demonstrate considerable plasticity and should be regarded - at least in the rabbit and presumably monkey - as relatively independent entities.

REFERENCES

1.Gromysz, H. and Karczewski, W.A. (1981). The effects of brainstem transections on respiratory activity in the rabbit. Acta Neurobiol.Exp. 41, 225-235

2.Gromysz, H. and Karczewski, W.A. (1982). Phrenic motoneurone activity in split-brainstem cats and monkeys. Respir.Physiol. 50, 51-61

3. Bainton, C.R., Kirkwood, P.A. and Sears, T.A. (1978), On the transmission of the stimulating effects of carbon dioxide to the muscles of respiration. J.Physiol., 280, 249-272

4.Gromysz, H. and Karczewski, W.A. (1984). The split-respiratory centre in the cat: responses to hypercapnia. <u>Respir. Physiol.</u>(in print).

5. St. John, W.M. (1983). Independent brain stem sites for ventilatory rhythmogenesis. <u>J. Appl. Physiol.</u> <u>55</u>, 433-439

6. Romaniuk, J.R. and Budzińska, K. (1982). Control of respiratory outputs in "split-respiratory centre" preparation. <u>Neurosci.</u> Suppl. 7, 180P

7. Janczewski, W.A. and Karczewski, W.A. (1984). Respiratory effects of pontine, medullary and spinal cord midline sections in the rabbit. <u>Respir. Physiol.</u>(in print).

8, Romaniuk, J.R. and Budzińska, K. (1984). Spinal respiratory reflexes in decerebrate and spinalized rabbits. S.E.P.C.R. Annual Meeting, Barcelona (Abstracts)p.162.

9

Conclusions on Respiratory Rhythmogenesis Drawn from Lesion and Cooling Experiments Predominantly in the Region of Ventrolateral Nucleus of Solitary Tract (vlNTS)

H. P. KOEPCHEN, D. KLUßENDORF, H. LAZAR, T. HUKUHARA and H. ABEL

Since the first descriptions by v. Baumgarten and associates (1, 2) it has been confirmed several times that the vlNTS contains respiration-related neurones (dorsal respiratory group (DRG) with predominantly inspiratory discharge patterns and different responses to lung stretch afferents and projections to spinal inspiratory motor neurones. Various functions for respiratory neurogenesis have been ascribed to these neurones : pacemaker for respiratory rhythmogenesis, transmission of pulmonary stretch afferents, mediation of drive to phrenic and intercostal I motoneurones, intracentral feedback for the termination of inspiration ("inspiratory off-switch") or simply an additional group of I premotor neurones. Since this group is located in a circumscribed region near the dorsal medullary surface, the possibility of separate exclusion and observation of the effects on respiratory pattern and drive offers itself. Older lesion experiments around the obex without knowledge of the existence and site of this nucleus yielded contradictory results. In 1974 Koepchen et al. (3) performed acute lesions restricted to the vlNTS region and observed apneustic prolongation of inspiration. This basic finding was confirmed and extended in consecutive analyses (4, 5, 6). Recent studies with different methods of lesioning either confirmed (7, 8) or contradicted (9) these results. With regard to the basic question whether the DRG plays a significant role in respiratory timing or solely transmits rhythmic drives generated somewhere else in medullary networks, the results of three different kinds of lesion or exclusion of the vlNTS region will be reported in detail.

1. __Method__ : sets of electrolytic lesions through the tip of a microelectrode previously recording activity from DRG.   2. __Method__ : gross lesions from the surface by suction, coagulation or mechanical destruction destroying the whole extent of the vlNTS region. 3. __Method__ : reversible cooling of the region by a cooling probe with 1 mm outer diameter in comparison with cooling at other sites within the medulla.

With all three methods the basic result was the same : generation of apneustic states in anaesthetized or decerebrated cats when the vlNTS region was involved and pulmonary stretch feedback was impaired or excluded. The mininum volume of damaged tissue in the vlNTS region to prolong inspiration was in the order of 0.05 mm^3. Much larger lesions including the dorsomedial NTS, the TS itself, dorsal vagal nucleus and hypoglossal nucleus, dorsal column nuclei, sensory trigeminal nucleus had no effect on respiratory timing.

Respiratory rhythm with severely altered pattern persisted even after total bilateral destruction of the entire NTS region. The "medullary apneusis" after lesion of during cooling shows the following characteristics : prolongation of T_I without distinct changes in T_E ; slight decrease of rate of rise of phrenic nerve activity ; large variations of T_I in the direction of prolongation with the lower limit at control T_I ; increase of V_T or phrenic peak discharge, respectively ; plateau phase of inspiratory activity or imperfect termination of phrenic bursts with decaying phases until near to zero and resuscitation of I activity without interposed E phase ; imperfect inhibition of retroambigual E activity during I plateau phase ; graded prolongation of T_I by graded cooling ; induction of apneusis by temporary exclusion of pulmonary inflation ; increase of phrenic amplitude and decrease of T_I by increasing P_{CO2} ; increase in CO_2 threshold tested by hyperventilation ; shortening or switch-off of prolonged phrenic bursts by afferent vagal stimulation, by unspecific afferent stimulation, by warming of contralateral vlNTS region or by elevation of body temperature ; tendency of restoration of normal rhythm beginning within minutes after lesion and completed after about 20 h and temporary undershoot of T_I during rewarming. The effects are easily reproducible and not caused by extraordinary changes in blood pressure.

The following <u>conclusions</u> are drawn : the vlNTS region contributes considerably to termination of inspiration and therewith to respiratory timing. The experiments confirm that active inhibition is involved in the off-switch process. The partial functions of mediation of central chemical drive and ramp generation are concerned to a minor degree, i.e. are partly inherent in the same structures or cannot be entirely separated locally by the methods applied. The similarity of the properties of the apneustic state to pontine apneusis raises the question whether connections from or to the NPBM or Koelliker-Fuse nucleus can be involved or responsible. This cannot yet be answered finally, but the CO_2 response in apneustic states after pontomedullary transection differs from that after lesion of vlNTS region. The vlNTS region is by no means the only structure with off-switch ability, but a site where this function can be strongly influenced without interfering with the expiratory timing demonstrating the existence of facultatively independent subfunctions of the oscillator.

Cooling at some other sites of the medulla can likewise influence T_I or other respiratory parameters separately, as e.g. phrenic amplitude without change in timing, T_E, i.e. inhibition or starting, respectively, of next inspiratory ramp. In apneic states local cooling can restore respiratory rhythm, demonstrating that certain types of expiratory apnoea can be produced by overactivity of inhibitory mechanisms representing the other extreme of imbalance of partial functions in contrast to inspiratory apneusis.

From the experiments the picture of a longitudinal inhomogeneous system emerges which normally is strongly coordinated but on the other hand has oscillatory potency in many of its parts. This explains the low vulnerability and astonishing tendency to restoration of rhythm after lesions.

References

1. v. BAUMGARTEN, R. and KANZOW, E. (1958) Arch. Ital. Biol. <u>96</u>, 361-373.

2. v. BAUMGARTEN, R., BALTHASAR, K. and KOEPCHEN, H.P. (1960) Pflügers Arch. <u>270</u>, 504-528.

3. KOEPCHEN, H.P., LAZAR, H. and BORCHET, J. (1974) Proc. IUPS, Vol XI, 8I

4. LAZAR, H., BORCHET, J., HUKUHARA, T. and KOEPCHEN, H.P. (1978) Pflügers Arch. <u>377</u>, R57

5. KOEPCHEN, H.P., LAZAR, H. and KLUßENDORF, D. (1981) Pflügers Arch. <u>389</u>, R53

6. KOEPCHEN, H.P., LAZAR, H., KLUßENDORF, D. and HUKUHARA, T. (1983) Proc. IUPS, Vol. XV, 245

7. BERGER, A.J. and COONEY, K.A. (1982) J. Appl. Physiol. <u>52</u>, 131-140

8. Ref. v. EULER, C. (1983) J. Appl. Physiol. <u>55</u>, 1647-1659

9. SPECK, D.F. and FELDMAN, J.L. (1982) J. Neurosci. <u>2</u>, 744-757

Research supported by Deutsche Forschungsgemeinschaft

10

Multiple Roles for the Inspiratory Premotor Neurones?

P. GAUTHIER and R. MONTEAU

INTRODUCTION

Two groups of central inspiratory neurones could be distinguished on the basis of antidromic stimulation tests (1) : i) interneurones (IN) whose axons lie entirely within the bulbo-pontine structures and ii) inspiratory bulbospinal neurones (IBSN) sending an axonal projection down to the spinal chord. The IBSN are the source of the drive to the spinal inspiratory motoneurones (2, 3) and constitute a pool of output premotor neurones.

The role of the IBSN as an output is now well established but few information is available concerning the possible others functions of these neurones within the respiratory network. In order to evaluate such a possibility, the activity of medullary IN and IBSN was compared to, during the inspiratory (I) on-switch obtained from mesencephalic electrical stimulation (4, 5). The interpretation of this study indicates that the IBSN could subserve the building up of the stimulus-evoked I activity by acting as a pool of (a) "input", (b) "memory storage" and (c) "onsetting" neurones.

MATERIAL AND METHODS

Data were analysed from experiments on adult cats (i) anaesthetized with an initial Alfatesine i.m. injection followed by an urethan-chloralose i.v. injection (100 and 20 mg/kg respectively) (ii) paralysed with gallamine triethiodide and artificially ventilated with room air in order to maintain the end tidal CO_2 of approximately 4.5 %.

As illustrated in Fig. 1, classical surgical procedures allow access to the mid-brain, the medulla, the spinal chord and vagus and phrenic nerves in order (a) to record C5 and/or C6 gross phrenic activity using bipolar silver wire electrodes and single inspiratory medullary units in the dorsal and ventral respiratory nuclei (DRN, VRN ; (1)) via tungsten microelectrodes (impedance 9-12 M at 1 KHz) (b) to stimulate (i) mesencephalic periaqueductal gray (PAG) and the adjacent reticular formation (RMF) (planes A1, A2) for the eliciting of the I on-switch effects (4, 5, 6) and (ii) spinal chord and vagus nerve for classification of medullary units : neurones were designated as bulbospinal if the criteria for antidromic invasion (7) were fulfilled upon delivery of stimuli to the spinal chord and as interneurones if either of spinal and ipsilateral vagal stimulations were ineffective in producing an antidromic activation. As previously reported (6) the type of recruitment of the neurones allowed to distinguish "late" neurones (spontaneous delay of recruitment from the onset of the phrenic activity greater than 10 % of the spontaneous I duration, Fig.3A) as opposed to the early recruited ones named "all" for those exhibiting a stable or an increasing firing rate (Fig. 2A).

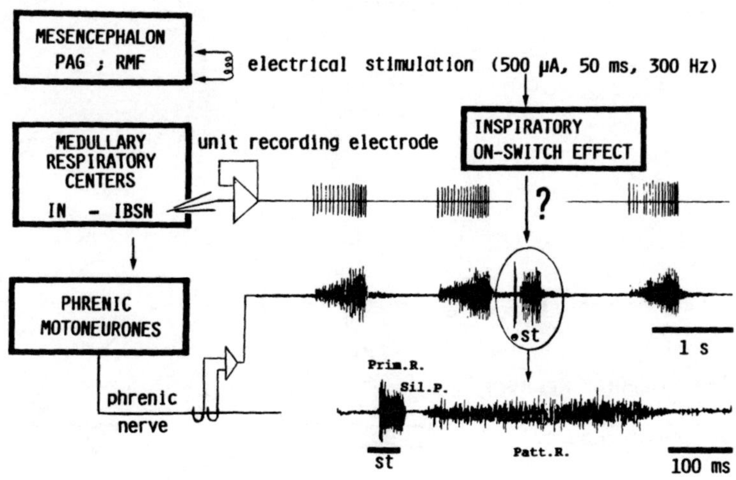

Figure 1 : <u>Schematic of the experimental set-up</u>. Abbreviations : PAG : Periaqueductal gray ; RMF : Reticular Mesencephalic Formation ; IN : Interneurones ; IBSN : Inspiratory Bulbospinal Neurones ; Prim.R. : Primary Response ; Sil.P. : Silent Phase ; Patt.R. : Patterned Response. For further explanations see text.

RESULTS AND INTERPRETATION

As illustrated in Fig. 1 the I on-switch effect evoked during the expiratory (E) phase from the mesencephalic stimulation (500 A, 50 ms, 300 Hz) consisted of a marked phrenic activation during the train (Primary Response : Prim.R., Fig. 1) followed by a long latency (30 to 100 ms from the end of the Prim.R. : Silent Phase, Sil.P., Fig. 1) long lasting discharge (300 to 1000 ms : Patterned Response : Patt.R., Fig. 1) which corresponded to the I on-switch. Since an apneustic Patt.R. can be still elicited during an apneustic breathing pattern obtained after a bilateral destruction of the pontine respiratory centers (4) the Patt.R. can be attributed to the on-switch of an "inspiratory medullary program". Thus although primarily due to an "artificial" stimulation, the Patt.R. is built up in such a way that it exhibits certain characteristics of the inspiratory medullary network organization.

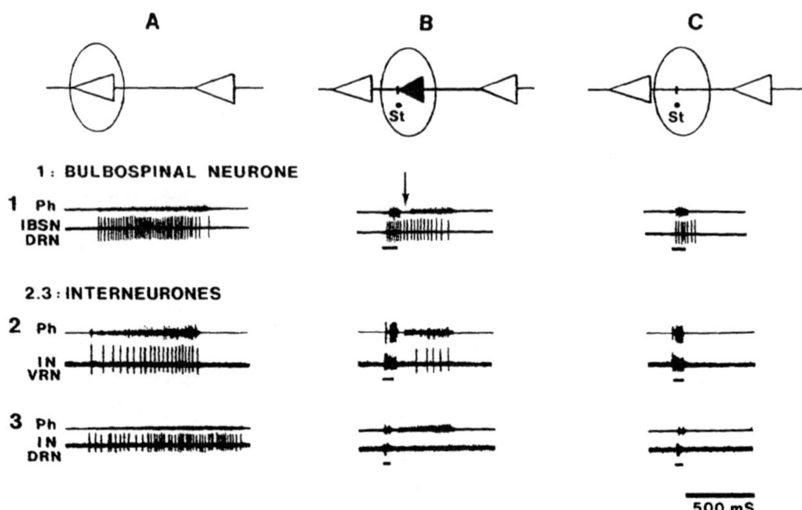

Figure 2 : <u>Inspiratory on-switch effect on I "all" medullary neurones</u>
Upper : diagram of the phrenic nerve activity. A : spontaneous activity, B : stimulus-evoked inspiration obtained from mesencephalic stimulation (st : 500 A, 50 ms, 300 Hz), C : ineffective stimulus. The activities of an inspiratory bulbospinal neurone (IBSN : 1) and two interneurones (IN : 2, 3) are illustrated in relation to the changes of phrenic (Ph) activity.

The activity of 75 medullary I-all (36 IBSN, 12 IN) and late (19 IBSN, 8 IN) neurones located in either the DRN or VRN were studied during the previously described changes of phrenic activity.

The most striking finding is that during the I on-switch the IBSN strongly discharged (Fig. 2, 3 : B1) whereas the IN were activated late or not at all (Fig. 2B : 2, 3 ; Fig. 3B : 2). Characteristics of the IBSNs activation were in agreement with the following assumptions :

1/ The IBSN : a pool of "input" neurones

The IBSN constitute the main point of impact of the mesencephalic stimulation : all of them (all and late) were activated during the stimulus train (Fig. 2 : B1 ; Fig. 3 : B1) and great majority of them (70 %) exhibited a short latency (12 +- 6 ms with respect to the first shock) burst (2 to 13 spikes, 134 +- 72 spikes/s mean frequency) of 42 +- 10 ms duration which was likely to produce the short-term phrenic activation (Prim.R., latency 20 +- 5 ms). In contrast the IN were not activated (Fig. 2B : 2,3 : Fig. 3B : 2). Thus, the IBSN constitute in these conditions not only "output" neurones projecting onto spinal motoneurones but also "input" neurones for the I on-switch mechanism.

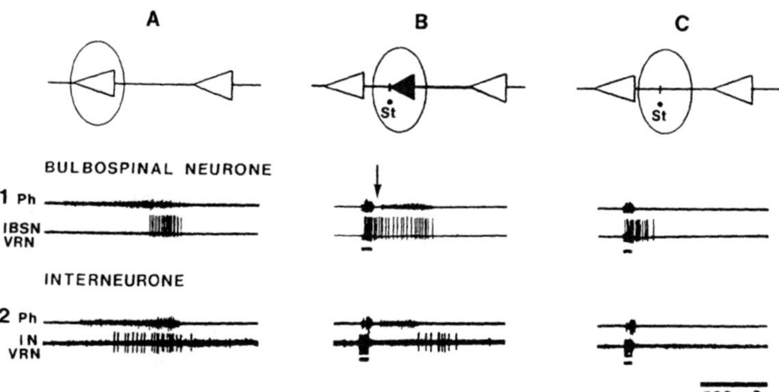

Figure 3 : <u>Inspiratory on-switch effect on I "late" medullary neurones</u>

Same ordering as in Fig. 2.

2/ The IBSN : a pool of "memory storage" neurones

The IBSN can continue to discharge after the initial activation evoked by the stimulation effect. About half (46 %) of them (both all and late) continued to discharge (78 +- 48 spikes/s) during the phrenic Sil.P. (arrow, Fig. 2B : 1 ; Fig. 3B : 1) which preceded the onset of the premature long phrenic burst (Patt.R.). This "after" long lasting discharge could also be observed for 22 % of the previous (46 %) IBSN even if a Patt.R. was not evoked in the mid E phase (Fig. 2C : 1 ; Fig. 3C : 1) although a Prim.R. was evoked. Thus, the IBSN seem to be able to keep a trace of an initial activation.

3/ The IBSN : a pool of "onsetting" neurones

The following results agree with the idea that the IBSN pattern discharge is likely to onset the activity of the IN and to induce the I on-switch phenomenon (Patt.R.) :

- Cumulative recruitment curves of IN and IBSN showed that the IBSN activation during the I on-switch was always stronger and earlier to that occurred during the spontaneous I phase (Fig. 4 : A and B1, 2) and to that occurred for the IN (Fig. 4B : 1, 2). For example about 50 % of the IBSN were already active before the Patt.R. whereas a similar percentage of active IN was only obtained 200 ms later (Fig. 4B : 1).

- The duration of the phase latency (Sil.P.) of the Patt.R. was related to the firing rate of the IBSNs : the stronger their frequency during this phase, the shorter its duration.

- During the Patt.R. and as compared to that observed during the spontaneous I phase all the IBSN fired with similar or higher mean discharge frequency whereas 80 % of the IN had a lower firing rate.

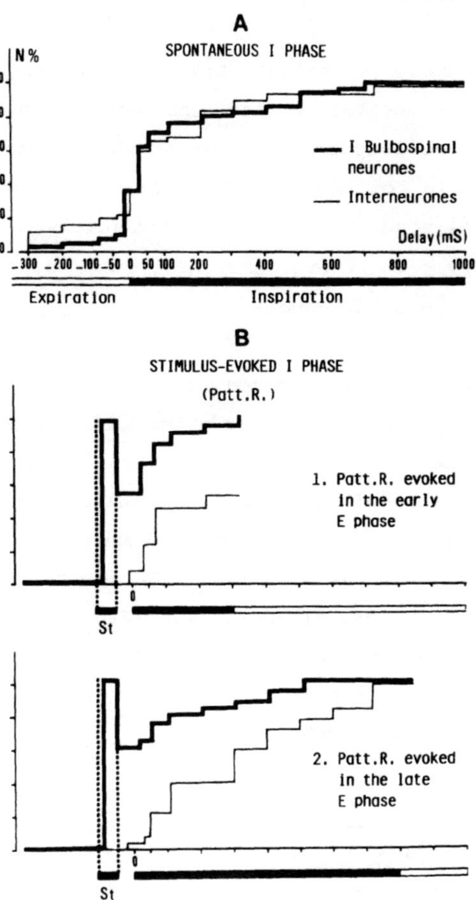

Figure 4 : <u>Quantitative study of the IBSN and IN recruitment delay</u>
Cumulative recruitment curves of inspiratory bulbospinal neurones (IBSNs) and interneurones (INs) during spontaneous I phase (A) and during stimulus-evoked inspirations (Patt.R.) (B) obtained in the early (B1) and the late (B2) expiratory phase. In each graph : abscissa : mean delay of recruitment of the neurones as related to the onset of the phrenic nerve discharge (time 0 ms) ; ordinate : number of active neurones firing at each time, expressed as the percentage (N %) of the total number of neurones in each group (IBSNs, INs). In A, note the similar recruitment of IBSNs and INs. In B, note in each case the early and strong recruitment of the IBSNs.

SUMMARY AND CONCLUSIONS

These results indicate that the IBSN highly influence the short-term phrenic activation and the generation of a premature phrenic burst onset induced from mesencephalic stimulation whereas the IN are a little, or not at all, involved in. Thus, at least in there experimental conditions the IBSN are involved in the generation of the respiratory pattern even though it runs counter to some recent data (8) obtained in other conditions.

The mechanisms involved in the I on-switch could be tentatively described as follows : the stimulus train mainly activates a pool of "driver" neurones comprising both "all" and "late" IBSN (Fig. 5 : arrow 6) so that their level of excitability increases in such a way that a postdischarge persists for some of them. This postdischarge would progressively increase the activity of the other IBSN by a system of mutual central excitation (Fig. 5 : 11, 12) as recently suggested by a cross-correlation study (12) until the IN were activated and that the resulting central I activity was sufficient for producing the I on-switch.

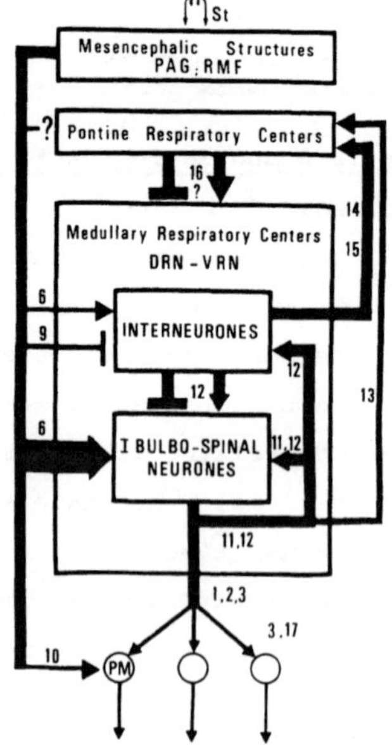

FIGURE 5 :

Activation of the inspiratory medullary neurones during the I on-switch mechanism and probable interactions.

Excitation

Inhibition

- numbers 1 to 17 correspond to references.

PM : Phrenic Motoneurones

1 BIANCHI A.L. (1971). Localisation et étude des neurones respiratoires bulbaires. Mise en jeu antidromique par stimulation spinale ou vagale. J. Physiol. (Paris), 63,5-40.

2 COHEN M.I., PIERCEY M.F., GOOTMAN P.M. and WOLOTSKY P. (1974). Synaptic connections between medullary inspiratory neurons and phrenic motoneurons as revealed by cross-correlation. Brain res., 81, 319-324.

3 HILAIRE G. and MONTEAU R. (1976). Connexions entre les neurones inspiratoires bulbaires et les motoneurones phréniques et intercostaux. J. Physiol. (Paris), 72, 987-1000.

4 GAUTHIER P., MONTEAU R. and DUSSARDIER M. (1983). Inspiratory on-switch evoked by stimulation of mesencephalic structures : a patterned response. Exp. Brain Res. 51, 261-270.

5 GAUTHIER P., MONTEAU R. and HILAIRE G. (1984). Inspiratory on-switch evoked by stimulation of the mesencephalon : activity of phrenic and laryngeal motoneurones. Exp. Brain Res., 55, 197-204.

6 GAUTHIER P. and MONTEAU R. (1984). Inspiratory on-switch evoked by mesencephalic stimulation : activity of medullary respiratory neurones. Exp. Brain Res., 56, 475-487.

7 BARILLOT J.C., BIANCHI A.L., DUSSARDIER M and GAUTHIER P. (1980). Study of the validity of the collision test. Application to the bulbo-spinal respiratory neurons. J. Physiol. (Paris), 76, 845-858.

8 FELDMAN J.L., McCRIMMON D.R. and SPECK D.F. (1984). Effect of synchronous activation of medullary inspiratory bulbo-spinal neurones on phrenic nerve discharge in cat. J. Physiol., 347, 241-254.

9 BASSAL M., BIANCHI A.L. and DUSSARDIER M. (1981). Effet de la stimulation des structures nerveuses centrales sur l'activité des neurones respiratoires chez le chat. J. Physiol. (Paris), 77, 779-795.

10 BASSAL M. and BIANCHI A.L. (1981). Effets de la stimulation des structures nerveuses centrales sur les activités respiratoires efférentes chez le chat. II. Réponse à la stimulation sous-corticale. J. Physiol. (Paris), 77, 759-777.

11 HILAIRE G. (1979). Contribution à l'étude du contrôle bulbaire de l'activité des motoneurones phréniques chez le chat. Thèse Doctorat ès-Sciences, Marseille, 260 p.

12 HILAIRE G., MONTEAU R. and BIANCHI A.L. (1984). A cross-correlation study of interactions among respiratory neurones of dorsal, ventral and retrofacial groups in cat medulla. Brain Res., 302, 19-31.

13 COHEN M.I. (1976). Synaptic relations between inspiratory neurons. In : Respiratory centres and afferent systems. B. DURON ed., INSERM, 4-6 mars 1976, 59, 19-29.

14 BIANCHI A.L. and ST JOHN W.M. (1981). Pontile axonal projections of medullary respiratory neurons. Respir. Physiol., 45, 167-188.

15 BIANCHI A.L. and BARILLOT J.C. (1982). Respiratory neurons in the region of the retrofacial nucleus : pontile, medullary, spinal and vagal projections. Neurosci. lett., 31, 277-282.

16 BIANCHI A.L. and ST JOHN W.M. (1982). Medullary axonal projections of respiratory neurons of pontile pneumotaxic center. Respir. Physiol., 48, 357-373.

17 HILAIRE G., GAUTHIER P. and MONTEAU R. (1983). Central respiratory drive and recruitment order of phrenic and inspiratory laryngeal motoneurones. Respir. Physiol., 51, 341-359.

11

Forebrain Control of the Respiratory Rhythm

R. M. HARPER, R. C. FRYSINGER, J. D. MARKS, J. X. ZHANG
and R. D. FROSTIG

INTRODUCTION

Respiratory rhythmogenesis occurs in the absence of forebrain structures, and may occur even in high spinal preparations (1). Such preparations, however, cannot maintain respiration appropriate for normal behavior. Normal respiratory patterning responds to thermoregulatory requirements as well as to momentary changes in arousal, affect, vocalization, and body position, and this responsivity depends on the integrity of suprapontine structures. Temperature regulation, for example, requires anterior hypothalamic regions, while arousal is partially controlled by midline thalamic areas as well as by limbic structures.

Role of Sleep States

Anesthesia or brain transection abolishes many of the responses modulated by suprapontine structures. Thus, intact, drug-free animals are most useful for studying suprapontine mechanisms involved in respiratory patterning. The intact drug-free preparation also allows the use of a natural physiological condition to alter respiratory patterning, i.e., different sleep-waking states.

Quiet sleep is characterized by regular, slow respiration, while REM sleep is associated with very great respiratory variability. Waking states are also associated with a great deal of variation in respiratory cycle timing due to moment to moment changes in behavior.

Sleep-waking states provide a means to separate certain forebrain and brain stem functions. Although quiet (QS) and active (REM) sleep states require the interaction of both forebrain and brain stem sites for full expression of each state (2,3), the largest proportion of REM sleep effects depends on the integrity of brain stem mechanisms, while forebrain structures underlie many of the characteristics of QS (4). Thus, during QS, synchronization of thalamic and cortical areas produces certain manifestations of that state, such as large amplitude, low frequency EEG waves, while phasic activation of excitatory - inhibitory sequences by pontine generators underlies the major characteristics of REM sleep.

REM sleep tends to functionally "dissociate" forebrain and brain stem areas. Although train stimulation of suprapontine structures, such as the central amygdala, elicits rapid elevation of blood pressure in the waking state (5), this effect is diminished in QS and nearly abolished during REM sleep (6). Similarly, stimulation of the orbital cortex typically has pronounced effects on respiratory patterns during waking or QS but profoundly diminished effects during REM (7).

Sleep-waking states thus provide the opportunity to study a variety of respiratory patterns and sources of change in respiratory cycle timing. We have been using this approach to investigate two suprapontine - brainstem systems which may alter respiratory rhythmogenesis.

<u>Cortical-Amygdaloid-Pontine Systems</u>

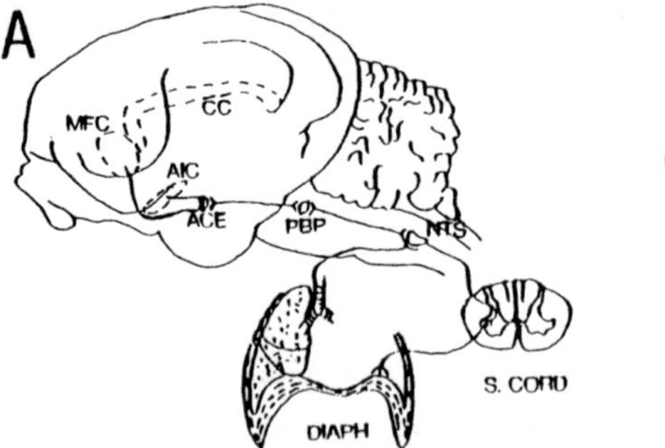
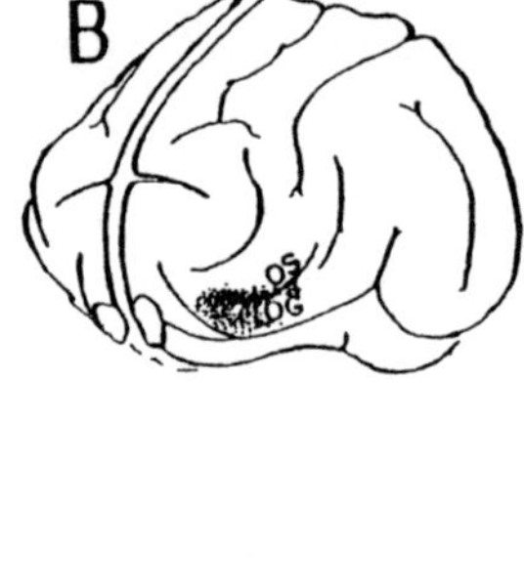

Figure 1. A. Simplified schematic representation of the principal anatomical relationships of the cortical, amygdaloid, and pontine structures comprising one respiratory related suprapontine system. ACE=central nucleus of the amygdala; AIC=agranular insular cortex; MFC=medial frontal cortex; PBP=parabrachial pons; NTS=nucleus of the solitary tract.
B. The orbital gyrus of the frontal cortex (stippled area) is a cortical area which can profoundly influence visceral and somatic (including respiratory) activity through projections to pontine reticular areas. OS=orbital sulcus; OG=orbital gyrus.

The first system consists of the medial frontal (infralimbic) cortex, central nucleus of the amygdala (ACE), and parabrachial pons. These structures have strong neuroanatomical interconnections, and are capable of integrating a variety of inputs from cortical and limbic regions which can exert, via the ACE, influence on respiratory phase switching areas in the parabrachial pons as well as the nucleus tractus solitarius in the medulla (8,9,10,11,12; Figure 1).

This limbic system has powerful influences on respiratory rhythmogenesis, but this control is extremely state-dependent. Train stimulation of the ACE, for example, will result in apneusis, together with electrocortical and behavioral arousal, and a rapid transient rise in blood pressure. Single pulse stimulation of the ACE at a rate slightly faster than the ongoing respiratory rate will entrain the respiratory rhythm, an effect which is abolished by QS or REM sleep (13).

Certain neurons in the ACE exhibit phasic, state-related discharge with the respiratory cycle. Typically, neurons will show a phasic discharge to the respiratory cycle during waking, but this relationship will be abolished during REM sleep. Other neurons phasically discharge during QS or REM sleep but not during waking (14). Cortical areas projecting to the region of the ACE also exhibit phasic discharge with the respiratory cycle, an effect which is also state-dependent (29). Finally, pontine "pneumotaxic" neurons discharge with the respiratory cycle in intact preparations (15,16).

The functional nature of this respiratory-related neuronal discharge is unclear. However, respiratory modulation of neuronal discharge at all levels of this system illustrates the state-dependent involvement of these structures in respiratory processing. This phasic discharge may provide excitatory input to the parabrachial pons which will then increase the bias to pontine reticular motor neurons. Such enhanced tone may contribute to inspiratory activation (17) and pathways for such an interaction exist (18).

There is evidence that at least part of the input of the cortico-amygdalo-pontine system is of an affective nature. Lesions of the ACE will block aversively conditioned bradycardia responses (19), and classically conditioned respiratory patterns are blocked by cooling of the ACE (20). There is, in addition, a marked electrocortical and behavioral arousal with ACE stimulation. Thus, affective stimuli which produce arousal can alter respiratory patterning through this limbic system. However, the system may be involved in a variety of arousal inducing stimuli other than affect. For example, the projections to the parabrachial pons from the ACE overlap chemoreceptor afferent projections from the upper airway (21). Sullivan and his colleagues (22) have described CO_2 receptors in the upper airway, and it is tempting to speculate that the ACE-parabrachial pons projections may participate in a CO_2 mediated arousal response.

<u>Orbital Frontal Cortex</u>

A second forebrain system involved in respiratory patterning consists of the orbital frontal cortex and its projections to motor and reticular areas of the brain stem (23,24; Figure 1B). This system has profound inhibitory actions on somatic motoneurons throughout the neuraxis (25). Bassal and Bianchi (26) have shown powerful effects on respiratory EMG and phasic discharge by stimulation of a number of frontal cortex areas in the anesthetized cat, including inhibitory effects from the orbital frontal cortex. Paired-pulse stimulation of the orbital frontal cortex results in a marked delay in the onset of inspiration, an effect which is dependent on timing of stimuli with respect to the respiratory cycle (7). This effect is most prominent in waking and QS and is virtually abolished in REM sleep.

CONCLUSIONS

We have described two suprapontine systems which exert control over respiratory rhythmogenesis. The first is a limbic system which is primarily concerned with "arousal". Train stimulation of this cortical-ACE-parabrachial pontine system results in apneusis, electrocortical arousal, and a rise in arterial pressure. The second is a cortical system with major inhibitory effects on respiration and on a variety of somatic musculature. This orbital frontal cortex system is generally inhibitory to all motor systems, and has major effects on blood pressure as well. These systems have a powerful state-related influence on respiratory rhythmogenesis as one component of a constellation of effects on CNS, autonomic and skeletomuscular systems.

The transition from waking to QS appears to involve a graded decrease in the influence of forebrain mechanisms on somatic and visceral control systems. Certain effects, such as entrainment of the respiratory cycle by single pulse ACE stimulation, are abolished during QS, while increased respiratory rate during such stimulation still observable. The effect of orbital cortex stimulation on respiratory timing and the elevation in blood pressure following train stimulation to the ACE are reduced during QS, but remain qualitatively similar to effects observed during the waking state.

During REM, the effects of train stimulation to the ACE or orbital cortex are profoundly reduced, implying a greatly diminished role of the forebrain during that state. In fact, respiratory pattern variability during REM sleep appears to be heavily dependent on phasic pontine activity. We have previously demonstrated, for example, (27) that pneumotaxic area neurons show a particularly strong interaction in synchrony with episodic REM events which are generated by pontine mechanisms.

Other suprapontine respiratory control mechanisms lose influence on respiratory patterning during REM sleep. The action of thermoregulatory mechanisms which play a major role in respiratory patterning is abolished during REM sleep (28). This phenomenon has particular importance for patterning of respiration in infants, whose sleep consists primarily of the REM state.

SUMMARY

There are several forebrain mechanisms which can alter respiratory rhythmogenesis. These forebrain mechanisms may operate by altering the overall level of tone to inspiratory or expiratory motoneurons; however, their contributions may not be selective to the respiratory musculature alone. One forebrain system, consisting of the infralimbic cortex, the ACE, and the parabrachial pons, appears to stimulate the inspiratory system, while the orbital cortex-reticular motor projections appear to operate by delaying inspiration. Both of these systems are greatly affected by sleep states. QS appears to reduce the effectiveness of forebrain input, while REM sleep appears to abolish these descending forebrain influences, allowing pontine phasic mechanisms to vary respiratory patterning.

REFERENCES

1. Aoki, M., Mori, S., Kawahara, K., Watanabe, H., and Ebata, N. (1980). Generation of spontaneous respiratory rhythm in high spinal cats. Brain Res., 202, 51-63.

2. Siegel, J.M., Tomaszewski, K.S., and Nienhuis, R. (1984). Behavioral states in the chronic medullary and mid-pontine cat. EEG Clin. Neurophysiol., In Press.

3. Siegel, J.M., Nienhuis, R., and Tomaszewski, K.S. (1984). REM sleep signs rostral to chronic transections at the pontomedullary junction. Neurosci. Lett., 45, 241-246.

4. Jouvet, M. (1965). Paradoxical sleep - a study of its nature and mechanisms. In: Akert, K., Bally, C., and Schade, J.P. (eds). Sleep Mechanisms. pp. 20-62. (Amsterdam: Elsevier).

5. Stock, G., Rupprecht, U., Stumpf, H., and Schlor, K.H. (1981). Cardiovascular changes during arousal elicited by stimulation of amygdala, hypothalamus and locus coeruleus. J. Auton. Nerv. Sys., 3, 503-510.

6. Frysinger, R.C., Marks, J.D., Trelease, R.B., Schechtman, V.L., and Harper, R.M. (1984). Sleep states attenuate the pressor response to central amygdala stimulation. Exp. Neurol., 83, 604-617.

7. Marks, J.D., and Harper, R.M. (1984). Orbital cortex stimulation alters respiratory cycle timing in the drug-free cat. Soc. Neurosci. Abs., 10, 708.

8. Hopkins, D.A., and Holstege, G. (1978). Amygdaloid projections to the mesencephalon, pons and medulla oblongata in the cat. Exp. Brain Res., 32, 529-547.

9. Otterson, O.P., and Ben-Ari, Y. (1979). Afferent connections to the amygdaloid complex of the rat and cat. J. Comp. Neurol., 187, 401-424.

10. Schwaber, J.S., Kapp, B.S., and Higgins, G. (1980). The origin and extent of direct amygdala projections to the region of the dorsal motor nucleus of the vagus and the nucleus of the solitary tract. Neurosci. Lett., 20, 15-20.

11. Otterson, O.P. (1981). Afferent connections to the amygdaloid complex of the rat with some observations in the cat. III. Afferents from the lower brain stem. J. Comp. Neurol., 202, 335-356.

12. Kapp, B.S., Schwaber, J.S., and Driscoll-Mendes, P.A. (1982). Insular and medial frontal cortex projections to the amygdaloid central nucleus in the rabbit. Soc. Neurosci. Abs., 8, 77.

13. Harper, R.M., Frysinger, R.C., Trelease, R.B., and Marks, J.D. (1984). State-dependent alteration of respiratory cycle timing by stimulation of the central nucleus of the amygdala. Brain Research, 306, 1-8.

14. Zhang, J.X., Harper, R.M., and Frysinger, R.C. Neuronal discharge patterns in the central nucleus of the amygdala during sleep-waking states. Soc. Neurosci. Abs., 9, 1163.

15. Sieck, G.C., and Harper, R.M. (1980). Discharge of neurons in the parabrachial pons related to the cardiac cycle: changes during different sleep-waking states. Brain Res., 199, 385-399.

16. Lydic, R., and Orem, J. (1979). Respiratory neurons of the pneumotaxic center during sleep and wakefulness. Neurosci. Lett., 15, 187-192.

17. Sears, T.A., Berger, A.J., and Phillipson, E.A. (1982). Reciprocal tonic activation of inspiratory and expiratory motoneurones by chemical drives. Nature, 299, 728-730.

18. Takeuchi, Y., Uemura, M., Matsuda, K., Matsushima, R., and Mizuno, N. (1980). Parabrachial nucleus neurons projecting to the lower brain stem and the spinal cord. A study in the cat by the Fink-Heimer and the horseradish peroxidase methods. Exp. Neurol., 70, 403-413.

19. Kapp, B.S., Gallagher, M., Frysinger, R.C., and Applegate, C.D. (1983). The amygdala, emotion and cardiovascular conditioning. In: Ben-Ari, Y. (ed). The Amygdaloid Complex. pp. 355-366. (Amsterdam: Elsevier/North-Holland Biomedical Press).

20. Zhang, J.X., and Harper, R.M. (1984). Cryogenic blockade of the central nucleus of the amygdala attenuates aversively conditioned cardiorespiratory responses. Soc. Neurosci. Abs., 10, 614.

21. Norgren, R. (1976). Taste pathways to hypothalamus and amygdala. J. Comp. Neurol., 166, 17-30.

22. Sullivan, C.E. (1984). Upper airway function and sleep apnea; relevance for unexpected death in infancy. In: Harper, R.M., and Hoffman, H. (eds). Proceedings of International Symposium: Santa Monica Sudden Infant Death Syndrome Conference. (New York: Spectrum), In Press.

23. Mizuno, N., Clemente, C.D., and Sauerland, E.K. (1969). Projections from the orbital gyrus in the cat. II. To telencephalic and diencephalic structures. J. Comp. Neurol., 136, 127-142.

24. Mizuno, N. Sauerland, E.K., and Clemente, C.D. (1968). Projections from the orbital gyrus in the cat. I. To brain stem structures. J. Comp. Neurol., 133, 463-475.

25. Sauerland, E.K., and Clemente, C.D. (1973). The role of the brain stem in orbital cortex induced inhibition of somatic reflexes. In: Pribram, K.H., and Luria, A.R. (eds). Psychophysiology of the Frontal Lobes. pp. 167-184. (New York: Academic Press).

26. Bassal, M., and Bianchi, A.L. (1981). Effets de la stimulation des structures nerveuses centrales sur les activités respiratoires efférents chez le chat. I. Reponses à la stimulation corticale. J. Physiol., Paris, 77, 741-757.

27. Harper, R.M., and Sieck, G.C. (1980). Discharge correlations between neurons in the nucleus parabrachialis medialis during sleep-waking states. Brain Res., 199, 343-358.

28. Glotzbach, S., and Heller, H.C. (1976). Central nervous regulation of body temperature during sleep. Science, 194, 537.

29. Frysinger, R.C., Harper, R.M., and Frostig, R.D. (1984). State-dependent discharge interactions of neurons in the infralimbic cortex and the central nucleus of the amygdala with the cardiac cycle. Soc. Neurosci. Abs., 10, 615.

ACKNOWLEDGEMENTS

     This research was supported by HL 22418-07 and AHA-GLAA 678-1G3. We wish to thank D. Taube, V. Schechtman, and R.K. Harper for their assistance in this research.

12

Influence of Caudal Raphe Complex on Phrenic Motoneurons

P. M. LALLEY

INTRODUCTION

Medullary neurons within the caudal raphe complex are involved in noci-
ception and the control of cardiovascular and other autonomic functions.
Raphe neurons may also contribute to respiratory regulation, since
breathing is altered dramatically during raphe stimulation (1). Certain
respiratory reflexes and the neurons which control them are inhibited by
stimulation of r.magnus (2). These observations, along with the knowledge that many raphe neurons are serotoninergic prompted the present
study. Several questions generated from the studies cited above are ad-
dressed: What influence has each of the three subnuclei of the caudal
raphe complex on phrenic motoneurons?; Are the responses to raphe stimulation due to direct (postsynaptic) influences on phrenic motoneurons?;
Are the responses attributable to serotoninergic mechanisms?

METHODS

Experiments were performed on cats anesthetized with chloralose-urethane
or pentobarbital. Several midcollicular decerebrate preparations were
also utilized. Respiratory neural discharges were recorded from the
phrenic nerve, from dissected filaments of the vagus nerve (central end)
and from single phrenic motoneurons (intracellular and extracellular recording). Bipolar coaxial electrodes were used to stimulate the three
major divisions of the caudal raphe complex. In pharmacological experiments, drugs were administered intravenously or by microelectrophoresis.

RESULTS

Stimulation at two sites within the caudal raphe complex (r. obscuris, r. magnus) produced inhibition of respiratory neural discharges recorded from phrenic and vagal motoneurons. The degree of inhibition varied with stimulus intensity (40-160 µA, 0.08 ms pulse duration) and frequency (12-100 Hz). Intracellular recording from phrenic motoneurons revealed that the depression of spontaneous discharges during stimulation is evidently not mediated by postsynaptic inhibition. Figure 1 illustrates inhibition of firing on two motoneurons (1A, 1C) during stimulation of r. obscuris.

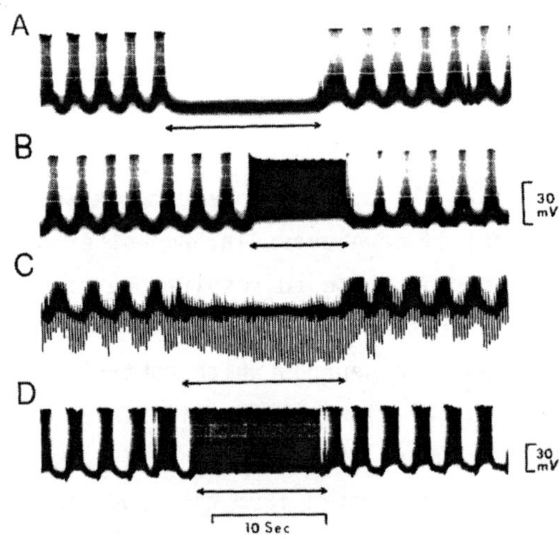

FIGURE 1   Intracellular records of responses in 2 neurons of the same
            cat to stimulation of r. obscuris and r. pallidus. Peak mem-
            brane potentials for the 2 neurons were -65 MV (A, B) and
            -70 MV (C, D).   A,C: Responses during stimulation of
            r. obscuris. In C, hyperpolarizing electrotonic potentials
            evoked by intracellular constant current pulses (6na, 20ms)
            are seen. B,D: Responses during stimulation of r. pallidus.
            Responses in 1B evoked by stimulation 5mm below the dorsal
            surface of the medulla; response in 1D produced during stimu-
            lation 4mm below surface in the same electrode track.

During stimulation, the membrane potential is held at a relatively hyperpolarized level. Hyperpolarization and depression of firing were accompanied by increased input resistance (1C). Stimulation of r. magnus also produces hyperpolarization and inhibition of cell firing, with increased input resistance. Intracellular chloride injection reversed

I.P.S.P.'s, leading to action potentials during the expiratory time (T_E). During stimulation of r. obscuris or r. magnus, C.R.D.P.'s, depolarizing I.P.S.P.'s, and action potentials were reduced or abolished. Thus the inhibition produced during raphe stimulation is evidently not exerted directly on phrenic motoneurons. The primary site of inhibition may be in the brainstem, or at more rostral supraspinal sites, since stimulation of r. obscuris or r. magnus also inhibited the firing of vagal respiratory motoneurons.

Stimulation of r. pallidus produced excitation of phrenic and vagal respiratory motoneurons. Sustained firing was evoked by 100 Hz stimulation at maximal intensity. Records 1B and 1D show responses from two phrenic motoneurons in the same experiment. Excitatory responses were more intense at more ventral points. At all points of stimulation within r. pallidus, depolarization and discharge of action potentials were observed, along with reduced input resistance. Other patterns of respiratory discharge occurred at lower intensities. Lowest effective intensities prolonged T_I and shortened T_E. These changes increased with intensity until responses similar to those shown in Figure 1 occurred, provided that the cats were paralyzed and artificially ventilated. In animals breathing spontaneously, expiratory pauses occurred during maximal stimulation, following an initial period of apneusis. The escape from apneusis is attributed to activation of expiratory neurons by chemoreceptors, since restoration of T_E also occurred in cats ventilated with 5% CO_2 or 15% O_2.

Single shocks or brief (5ms) trains were used to determine onset latencies for responses evoked in r. obscuris, r. magnus or r. pallidus. The latency for the response to r. pallidus stimulation was 4-6ms. This response was sometimes preceded at higher intensities during inspiration by a weak discharge with a 3-ms latency. The early response is attributed to stimulation of bulbospinal inspiratory axons. Similar latencies (4-8ms) for the inhibitory responses to stimulation of r. magnus or r. obscuris were recorded from phrenic motoneurons and vagal motoneurons.

In pharmacological experiments, serotonin (5HT) receptor blockers reduced both inhibition and excitation, while inhibition of 5HT uptake enhanced the responses. However, the effect of receptor blockade was overridden by increasing stimulus intensity, and the changes in responses produced by the two drug treatments were moderate. Likewise, reserpinization produced only a moderate reduction in the responses to raphe stimulation. In other experiments, microelectrophoretic application of 5HT to phrenic motoneurons produced very small increases in spontaneous firing frequency.

CONCLUSIONS

Dramatic changes in firing patterns of respiratory neurons occur during stimulation of the caudal raphe complex. The responses to stimulation of r.magnus and r.obscuris are identical with respect to effect (inhibition of firing) and the associated increase in input resistance. Stimulation of r.pallidus, however, depolarizes phrenic motoneurons. The greater responses at more ventral sites in r.pallidus could be due to factors associated exclusively with r.pallidus. However, an alternative explanation is that the relatively smaller response at more dorsal sites is related to stimulus spread to r.obscuris.

Chemoreceptor stimulation during hypercapnea or hypoxia overcame the effects of stimulation at all sites in the raphe complex. Escape from the effects of raphe stimulation occurred as quickly as 20 seconds after the onset of administration of 5% CO_2 or 15% O_2; therefore, it is assumed that peripheral chemoreceptors are involved, at least in part.

Serotonin appears to play a role in the responses to raphe stimulation; however, other neurotransmitters must be involved. With regard to serotoninergic neurons, the relevant axon terminals are most likely to be in the medulla. Since 5HT is released from nonmyelinated and finely myelinated axons, the relatively short conduction distance from caudal raphe regions to dorsal and ventral respiratory neurons would explain the observed latencies, which are too early to allow for a nonmyelinated raphe-spinal conduction pathway.

REFERENCES

1. Pitts, R.F., Magoun, H.W. and Ranson, S.W. (1939). Localization of the medullary respiratory centers in the cat. Amer. J. Physiol. <u>126</u>: 673-688.

2. Sessle, B.J., Ball, G.J. and Lucier, G.E. (1981). Suppressive influences from periaqueductal gray and nucleus raphe magnus on respiratory and related reflex activities and on solitary tract neurons, and effect of naloxone. Brain Res. <u>216</u>: 145-161.

13

Generation of a Rhythmic Neuronal Discharge in the Hippocampus

R. MILES, R. K. S. WONG and R. D. TRAUB

When GABA-dependent inhibition is blocked the discharge of hippo-campal pyramidal cells becomes synchronized. This bursting activity of a large neuronal population occurs rhythmically with a period of several seconds. A group of cells which act as the pacemaker for rhythm generation has been identified. Here we consider intrinsic neuronal properties and synaptic interactions occurring within the pacemaker region. The influence of single cells on this synchronized population discharge is described. Finally we ask whether factors underlying this behaviour may contribute to the generation of the synchronized discharge of respiratory neurones in the mammalian CNS.

On exposure to penicillin or picrotoxin, pyramidal cells in hippo-campal slices maintained in vitro exhibit a rhythmic population dis-charge of period 2-8 s. Intracellularly it consists of a burst of activity of duration 50-300 ms, synchronous in all cells to within 20 ms, which is succeeded by a prolonged afterhyperpolarization. Pyramidal cells in different areas of the hippocampus appear to play functionally different roles in the genesis of this activity. Thus, the synchronous discharge of CA1 pyramidal cells normally depends on a coherent excitatory synaptic input. This input derives from axon collaterals of CA2-3 pyramidal cells which act as the pacemaker in the generation of the rhythm.

Within the pacemaker region both intrinsic neuronal conductances and the nature of local synaptic connections are important in the generation of a synchronized discharge. As well as fast Na^+ and K^+ currents involved in the generation of a single action potential, CA3 pyramidal cells possess both inward and outward currents of slower time course. A slow inward current is carried by Ca^{++} [1] and possibly also by Na^+. Several outward K^+ currents, including one dependent on the level of intracellular Ca^{++} [2], seem to exist. These currents are responsible for the typical burst firing pattern of CA3 pyramidal cells. Bursts consist of several action potentials superimposed on a slow depolarization of duration 20-50 ms. They are followed by a pronounced after-hyperpolarization and normally occur asynchronously in different cells.

Local excitatory connections exist between CA3 pyramidal cells [3]. Such synapses are scarce (seen in about 5% of simultaneous recordings) but powerful. The unitary event is of amplitude about 1 mV and duration about 50 ms [4]. These synapses potentiate when repeatedly activated at intervals shorter than about 100 ms. Since the interval between action potentials in a burst is typically 10-15 ms this firing pattern leads to facilitation of synaptic events and may result in an obligatory spread of bursting between coupled pyramidal cells.

However synaptic inhibition normally prevents the spread of activity in the pyramidal cell population. When afferent fibres are stimulated the resulting epsp is succeeded by a di-synaptic ipsp and burst firing does not occur. The feedback nature of some hippocampal inhibition also results in an effective regulation of spontaneous pyramidal cell activity [5]. Synapses made by a pyramidal cell onto an inhibitory cell are sometimes strong enough to cause it to fire. The unitary ipsp is of amplitude 1-3 mV and if a recurrent synapse is made onto the same pyramidal cell its latency is such that a burst does not develop.

If activity does spread within the pyramidal cell population through local excitatory synapses it should be possible to affect the synchronized rhythm by activating a small group of cells. Figure 1 shows that in some cases a single cell could influence rhythmic population discharge. About 30% of neurones tested could initiate or partially entrain synchronized activity, induced by exposure to picrotoxin, in small segments of the CA3 pacemaker region containing about 1000 pyramidal cells [6]. A 50-200 ms delay often elapsed between the activation of a single cell and the resulting population discharge (Fig. 1b). Computer simulations suggest the delay reflects time taken for activity to spread from one cell to other cells in a cascade with the number of active cells increasing until the whole population discharges simultaneously [7]. Even though excitatory connections are very sparse, synchronization will result as long as each cell receives some excitatory input.

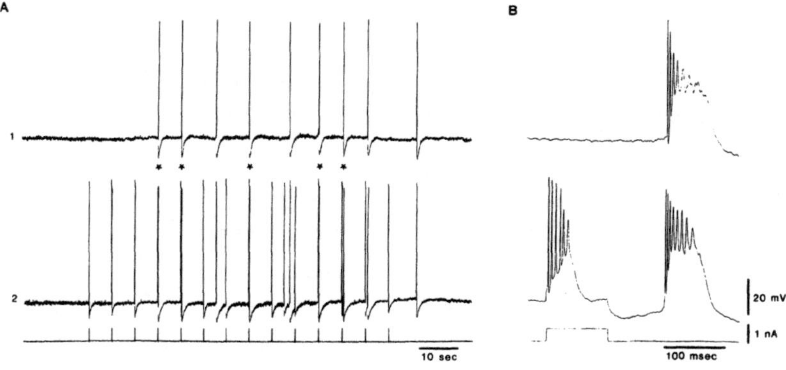

Figure 1. a. A single cell can initiate synchronized rhythmic activity in a segment of the CA3 hippocampal region in the presence of picrotoxin (100 µM). Traces are, from above, intracellular recordings from cells 1 and 2 and current injected into cell 2. b. Activation of cell 2 was followed with latency 50-200 msec by a simultaneous burst in cells 1 and 2 (as at stars in Fig. 1a). Note that these cells were not monosynaptically connected.

This synchronized activity observed in the hippocampus in vitro has many features in common with a simple form of epilepsy. Can our analysis of this rhythmic neuronal activity be applied to the generation of the respiratory rhythm? Probably not, since several different strategies seem likely to be employed in the generation of synchronized rhythms in the mammalian CNS. Nevertheless it may be useful to ask two questions about intrinsic cellular properties and synaptic connections within the group or groups of pacemaker neurones for respiration. Firstly, do these cells possess intrinsic currents which can partly or completely play the role, often ascribed to synaptic inhibition, of terminating a burst of action potentials? Secondly, are local excitatory synapses, which apparently need not be very densely distributed, responsible for synchronization of the pacemaker population as proposed by Salmoiraghi and Burns [8]?

This work was supported by NS 18464.

REFERENCES

1.   Johnston, D., Hablitz, J. J. and Wilson, W. A. (1980). Voltage clamp discloses slow inward current in hippocampal burst firing neurones. Nature, 286, 391-393.

2.   Hotson, J. R. and Prince, D. A. (1980). A calcium activated hyperpolarization follows repetitive firing in hippocampal neurons. J. Neurophysiol., 43, 409-419.

3.   MacVicar, B. A. and Dudek, F. E. (1980). Local synaptic circuits in rat hippocampus: interactions between pyramidal cells. Brain Res., 184, 220-223.

4.   Miles, R. and Wong, R. K. S. (1983). Properties of recurrent excitation in the CA3 region of the hippocampus. Neurosci. Abst., 9, 909.

5.   Miles, R. and Wong, R. K. S. (1984). Unitary inhibitory synaptic potentials in the guinea-pig hippocampus in vitro. J. Physiol., (in the press).

6.   Miles, R. and Wong, R. K. S. (1983). Single neurones can initiate synchronized population discharge in the hippocampus. Nature, 306, 371-373.

7.   Traub, R. D. and Wong, R. K. S. (1982). Cellular mechanism of neuronal synchronization in epilepsy. Science, 216, 745-747.

8.   Burns, B. D. and Salmoiraghi, G. C. (1960). Repetitive firing of respiratory neurones during their burst activity. J. Neurophysiol., 23, 27-46.

Section 2
Neuronal Organization of Respiratory Centers

14

Interconnective Pathways Between Respiratory Groups of Neurons: Results from Electrophysiological Experiments as Opposed to Anatomical Tracing Methods

A. L. BIANCHI

Neuroanatomical tracing methods have demonstrated many connections between the different nuclei or subnuclei in the brainstem where are located the neurons involved in respiratory rhythmogenesis. Thus, following injection of horseradish peroxidase (HRP) into the medullary regions which include the dorsal (DRN) and ventral (VRN) respiratory nuclei (1,2,3) or into the spinal cord (4,5) labelled cells are found in both the parabrachialis (NPBM) and Kölliker-Fuse nuclei (KF), that is within the "pneumotaxic center". Conversely, following injections of HRP into the nuclei of the pneumotaxic center, numerous cells are labelled in the area of the VRN and DRN (6,7). Similarly, after injections of radioactive aminoacids into area of the VRN and DRN labelled axons are seen projecting to the NPBM (7). Futhermore, studies of retrograde axonal transport of HRP have shown that efferent projections from cells in proximity to the retrofacial nucleus terminate in both the DRN (8), and the VRN (2). However, these neuroanatomical studies do not establish that the cells which are labelled are, in fact, respiratory neurons (see also 9) Electrophysiological evidence is necessary to confirm that these projections belong to the respiratory network of neurons responsible for the production of ventilatory movements. In this context the study by BIANCHI and St JOHN (10) indicates that, within the areas of DRN and VRN both respiratory and non-respiratory neurons have axons which project to the same area of the rostral pons. Other studies by

the same authors (11) have shown that within the area of the pneumotaxic center, both respiratory and non-respiratory neurons have axons
which project to the same area of the medulla. These results demonstrate that there are a limited number of respiratory neurons which
directly communicate in both directions between pons and medulla.
Non-respiratory neurons or non-active pontobulbar or pontospinal as
well as bulbopontine neurons cannot acquire a respiratory modulated
pattern of activity during local application of excitatory amino
acids (BIANCHI, DENAVIT-SAUBIE and MORIN-SURUN, results not published), and there are neurons serving a wide variety of physiological functions, including cardiovascular, gustatory and swallowing,
located in those same areas of the medulla which contain a high
concentration of respiratory unit activities. Therefore, it is likely
that a large number of labelled cells in neuroanatomical studies of
interconnections between central respiratory nuclei in the brainstem
do not belong primarily to the network of respiratory neurons.

By using the technique of antidromic mapping of the axons
of respiratory neurons it is possible to determine the relationships
between anatomy and function (12, 13, 14). Recently, it has been
shown that some expiratory neurons of the region of the retrofacial
nucleus (RFN) send their axons to the DRN or to both the DRN and spinal cord (15, 16). On the other hand, some inspiratory neurons of the
RFN project to the VRN (16) (see Table 1). However, the results of
LIPSKI and MERRILL (15) and FEDORKO and MERRILL (17) differ from
those of BIANCHI and BARILLOT (16) in regard to the nature of respiratory neurons recorded in the RFN and the proportion of their axonal
projections to the respiratory nuclei in the medulla and the spinal
cord. Such differences in the results can be easily explained mainly
by experimental technique. MERRILL's group used barbiturate-
anesthetized cats while decerebrated cats were used by BIANCHI and
BARILLOT (16). Indeed, in these latter experiments it was possible to
record from both inspiratory and expiratory neurons as close to each
other as 100 or 200 µm (Fig. 1). As shown in Figure 2 the frontal
plane anatomical reconstruction of the sites of recordings from respiratory neurons in the RFN indicates that the expiratory neurons are
intermingled with inspiratory and phase-spanning neurons. The larger
number of neurons (16) recorded in this region by BIANCHI and

BARILLOT (16) as opposed to others (15, 17) could also explain the smaller proportion of projections noted in their study. This is a more likely explanation than technical problems as proposed by FEDORKO and MERRILL (17), since, similar monopolar microstimulations with tungsten microelectrodes inserted within the medulla were used in both studies.

| | TYPES OF RESPIRATORY NEURONS (n=215) | | |
SITES OF PROJECTION	INSPIRATORY (n=99)*	EXPIRATORY (n=82)*	PH.SPANN. OR TONIC (n=34)*
ipsilateral VRN	1	4	1
controlateral VRN	22	3	2
ipsilateral DRN	2	3	2
controlateral DRN	1	12	0
ipsilateral sp. cord	0	0	3
controlateral sp. cord	4	2	0
ipsilat. VRN + sp.cord	0	0	1
controlat. VRN + sp.cord	1	4	0
cont.DRN + cont.sp.cord	0	4	0
cont.DRN + ipsil.sp.cord	0	3	0
ipsil.DRN + cont.sp.cord	0	1	0
ipsilateral rostral pons	1	1	3
vagus nerve	8	2	0
NAA	60	46	25

Table 1 : Functional types of respiratory neurons recorded in the region of the retrofacial nucleus (the so-called Bötzinger complex).
Abbreviations : VRN, ventral respiratory nucleus ; DRN, dorsal respiratory nucleus ; NAA, not antidromically activated. *, this number is less than the total number of sites of projection plus the NAA units because one inspiratory neuron and one expiratory neuron had ipsilateral projections in both the VRN and DRN, one expiratory neuron had controlateral projections in the VRN, DRN and spinal cord, one expiratory neuron had projections in the controlateral DRN and in both the ipsilateral and controlateral spinal cord, one phase-spanning neuron had projections in both the controlateral VRN and ipsilateral DRN and one phase-spanning neuron had ipsilateral projections in both the VRN and DRN (from BIANCHI and BARILLOT, results not published).

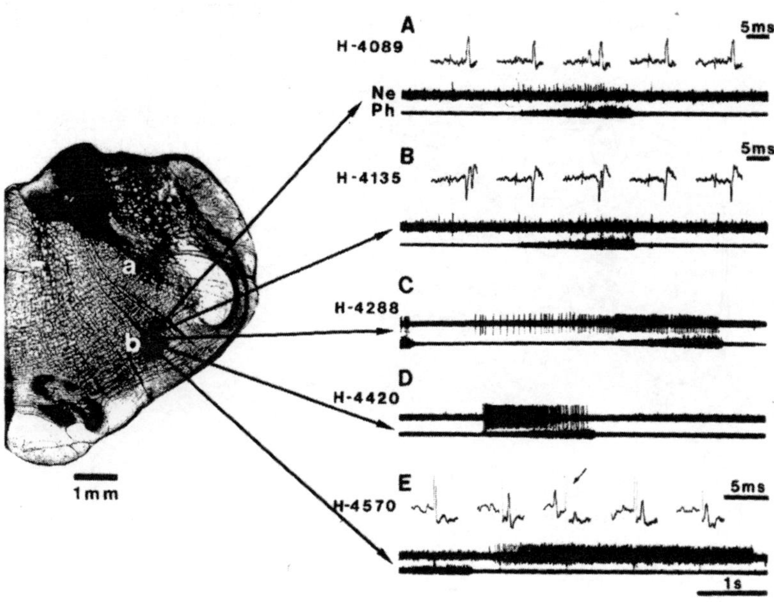

FIGURE 1 Example of unitary respiratory activity of the RFN recorded
successively in the same track at very close sites. Location of each
neuronal activity is shown on frontal section at 4 mm rostral to the
obex (a : control electrocoagulation 2 mm above the lowest point   b
were the unitary activity shown in E was recorded and where an electrocoagulation (500 µA, 8 µsec) has been done). H - indicates the
depth in µm from the dorsal surface of the medulla at which the unitary neuronal activities were recorded. These are shown on the five
panels on the right. We recorded successively (A) an inspiratory activity, and at the same site an antidromic potential evoked by
stimulation of the vagus nerve. This potential did not show any
relationship to the unitary activity. (B) antidromic action potential
evoked by stimulation of the vagus nerve while no spontaneous activity was recorded. (C) phase-spanning expiratory-inspiratory discharging neurons not antidromically activated   (D) early inspiratory
discharging neuron exhibiting a decreasing pattern of discharge, not
antidromically activated. (E) expiratory neuron exhibiting an
augmenting pattern of discharge : this neuron could be antidromically
activated by stimulation of the controlateral DRN (arrow indicates
collision between orthodromic and antidromic action potentials. In
each panel, lower tracing : phrenic nerve activity ; other upper
tracings : unitary activity recorded at two time bases in A, B and E
(Bar above upper tracings indicates 5.0 msec ; bar below lower pair
of tracings in E designates 1.0 sec (to be used also for A, B, C and
D). By projecting the artefact of stimulation of each upper tracing
upon the middle tracings, the time may be determined at which stimuli
were delivered (from BIANCHI and BARILLOT, results not published).

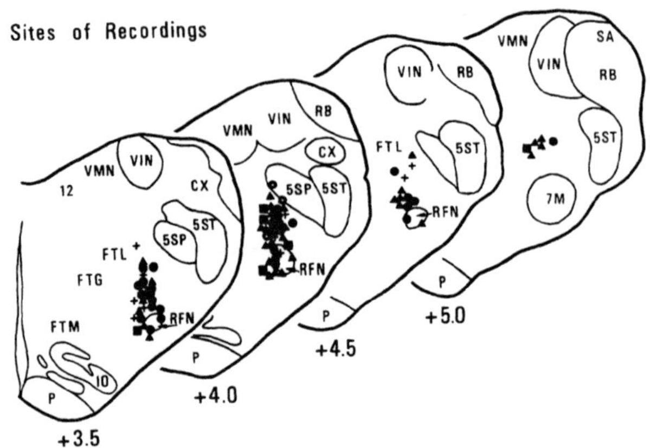

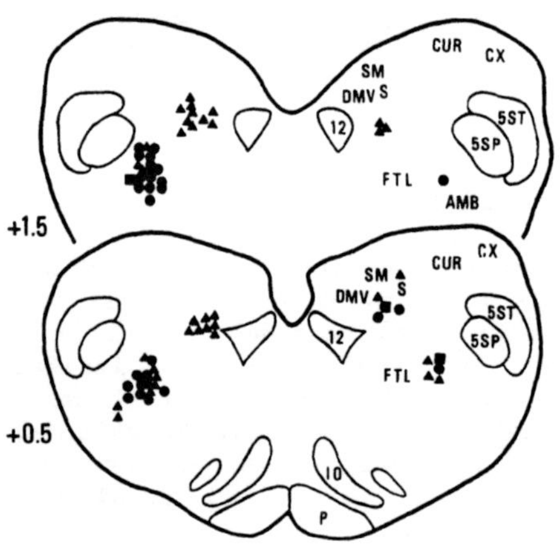

FIGURE 2 Loci of recordings of activities of respiratory neurons in the region of the retrofacial nucleus and of medullary-microstimulation. The sites at which activities were recorded either in the right or in the left side of the medulla from inspiratory neurons (●), expiratory neurons (▲), phase-spanning or tonically active neurons (■), and vagal motoneurons (+) have been projected on the closest right half-frontal sections to those illustrated here. The distance in mm is indicated by reference to the obex. Sites of stimulations are designated by the same symbols as used for the sites of recordings. They indicate the loci at which stimuli applied either ipsilaterally or controlaterally regarding the sites of recordings, elicited action potentials in the respiratory neurons recorded in the RFN. (from BIANCHI and BARILLOT, results not published).

Furthermore, the manner in which arrays of thin concentric bipolar electrodes (contact diameter, center 0.1 mm, outer : 0.25 mm) were positioned bilaterally and moved ventral and dorsal into the spinal cord could not fail to reveal spinal axons within the lateral funiculus (further description of the method in (2)).

By using cross-correlation analysis between pairs of respiratory unit activities, and spike-triggered averaging of post-synaptic noise, it has been possible to tentatively determine the functional nature of these interconnections. For example : I) the expiratory inhibition of the inspiratory neurons and phrenic motoneurons could be due to the expiratory neurons of the RFN (18, 19) ; II) positive correlations involved medullary respiratory pairs of neurons in RFN, VRN and DRN (20) as well as pontine medullary respiratory neuron pairs (21) ; III) a high incidence of correlation occurred between EI neurons of the RFN and membrane potential of post-inspiratory neurons of the VRN (22). In the results of HILAIRE et al. (20) cross-correlograms clearly indicated a high incidence of narrow peaks which could be interpretated as an excitatory influence of inspiratory bulbospinal neurons of the DRN upon the inspiratory neurons of the RFN. As a confirmation of the antidromic mapping study (16), HILAIRE et al. (20) have also shown in their cross-correlation study that narrow peaks could be obtained between inspiratory neurons of the RFN and the contralateral VRN which might correspond to an excitation of the former upon the latter. From these results and those of others (19, 21, 22), we have summarized the main afferent and efferent connections that the respiratory neurons in the RFN make with other groups of respiratory neurons in the brainstem (Fig. 3). We assumed that inspiratory bulbospinal neurons of the DRN could send axonal collaterals toward or terminate upon the inspiratory neurons of the RFN, which in turn could excitate the inspiratory bulbospinal neurons of the VRN. Inhibition of the activity of inspiratory bulbo-spinal neurons in both DRN and VRN, as well as inhibition of the inspiratory motoneurons, could be due to the expiratory neurons of the RFN (18, 19). Hence, it is likely that the neurons in the RFN play an important function in the genesis of respiration and intervene as a central relay within the respiratory center. Furthermore, there are a limited number of bulbopontile and pontobulbar respiratory neurons revealed by antidromic mapping experiments (10, 11), and there is evidence that the pneumotaxic center in the rostral pons strongly influences the activities of medullary respiratory neurons. Therefore we could assume, as speculated elsewhere (20), that the neurons in the RFN could be involved in the pathways by which inputs from the rostral pons reach the bulbospinal neurons, and inputs from the DRN and VRN reach the pneumotaxic area.

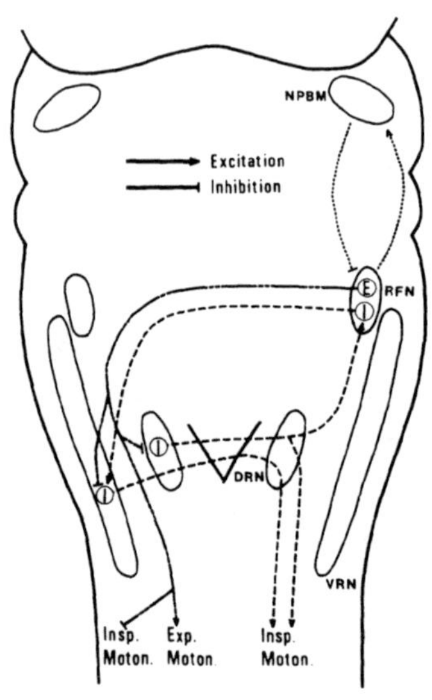

FIGURE 3 Dorsal view of the brainstem and schematic diagrams of the connections of the expiratory neurons (E) of the RFN and of inspiratory neurons (I) of the RFN, VRN and DRN as drawn from antidromic mapping experiments and cross-correlation studies (Reference in the text). Abbreviations : DRN and VRN, dorsal and ventral respiratory nuclei ; RFN, region of the retrofacial nucleus ; NPBM, parabrachial and Kölliker-Fuse nuclei.

In conclusion, electrophysiological studies show intermingling of respiratory and non-respiratory neurons within the well-known respiratory areas in the brainstem, with axonal projections terminating in the same areas. Therefore neuroanatomical studies with substances applied extracellularly cannot definitively establish interactions among respiratory neurons of the brainstem. Similarly neuromediator or neuromodulator neuronal systems observed in the brainstem by histofluorescence or immunoreaction cannot definitively establish that respiratory neurons are involved in these systems. Rather, such neuroanatomical as well as neuropharmacological studies only indicate the possibility of such interconnections or of the presence of neuromediator in the respiratory network : findings to be confirmed by electrophysiological techniques.

REFERENCES

1 DENAVIT-SAUBIE, M. and RICHE, D. (1977) Descending input from the pneumotaxic system to the lateral respiratory nucleus of the medulla. An anatomical study with the horseradish peroxidase technique. Neurosci. Lett., 6, 121-126.

2 BYSTRZYCKA, E.K. (1980). Afferent projections to the dorsal and ventral respiratory nuclei in the medulla oblongata of the cat studied by the horseradish peroxidase technique. Brain Res., 185, 59-66.

3 KALIA, M. (1981). Anatomical organization of central respiratory neurones. Ann.Rev.Physiol., 43, 105-120.

4 KYUPERS, H.G.J.M. and MAISKY, V.A. (1975). Retrograde axonal transport of horseradish peroxidase from the spinal cord to brain stem cell groups in the cat. Neurosci.Lett., 1, 9-14.

5 TAKEUCHI, Y., UEMURA, M., MATSUDA, K., MATSUSCHIMA, R. and MIZUNO, N. (1980). Parabrachial nucleus neurons projecting to the lower brain stem and the spinal cord. A study in the cat by the Fink-Heimer and the horseradish peroxidase methods. Exp.Neurol., 70, 403-412.

6 KALIA, M. (1977). Neuroanatomical organization of the respiratory centers. Fed.Proc., 36, 2405-2411.

7 KING, G.W. (1980). Topology of ascending brainstem projections to nucleus parabrachialis in the cat. J.Comp.neurol., 191, 615-638.

8 KALIA, M., FELDMAN, J.L. and COHEN, M.I. (1979). Afferent projection to the inspiratory neuronal region of the ventrolateral nucleus of the tractus solitarius in the cat. Brain Res., 171, 135-141.

9 COHEN, M.I. (1979). Neurogenesis of respiratory rhythm in the mammal. Physiol.Rev., 59, 1105-1173.

10 BIANCHI, A.L. and St JOHN, W.M. (1981). Pontile axonal projections of medullary respiratory neurons. Respir.Physiol., 45, 167-183.

11 BIANCHI, A.L. and St JOHN, W.M. (1982). Medullary axonal projections of respiratory neurons of pontile pneumotaxic center. Respir.Physiol., 48, 357-373.

12 BIANCHI, A.L. (1971). Localisation et étude des neurones respiratoires bulbaires. Mise en jeu antidromique par stimulation spinale ou vagale. J.Physiol. (Paris), 63, 5-40.

13 BIANCHI, A.L. (1974). Modalités de décharge et propriétés anatomo-fonctionnelles des neurones respiratoires bulbaires. J.Physiol. (Paris), 68, 555-587.

14 MERRILL, E.G. (1974). Finding a respiratory function for the medullary respiratory neurones. In Bellairs, R. and Gray, E.G. (eds.). Essays on the nervous system, pp. 451-486 (Oxford, Clarendon Press).

15 LIPSKI, J. and MERRILL, E.G. (1980). Electrophysiological demonstration of the projection from expiratory neurons in rostral medulla to contralateral dorsal respiratory group. Brain Res., 197, 521-524.

16 BIANCHI, A.L. and BARILLOT, J.C. (1982). Respiratory neurons in the region of the retrofacial nucleus : pontile, medullary, spinal and vagal projections. Neurosci. Lett., 31, 277-282.

17 FEDORKO, L. and MERRILL, E.G. (1984). Axonal projections from the rostral expiratory neurones of the bötzinger complex to medulla and spinal cord in the cat. J. Physiol., 350, 487-496.

18 MERRILL, E.G., LIPSKI, J., KUBIN, L. and FEDORKO, L. (1983). Origin of the expiratory inhibition of nucleus tractus solitarius inspiratory neurones. Brain Res., 263, 43-50.

19 FEDORKO, L. and MERRILL, E.G. (1984). Bötzinger projections to brain stem and spinal cord. This volume.

20 HILAIRE, G., MONTEAU, R. and BIANCHI, A.L. (1984). A cross-correlation study of interactions among respiratory neurons of dorsal, ventral and retrofacial groups in cat medulla. Brain Res., 302, 19-31.

21 LINDSEY, B.G., SEGERS, L.S. and SHANNON, R. (1984). Interactions among medullary and pontine respiratory neurons. This volume.

22 REMMERS, J.E., TAKEDA, R., SCHULTZ, S.A. and HAJI, A. (1984). Relationship of membrane potential of ventral respiratory group neurons to action potentials of retro-facial respiratory units. This volume.

15

Relationship of Membrane Potential of Ventral Respiratory Group Neurons to Action Potentials of Retro-Facial Respiratory Units

J. E. REMMERS, R. TAKEDA, S. A. SCHULTZ and A. HAJI

Early inspiratory (EI) neurons of the retro-facial nucleus (RFN) project contralaterally to the vicinity of the bulbar ventral respiratory group (VRG) (1). Post-inspiratory (PI) neurons of the VRG receive prominent inhibitory post-synaptic potentials (IPSPs) during inspiration, and their membrane potential is often maximally hyperpolarized early in inspiration. We speculate that during inspiration PI neurons receive IPSPs from EI neurons, some of which may be located in the contralateral RFN. To examine this possibility, we simultaneously recorded extracellular action potentials (APs) of respiratory units in the RFN and membrane potential (MP) of VRG respiratory neurons, the former with metal microelectrodes (50 M ohms) and the latter with glass micropipettes (10-20 M ohms) filled with 3 M KCl. Decerebrate cats were paralyzed and ventilated by a phrenic driven servo-respirator. All neurons were tested for peripheral axons by stimulating the ipsilateral vagus and superior laryngeal nerves, as well as the spinal cord at C2. Stable, simultaneous recordings of MP and of APs were obtained in 82 instances. In addition to EI and PI neurons, augmenting inspiratory (AI) and expiratory (AE) neurons were recorded in both locations. The pre- and post-AP time average of MP of the VRG neuron was calculated using the RFN spike as a trigger. Similar histograms were constructed using the onset or rapid downstroke of the averaged phrenic neurogram as a reference event.

Seventeen post-RFN spike histograms, derived from 82 stable record-

ings, revealed hyperpolarizing waves lasting 8-10 msec. These began
1-2 msec after or before the RFN spike and appeared to be
incorporated in a high frequency oscillation (HFO). This HFO was
also demonstrated by histograms triggered by the onset of I or E
phases.

An example of this combination is illustrated in Figure 1. MP
of a PI neuron in the VRG was averaged following phrenic onset
(Insp. trigger, lower panel) and following the spike of the EI unit
(RFN trigger, upper panel). The former reveals a symmetrical wave
of MP (12 msec period) characteristic of the membrane potential HFO
(2), and the latter shows an 8-10 msec hyperpolarization following
the EI spike. Reversal of these waves was observed after
iontophoresis of chloride ions, suggesting IPSPs contribute to
the waves.

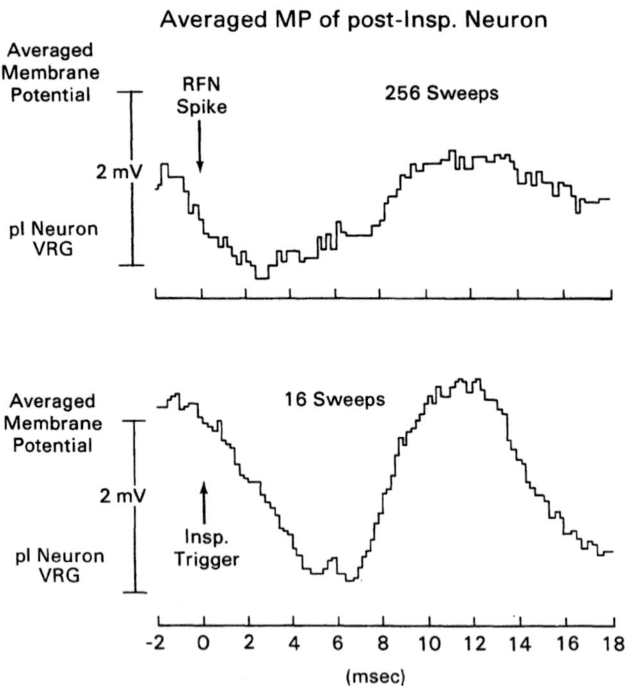

Figure 1. Upper panel: An example of a hyperpolarizing wave in the
           MP of a PI neuron averaged with respect to the spike of an
           early inspiratory neuron in the RFN. Lower panel: MP of
           the same neuron was averaged using onset of phrenic dis-
           charge as the reference event.

That the firing of RFN neurons occurs non-randomly with respect to
the HFO in MP of the VRG neuron is shown for this neuronal pair by
the distribution histogram shown in Figure 2. The occurrence of APs
in the RFN unit is registered with respect to time preceding and following peak depolarization in the HFO. APs occur preferentially
during an interval beginning 2 msec before and extending to 6 msec
after peak depolarization, indicating that the EI unit fires principally during the repolarizing portion of the HFO in the PI neuron.

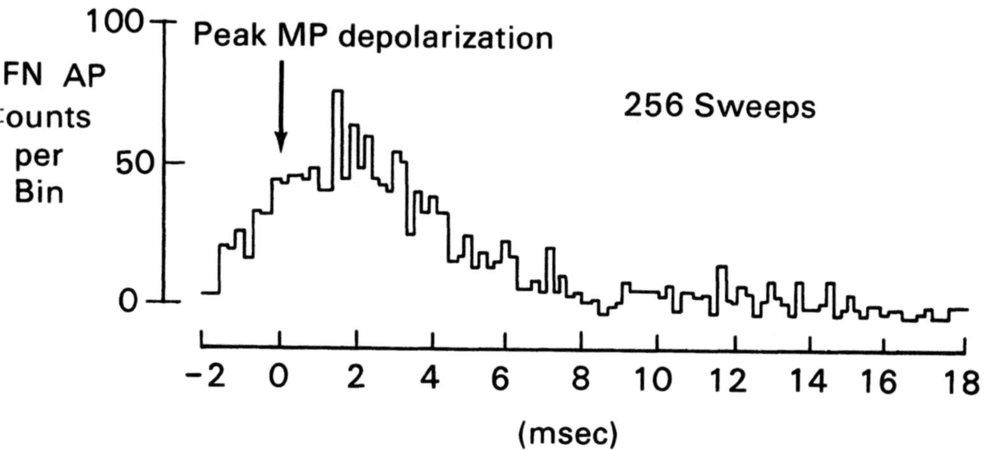

Figure 2.   The distribution of spikes from the RFN unit shown in
Figure 1 using peak depolarization of the VRG as the
reference event.

The distribution of these correlations amongst 4 categories of
respiratory neurons is shown in Figure 3. The highest incidence of
correlation occurred between EI spike of the RFN and membrane
potential of PI neurons of the VRG. Because of the large contribution
of HFO to the wave forms such as shown in Figure 1, the results
cannot be interpreted as revealing a synaptic projection from EI units

of the RFN to PI neurons of the VRG. Rather, they document the
phase relation between the former and the latter.

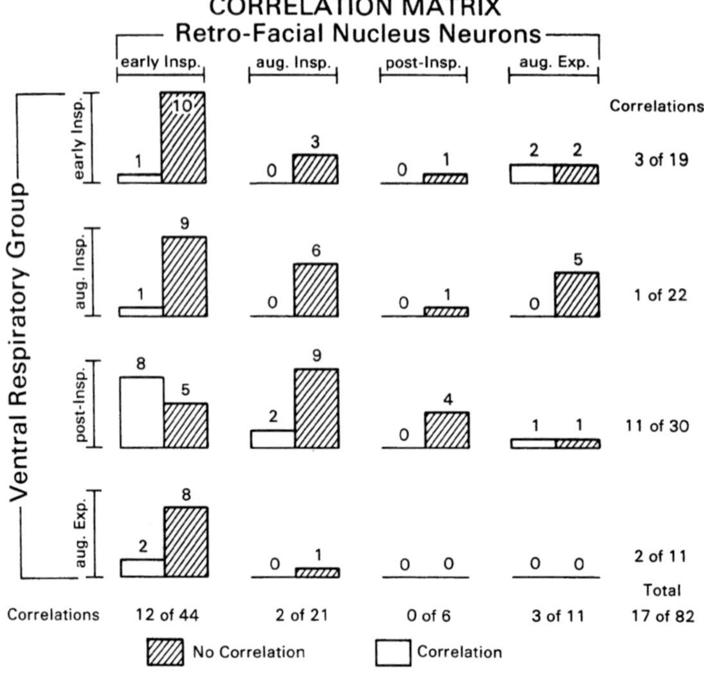

Figure 3. A correlation matrix providing the results of post-RFN
spike averages of MP of VRG respiratory neurons. Four
types of neurons are identified in the RFN and VRG
recording sites. Hatched bar indicates that no wave
was observed in the averaged MP; open bar indicates
that a hyperpolarizing wave appeared.

REFERENCES

1. BIANCHI, A.L. and BARILLOT, J.C. (1982). Respiratory neurons in
   the region of the retrofacial nucleus : pontile, medullary, spinal
   and vagal projections. Neurosci. Lett., 31, 277-282.

2. MITCHELL, R.A. and HERBERT, D.A. (1974). Synchronized high
   frequency synaptic potentials in medullary respiratory neurons.
   Brain Res., 75, 350-355.

16

Role of the Dorsal and Ventral Respiratory Groups: Generation of Spatiotemporal Patterns of Motoneuronal Activity

J. L. FELDMAN, D. R. McCRIMMON and D. F. SPECK

Airflow is produced by synergistic movements of the diaphragm, intercostal and abdominal muscles. The complex respiratory mechanics require an exquisitely coordinated spatiotemporal pattern of motoneuronal activity to achieve appropriate respiratory movement. This coordination could be achieved by: 1) a homogeneously distributed descending respiratory drive that is sculpted by segmental reflexes and segmentally distributed tonic drives, or 2) a differential distribution of the descending drive to the different motoneuron pools, where it could be further modified by segmental reflexes or tonic drives.

The likelihood of an inhomogeneous respiratory drive to motoneurons is suggested by the fact that at least two distinct brainstem groups, the DRG and VRG, contain respiratory modulated bulbospinal premotor neurons. Different functional roles for DRG and VRG neurons is suggested by differences in:

1) Afferent inputs. DRG neurons are located adjacent to the solitary tract. Some of these neurons receive monosynaptic inputs from respiratory related visceral afferents in the tract (1,2,3,4).

Similar direct inputs to VRG neurons have not been observed. Thus, DRG bulbospinal neurons receiving monosynaptic afferent inputs represent a more direct reflex pathway to motoneurons than VRG bulbospinal neurons. This suggests that the DRG and VRG have different roles in the mediation of reflex responses.

2) Descending projections. Individual DRG-VRG inspiratory bulbospinal neurons have limited and specific projections to phrenic and intercostal motoneurons (5,6). This suggests that the differential control of the diaphragm and intercostal muscles arises, in part, from a topological organization of descending DRG-VRG inputs. We have studied descending projections from the VRG using both anatomical and electrophysiological techniques. To examine the spinal connections from inspiratory neuronal regions of the VRG, anterograde transport of tritiated amino acids was used (7). Small volumes (20-50 nl) of tritiated amino acids were pressure ejected into the VRG through a micropipette which permitted simultaneous recording of respiratory-modulated activity. Standard processing procedures for autoradiography were used. Transport of tritiated amino acids revealed a marked bilateral projection to lamina IX at the C4-C6 level and a primarily contralateral projection to lamina VIII and IX in the thoracic spinal cord. Distinct descending pathways to the phrenic motor nucleus were observed in the lateral funiculi and in the ventral funiculi; descending fibers to the intercostal motoneurons in the spinal cord appeared restricted to the ventral funiculus. Labelling of axon terminals in both the cervical and thoracic cord was confined to ventral horn regions which contain primarily motoneurons. These results suggest the hypothesis that spinal interneurons in lamina I-VII are not important in processing the descending respiratory drive

from the brainstem to spinal motoneurons.

The almost exclusive C4-C6 projection to the phrenic motor nucleus (as well as to the T1-T12 intercostal motor nuclei) seen in these studies strongly suggests a monosynaptic projection from the brainstem premotor neurons to these motoneurons. In most limb motor systems (e.g. locomotion), the supraspinal input to the spinal motoneurons is mediated by spinal interneurons which integrate afferent and intrinsic spinal inputs with the descending command. These interneurons are commonly located in lamina V-VII (8). Thus, in limb motor systems, motoneuronal discharge produces movements that activate mainly somatic afferents. These afferents enter the spinal cord within proximal segments and can make direct contact with the spinal interneurons which also receive descending commands. In contrast, respiratory movements activate both somatic and visceral afferents. The latter enter the nervous system via cranial nerves and are especially important in the control of respiratory pattern. In the brainstem these afferents can make monosynaptic connections with the interneurons which conduct the descending commands **directly** to spinal respiratory motoneurons. Although both limb and respiratory muscles utilize interneurons to integrate the centrally generated pattern with afferent feedback, these systems differ in the location of (immediately) pre-motor interneurons (spinal cord vs. brainstem).

In order to determine the distribution of outputs from individual neurons within the DRG-VRG, we have used the technique of unit-nerve cross-correlation analysis (9). Discharges of single DRG-VRG neurons were recorded with tungsten microelectrodes. Simultaneous recordings were also made of the ipsi- and contra-lateral phrenic nerves and a contralateral rostral (T2-4), intermediate (T6-T7) and caudal

(T10-T11) external intercostal nerve. Preliminary analysis indicates

that each medullary inspiratory neuron has a distinct projection to

these nerves. Some neurons exhibit correlations suggestive of a

direct input with only one nerve while others appear to project to two

or more segments. Correlations from some neurons suggest that an

individual unit may project to both the phrenic and intercostal

nerves.

The organization of descending projections may indicate a mechanism

for preprogramming of the descending respiratory drive as a

feedforward mechanism for control of movement. This bulbospinal

control is modified by cranial nerve afferent activity with subsequent

fine tuning by segmental and suprasegmental reflexes. [Supported by

N.I.H. Grant HL-23820. D.R.M. is a P.B. Francis Fellow of the

Puritan-Bennett Foundation.]

REFERENCES
1). Berger, A.J. and Averill, D.B. (1983). Projections of single
pulmonary stretch receptors to the solitary tract region. _J.
Neurophysiol_. 49: 819-830.
2). Camerer, H., Richter, D.W., Rohrig, N. and Meesmanan, M.
(1979). Lung stretch receptor inputs to R(beta)-neurones: a model
for "respiratory gating". In: _Central Nervous Control Mechanisms in
Breathing_ ed. by C. von Euler and H. Langercrantz,. Oxford:
Pergamon Press, p. 261-266.
3). Donoghue, S., Felder, Jordan, D. and Spyer, K.M. (1984). The
central projections of carotid baroceptors and chemoreceptors in the
cat: a neurophysiological study. _J. Physiol._ (London) 347:
397-409.
4). Donoghue, S., Garcia, M. Jordan, D. and Spyer, K.M. (1982).
Identification and brain-stem projections of aortic baroceptor
afferent neurons in nodose ganglia of cats and rabbits. _J. Physiol._
(London) 322: 337-352.
5). Hilaire, G. and Monteau, R. (1976). Connexions entre les
neurones inspiratoire bulbaires et les motoneurones phreniques et
intercostaux. _J. Physiol.,_ Paris 72: 987-1000.
6). Kirkwood, P.A., Sears, T.A. and Stagg, D. (1974). Synchronized
firing of respiratory motoneurones during spontaneous breathing in the
anesthetized cat. _J. Physiol._ (London) 239: 11-13.
7). Feldman, J.L. and Speck, D.F. (1983). Efferent connections of
the ventral respiratory group in the medulla of cats: An
Autoradiographic Study. _Soc. Neurosci. Abs_. 9: 362.

8). Nyberg-Hansen, R. (1956). Sites and mode of termination of reiculo-spinal fibers in cat.   An experimental study with silver implegnation methods. J. Comp. Neurol. 124: 71-100.
9).   McCrimmon, D.R., Speck, D.F. and Feldman, J.L.   (1983). Projections of medullary inspiratory neurons to phrenic and intercostal motor nerves assessed by cross-correlation analysis. Soc. Neurosci. Abs. 9: 362.

Role of the Dorsal and Ventral Respiratory Groups: Generation of Respiratory Rhythm

D. F. SPECK, D. R. McCRIMMON and J. L. FELDMAN

Since the pattern of respiratory motoneuronal activity in the paralyzed, decerebrate, decerebellate animal is similar to that of the intact mammal, one concludes that a respiratory pattern generator is contained within the brainstem and spinal cord . The presence of a respiratory oscillator in the isolated brainstem is demonstrated by the observation that transection at the spino-medullary junction abolishes respiratory movements of all the respiratory muscles innervated by spinal motoneurons (diaphragm, intercostals, abdominal and accessory muscles), while the periodic respiratory activity of cranial motoneurons (laryngeal, pharyngeal, facial, trigeminal, and hypoglossal) persists (1,2,3).

The dorsal and ventral respiratory groups (DRG and VRG) have been proposed as sites for respiratory rhythm generation. However, three studies published by this laboratory make a strong case that structures other than the DRG and VRG are involved in respiratory rhythmogenesis. First, cross-correlation analysis of the interactions between neurons in the DRG and VRG showed that approximately 17% of the neuronal pairs had correlated activity (4). The correlation characteristics suggested that DRG-VRG neurons were driven by common inputs. Very few (3/366) correlation troughs indicative of unidirectional inhibitory interactions were observed. The absence of inhibitory interactions hypothesized to underlie rhythm generation and the predominance of shared inputs indicate that respiratory rhythm generating mechanisms may lie outside the DRG-VRG.

In a second study (5), successive lesions within the DRG and VRG of vagotomized cats did not consistently alter respiratory rhythm. Small lesions within these areas produced modest reductions in the magnitude of the phrenic nerve discharge; however successively larger lesions clearly decreased respiratory neural outflow in both phrenic and recurrent laryngeal nerves. Extensive bilateral destruction of the DRG-VRG did not change the average duration of inspiration although there was an increase in the variance of respiratory timing and amplitude. These results suggest that the DRG-VRG is not necessary for respiratory rhythmogenesis although these areas do play an important role in determining the depth and stability of respiratory outflow.

Although DRG-VRG neurons are unlikely to be necessary for the production of respiratory rhythmicity, the microlesion study was not designed to determine if DRG-VRG bulbospinal neuronal activity affects respiratory rhythm. To test this possibility we used a paradigm based on the experimental observation in invertebrates that synchronous activation of a component of a rhythm-generating circuit will phase shift or reset the rhythm. Although it is unlikely that perturbation of a single neuron in the cat central respiratory pattern generator would have a measureable effect on respiratory neural outflow, a synchronized perturbation of a large portion of the generator should alter the output pattern. By taking advantage of the concentration of descending inspiratory premotor axons in the upper cervical spinal cord, it was possible to synchronously activate large numbers of DRG-VRG bulbospinal neurons (6). Such antidromic activation should produce orthodromic spikes in medullary axon collaterals of bulbospinal neurons. If such collaterals exist and project to neurons in the circuits generating either respiratory timing or the augmenting inspiratory discharge, then a discernible change in pattern should result from synchronous activation. Bilateral spinal cord stimulation was performed at frequencies between 10-300 Hz and at intensities up to 150 uA. Continuous trains of stimuli or gated trains delivered every fourth inspiratory or expiratory cycle had little or no effect on the timing of respiratory cycles. Similarly, brief trains of bilateral spinal cord stimulation delivered at various delays after the onset of inspiration had only a transient effect (less than 60 msec) on the pattern of phrenic nerve discharge. From this study we conclude that the bulbospinal inspiratory neurons are not responsible for the generation of respiratory timing signals and play, at most, a limited role in the generation of the augmenting central inspiratory activity.

Results of these three studies are not consonant with the hypothesis that the DRG or VRG are necessary for respiratory rhythmogenesis. However, it is likely that the neurons in these regions play a very important role in the production of respiratory pattern by integrating respiratory-related inputs and controlling the spatiotemporal pattern of activity in spinal respiratory motoneurons.

Certainly, their location in the brainstem makes them ideally suited to integrate chemoreceptor and visceral mechanoreceptor cranial nerve afferent activity and respiratory rhythmic inputs to produce the appropriate spatiotemporal pattern of descending input to spinal respiratory motoneurons. We are currently investigating both the output pathways from the DRG-VRG and the role of the DRG-VRG in the processing of visceral afferent information. Our studies concerning pulmonary stretch receptor and superior laryngeal nerve inputs to the DRG (7) are included as a separate abstract and will be presented as a poster. Organization of the descending projections from the DRG and VRG will be discussed in the following lecture.

[Supported by N.I.H. Grant HL-23820. D.R.M. is a P.B. Francis Fellow of the Puritan-Bennett Foundation. The current address for D.F. Speck is: Department of Physiology and Biophysics, University of Kentucky, Lexington, KY.]

REFERENCES
1). BINET, L., STRUMZA, M.V. and STRUMZA-POUTONNET, J.M. (1953). Sur la "respiration" experimentalle medullaire. J. Physiol. (London) 45: 41-43.
2). HUKAHARA, T., Jr. (1976). Functional organization of brain stem respiratory neurons and its afferences. In: Respiratory Centres and Afferent Systems ed. by B. Duron, Paris: INSERM, p. 41-53.
3). ST. JOHN, W.M., BARTLETT, D. Jr., KNUTH, K.V. and HWANG, J. (1981). Brain stem genesis of automatic ventilatory patterns independent of spinal mechanisms. J. Appl. Physiol. 51: 204-210.
4). FELDMAN, J.L. and SPECK, D.F. (1983). Interactions among inspiratory neurons in dorsal and ventral respiratory groups in cat medulla. J. Neurophysiol. 49: 472.
5). SPECK, D.F. and FELDMAN, J.L. (1982). The effects of microstimulation and microlesions in the ventral and dorsal respiratory groups in cat. J. Neurosci. 2: 744-757.
6). FELDMAN, J.L., McCRIMMON, D.R. and SPECK, DF. (1984). Effect of synchronous activation of medullary inspiratory bulbo-spinal neurones on phrenic nerve discharge in cat. J. Physiol. (London) 347: 241-254.
7). McCRIMMON, D.R., SPECK, D.F. and FELDMAN, J.L. (1984). Role of the dorsal respiratory group (DRG) in processing vagal and superior laryngeal nerve afferent in cat. Soc. Neurosci. Abs. 10: In Press.

18

Anatomical and Functional Correlates of Dorsal Respiratory Group Neurons

A. J. BERGER, D. B. AVERILL, W. E. CAMERON and T. E. DICK

INTRODUCTION

One reason for investigating brain stem respiratory neurons is a belief that the role of the central nervous system in respiration can be understood if both properties and interactions between individual neurons are known. Implicit in this approach is the hypothesis that the behavior of the respiratory system can be synthesized based upon a neurophysiologic and morphologic understanding of its neuronal elements and their interconnections.

The dorsal respiratory group (DRG) located within the ventrolateral nucleus of the tractus solitarius (vl-NTS) constitutes an aggregation of respiratory neurons that provide drive to phrenic motoneurons and receive inputs from slowly adapting pulmonary stretch receptors (PSRs). Thus these neurons may have a role in the variety of respiratory reflex behaviors resulting from lung inflation-derived and PSR-mediated input. In the anesthetized, paralyzed, artificially ventilated cat, we have studied both function and form of DRG neurons, using inputs from single PSRs as a common thread in these studies.

NEUROPHYSIOLOGIC STUDIES

Extracellular Spike-Triggered Averaging (STA) to Determine the Projection of Single PSRs to the Solitary Tract Region

We mapped the projection of single PSRs to the medulla of cat using STA of extracellular field potentials [1]. Natural spike activity of single PSRs was recorded in-continuity in the nodose ganglion, and this activity provided the trigger for an averaging computer.

Within the medulla, we observed three forms of electrical potentials indicative of PSR central projections: 1) Axonal potentials were recorded when the brain stem electrode was in the vicinity of the afferent axon. 2) Terminal potentials were recorded when the electrode was adjacent to the terminations of the afferent axon. 3) Focal synaptic potentials were recorded when the electrode was near postsynaptic neurons receiving input from the PSR. The central projections of eleven PSRs were mapped, and maxima of terminal potentials in a region 1 mm rostral to the obex were observed in: the medial-NTS (6 PSRs), the vl-NTS (3 PSRs), and the area just dorsolateral to the solitary tract (2 PSRs). Maxima of focal synaptic potentials were recorded in the medial-NTS (2 PSRs), in the vl-NTS (2 PSRs), and in the area just dorsolateral to the solitary tract (1 PSR).

These results showed the presence of both pre- and postsynaptic effects of single PSRs in three different regions about the solitary tract, including the vl-NTS. But this latter region accounted for only a fraction of the input from the PSRs studied. These results are consistent with those obtained with microstimulation by Donoghue et al. [2].

Cross-correlation between PSRs and DRG Neurons
The results of the extracellular STA and microstimulation studies suggest the possibility of monosynaptic connections between PSRs and DRG neurons. To investigate this, cross-correlational analysis of activity from single PSRs and single cells in the DRG was employed [3]. Dorsal respiratory group cells were classified using cycle-triggered histograms derived from the cycle-triggered pumping paradigm [4]. Differences in a cell's firing pattern during lung inflation and that observed when withholding inflation during central I activity were used to classify a cell's behavior. The discharge of Iα neurons is inhibited by lung inflation, whereas the discharge of Iβ neurons is facilitated by lung inflation. Also investigated was a third DRG neuronal type, the P-cell, which is excited by lung inflation, irrespective of central respiratory drive [5].

Activity from 92 DRG neurons and 30 PSRs was recorded. Of the 92 neurons, 23 were classified as P-cells, 19 as Iα, 44 as Iβ neurons, and the remaining 6 I-cells could not be classified because of insufficient testing. Cross-correlation analysis was performed on a total of 92 pairs of units, each pair consisting of a PSR and a DRG neuron. The PSRs studied were only those that exhibited evidence of terminal potentials, with or without focal synaptic potentials in the DRG region. Examples of positive cross-correlation histograms for both P-cells and Iβ neurons are shown in Figure 1. We

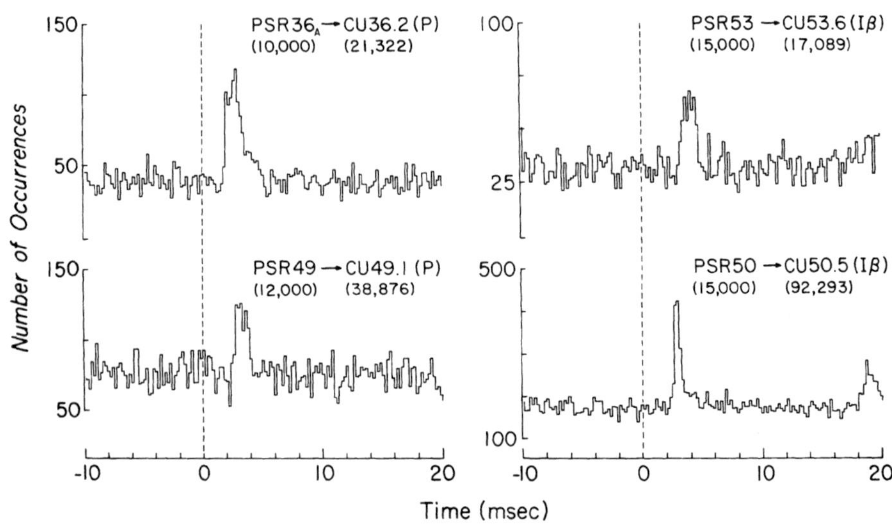

FIGURE 1   Four examples of positive cross-correlation histograms between single PSRs and single DRG P-cells (left panels) and Iβ neurons (right panels). The short latency peak in each correlogram indicates monosynaptic excitation from each PSR. Bin width for correlograms was 0.1 ms, and the number of reference (PSR trigger spikes) and correlated events (DRG neuron spikes) indicated in each panel.

observed that 26% of the P-cells studied received monosynaptic excitatory inputs from the PSRs recorded in this investigation. This was revealed by a sharp peak observed at short latency in the cross-correlation histogram. Twenty percent of Iβ neurons (9/44) had cross-correlation histograms also consistent with monosynaptic excitation from PSRs. Both early onset (7/32) and late onset (2/12) Iβ neurons exhibited synaptic input from PSRs. We did not observe, through the use of the cross-correlation method, any evidence for monosynaptic excitation of Iα neurons by PSRs.

Thus through the use of cross-correlation, these studies established that individual PSRs make monosynaptic excitatory connections in approximately the same proportion to P-cells and to Iβ neurons. Our inability to demonstrate monosynaptic connectivity between PSR afferents and single Iα neurons may be a consequence either of insensitivity of the cross-correlation method [6], or because such connectivity does not exist.

The results of the cross-correlation study also bear upon the issue of whether PSRs exert a facilitatory effect upon phrenic inspiratory discharge

[7]. This may be mediated by a pathway involving PSR monosynaptic excitation of Iβ neurons, and Iβ neuron monosynaptic excitation of phrenic motoneurons [8].   In this cross-correlational study, all the neural elements
that may subserve this facilitatory reflex were recorded simultaneously. In
5 of 9 cases studied where cross-correlation showed a PSR to monosynaptically excite a given Iβ neuron, the Iβ neuron was used to also provide a
trigger for STA of contralateral C5 phrenic nerve activity.   In these 5
cases, a delayed peak in the STA record suggested monosynaptic excitatory
projections to the phrenic nerve from these Iβ neurons. The overall results
demonstrated directly that PSRs monosynaptically project to Iβ neurons, and
that these latter cells can be shown to monosynaptically excite phrenic
motoneurons. We believe that these results provide a demonstration of the
components responsible for the lung volume-derived, PSR-mediated facilitation of phrenic motoneurons.

Intracellular STA between PSRs and DRG Neurons

To further study the connectivity between PSRs and DRG neurons, we
simultaneously recorded activity from PSRs, and intracellularly from DRG
neurons. The membrane potential of nonspiking DRG respiratory neurons was
averaged using the PSR spike as the trigger event. Figure 2A shows a
unitary EPSP in an Iβ neuron, occurring at short latency from the trigger

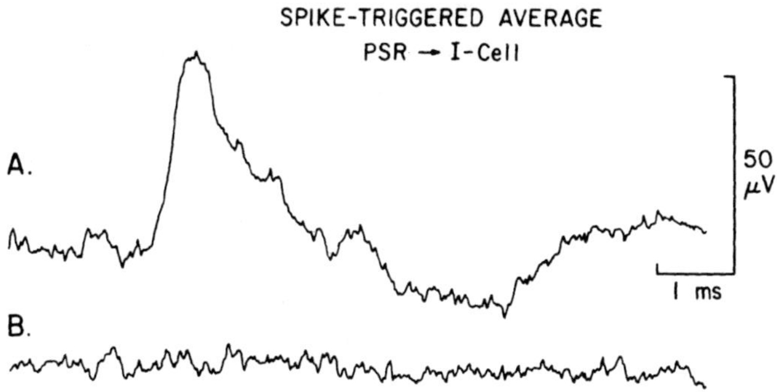

FIGURE 2   Spike-triggered average (STA) of membrane potential of Iβ neuron
derived from 1000 PSR triggers.   STA obtained when cell was not
firing.
A: Monosynaptic unitary EPSP derived from the PSR.
B: Extracellular average obtained immediately after loss of
impalement.   Calibration same in A and B.

(approx. 2 ms). The amplitude of approximately 50 µV, the latency, and the rise-time are appropriate for monosynaptic excitation of this Iβ neuron from the single PSR [9,10]. The record shown in Figure 2B is an extracellular control average; it is flat and indicates that the wave observed in Figure 2A was due to a transmembrane event. Clear evidence for the presence of PSR-derived, unitary monosynaptic EPSPs has been observed in both Iβ neurons and P-cells. As yet, however, EPSPs have not been observed in Iα neurons. Characterization of the unitary EPSPs (n=6 cells) revealed the following range of variables: amplitude, 15-239 µV; latency, 1.25-2.0 ms; 10-90% rise-time, 0.23-0.63 ms; and half-width, 0.66-1.38 ms.

NEUROANATOMICAL STUDIES

DRG Neuronal Morphology

Horseradish peroxidase (HRP) was iontophoretically injected into DRG inspiratory cells that were classified as Iα or Iβ based upon their response lung inflation. The purpose was to determine whether these two DRG neuronal types were morphologically distinct. We described previously the general dendritic morphology and initial axonal trajectory of a mixed population of DRG I-cells [11].

The morphology of 7 I-neurons of each type was analyzed in transverse (2 Iα, 4 Iβ) and horizontal planes of section (5 Iα, 3 Iβ). Geometric variables compared included: somal diameters, surface areas, and perimeters; maximal dendritic lengths and number of primary dendrites; axonal and initial segment diameters. No discernible differences were revealed and the dendritic arbors of these cells were similar. The main dendrites were oriented approximately rostrally and caudally along an axis parallel to the longitudinal axis of the solitary tract, and the dendritic arbors remained almost exclusively within the vl-NTS. Both cell types had dendritic spines and appendages. Their axonal trajectories were similar. The axons proceeded ventrally from the cell body, eventually crossing the medullary midline rostral to the soma. Ipsilateral and ventral to the cell bodies of both types, axonal bifurcations were observed. From these results, it can be concluded that Iα and Iβ neurons do not constitute two morphologically distinct classes of neurons. Their differing responses to lung inflation must arise from differences in afferent connectivity.

Distribution of PSR Central Terminations in the vl-NTS

Because only a small proportion of PSRs terminate in the vl-NTS and the dendrites of DRG I-cells remain almost exclusively within this sub-nucleus, it is of interest to determine the location of PSR terminals at this site. In the course of our studies of DRG I-cell morphology, we were able to inject a PSR central axon residing within the solitary tract region. The PSR axon was identified physiologically by its characteristic firing pattern -- that is, synchronous firing with lung inflations and an absence of a slow wave of depolarizing membrane potential upon which the action potentials were generated -- and ipsilateral vagal electrical stimulation resulted in generation of an action potential without the presence of an EPSP. Following iontophoresis of HRP into the axon, the axon was recovered by the histological procedures [11]. Axonal collaterals and presumed presynaptic terminals (HRP-filled swellings on the fine axon collaterals) were visualized. Figure 3 shows the location of 54 presynaptic terminals (solid circles); all were located within the vl-NTS. Synaptic terminals were oriented along a longitudinal column in the horizontal plane, and lay in a

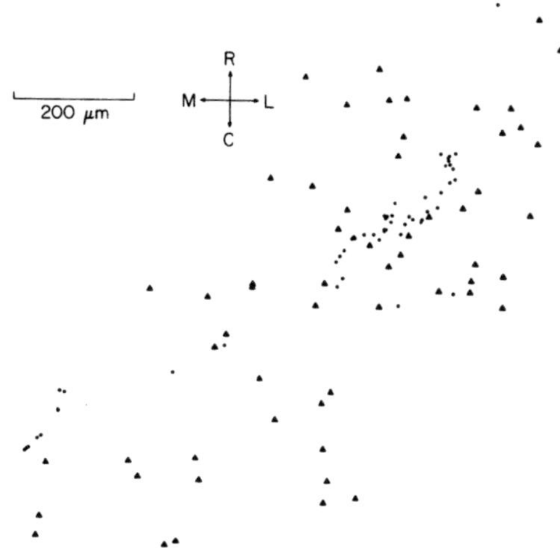

FIGURE 3 Horizontal reconstruction of presynaptic terminals (solid circles) of HRP-filled PSR axon in vl-NTS. Solid triangles indicate location of neuronal cell bodies within this nucleus. Reconstruction derived from six serial 100-µm thick sections beginning just ventral to tractus solitarius and centered just rostral to the obex. Abbreviations: C-caudal; L-lateral; M-medial; R-rostral.

plane parallel to the longitudinal axis of the solitary tract in a region extending from the obex to approximately 1 mm rostral to the obex. The 58 solid triangles shown in Figure 3 are the locations of Nissl-stained cells also within the vl-NTS. These cells probably include both I- and P-cells of the DRG. Therefore, PSR axonal collaterals and presumed presynaptic terminals travel parallel to the dendrites of DRG I-neurons. Thus synaptic contact between these elements appears to be highly efficient.

SUMMARY

The DRG is composed of two types of I-neurons that appear to be morphologically indistinct and may exhibit a continuum of behaviors with respect to their response to lung inflation [3,8,12]. These differing responses are most easily explained by differences in connectivity from PSRs. PSRs do make monosynaptic excitatory contacts with some DRG cells, and the anatomy of the dendritic trees of DRG I-neurons is appropriate for efficient receipt of these contacts. It appears that facilitation of phrenic motoneurons induced by modest lung inflation may be mediated by a disynaptic pathway involving Iβ neurons. Other lung volume-dependent and PSR-mediated reflexes, such as expiratory prolongation with increased functional residual capacity and the shortening of inspiration with tidal volume increases, may involve DRG P-cells and/or other relay sites in the solitary tract region where a majority of PSRs terminate.

ACKNOWLEDGMENTS

This work was supported by USPHS grant NS 14857, and NIH postdoctoral fellowships to W.E.C. and D.B.A.

REFERENCES

1. Berger, A.J. and Averill, D.B. (1983). Projection of single pulmonary stretch receptors to solitary tract region. J. Neurophysiol., 49, 819-302
2. Donoghue, S., Garcia, M., Jordan, D. and Spyer, K.M. (1982). The brainstem projections of pulmonary stretch afferent neurones in cats and rabbits. J. Physiol. London, 322, 353-63
3. Averill, D.B., Cameron, W.E. and Berger, A.J. (1984). Monosynaptic excitation of dorsal medullary respiratory neurons by slowly adapting pulmonary stretch receptors. J. Neurophysiol., 52, 771-85
4. Cohen, M.I. and Feldman, J.L. (1977). Models of respiratory phase-switching. Fed. Proc., 36, 2367-74

5. Berger, A.J. (1977). Dorsal respiratory group neurons in the medulla of cat: Spinal projections, responses to lung inflation and superior laryngeal nerve stimulation. Brain Res., 135, 231-54

6. Cope, T.C., Matsumura, M. and Fetz, E.E. (1982). Effects of single fiber Ia-EPSPs on firing probability of alpha-motoneurons. Neurosci. Abst., 8, 448

7. DiMarco, A.F., von Euler, C., Romaniuk, J.R. and Yamamoto Y. (1981). Positive feedback facilitation of external intercostal and phrenic inspiratory activity by pumonary stretch receptors. Acta Physiol. Scand., 113, 375-86

8. Lipski, J., Kubin, L. and Jodkowski, J. (1983). Synaptic action of Rβ neurons on phrenic motoneurons studied with spike-triggered averaging. Brain Res., 288, 105-18

9. Mendell, L.M. and Henneman, E. (1971). Terminals of single Ia fibers: Location, density and distribution within a pool of 300 homonymous motoneurons. J. Neurophysiol., 34, 171-87

10. Fetz, E.E. and Gustafsson, B. (1983). Relation between shapes of postsynaptic potentials and changes in firing probability of cat motoneurons. J. Physiol. London, 341, 387-410

11. Berger, A.J., Averill, D.B. and Cameron, W.E. (1984).   Morphology of inspiratory neurons located in the ventrolateral nucleus of the tractus solitarius of the cat. J. Comp. Neurol., 224, 60-70

12. Cohen, M.I. and Feldman, J.L. (1984).   Discharge properties of dorsal medullary inspiratory neurons: relation to pulmonary afferents and phrenic efferent discharge. J. Neurophysiol., 51, 753-76

19

Where are the Inspiratory and Expiratory Off-Switch Neurons?

M. I. COHEN, J. L. FELDMAN and D. SOMMER

<u>Inspiratory-Expiratory Phase-Switching</u>

The mechanisms leading to the abrupt termination of inspiration (I) have preoccupied many investigators. The discovery by Baumgarten & Kanzow (1958) of the "R-beta" neurons (I neurons that are also excited by lung inflation, later called "I-beta") led to the hypothesis that these neurons are inhibitory interneurons in the Breuer-Hering reflex arc and are responsible for the termination of I. The subsequent discovery of late-onset I-beta neurons (Cohen & Feldman, 1977; Baker & Remmers, 1980; Marino <u>et al</u>., 1981) resulted in a modified hypothesis stating that these late-onset neurons are responsible for the I off-switch. This hypothesis was based on the observations that: (a) such neurons start firing a short time before the end of I; (b) they are excited by lung inflation, which at the same time shortens the I phase; (c) many of them have the same peak frequency at the end of short I phases (when inflation is applied) and at the end of long I phases (when inflation is not applied).

In a study of dorsal respiratory group (DRG) I neurons (Cohen & Feldman, 1984), we related two properties of each neuron's discharge: (a) response to inflation; (b) pattern of crosscorrelation to phrenic nerve discharge.

Experiments were conducted in decerebrate, paralyzed cats that were ventilated with a cycle-triggered pump system (Feldman & Gautier, 1976), in which lung inflation is applied during the time of central I activity, monitored by phrenic discharge. The influence of lung afferent input was evaluated by withholding inflation during one I phase. This produced several types of response of DRG I neurons, which were classified as: <u>inflation(+)</u>, excited by inflation; <u>inflation(0)</u>, not affected by inflation; <u>inflation(-)</u>, depressed by inflation. The inflation(+) neurons, which correspond to the earlier category of I-beta neurons, fell into two groups, early-onset and late-onset.

For each neuron, the crosscorrelation histograms (CCH) of spike activity to ipsilateral and contralateral phrenic discharge were computed. Many of the CCHs had high-frequency oscillations (HFO); and in addition there was often a sharp short-latency (ca. 2 msec) peak superimposed on the HFO in the unit-(contralateral phrenic) CCH. On the basis of accepted interpretations (Kirkwood, 1979), we conclude that such correlation peaks indicate monosynaptic excitation of some contralateral phrenic motoneurons by the DRG neuron.

The distribution of these correlations among the neuron groups classified by inflation response was not uniform. The majority of the inflation(+) and inflation(0) neurons had correlation peaks indicating that they monosynaptically excited contralateral phrenic motoneurons; and there was no difference between these two groups with respect to incidence of the peaks. Thus, though the two types of neuron differed with respect to inflation response, they were indistinguishable with respect to unit-phrenic crosscorrelations. Moreover, about half of the late-onset inflation(+) neurons had such peaks. Similar results have been found (Lipski _et al._, 1983; Fedorko _et al._, 1983) using spike-triggered averages of phrenic motoneuron membrane potentials, which showed that monosynaptic EPSPs were produced in phrenic motoneurons by both alpha and beta DRG neurons.

These results cast doubt on the hypothesis that the I-beta neurons are responsible for termination of I. It is hard to imagine that neurons whose action is excitatory to I (phrenic) motoneurons would also have an I-inhibitory action (termination of I). Yet, in our sample half of the late-onset inflation(+) neurons did not have correlation peaks. It is therefore conceivable that these neurons actually comprise several distinct populations: some may be involved in I termination, while others may be involved in transmission of excitation from pulmonary afferents to phrenic motoneurons.

Expiratory-Inspiratory Phase-Switching

The mechanisms of the expiratory (E) off-switch have been relatively little studied. Our recordings from both rostral and caudal medullary E neurons having augmenting patterns show that the activity of these neurons terminates abruptly about 50-100 msec before the onset of phrenic discharge. It is known that the rostral (Botzinger) E neurons inhibit DRG I neurons (Merrill _et al._, 1983) and it is likely that this action is also exerted on other types of I neuron. Therefore, the onset of I discharge could be due to disinhibition when E neuron activity stops. However, one would still have to explain the mechanism of termination of E activity.

Some relevant information was obtained in experiments where the lungs were kept inflated during the E phase. When the inflation, which had lengthened E phase duration, was terminated, there was usually a delay of several hundred milliseconds before E neuron activity started to fall. This delay is interpreted as the mobilization time for the E-I switching process. Since the discharges of some expiratory-inspiratory neurons increase markedly at this time, these neurons may comprise the putative E off-switch population.

References

Baker, J. P., Jr. & Remmers, J. E. (1980). <u>Brain Res.</u> 200, 331-340.

Baumgarten, R. von & Kanzow, E. (1958). <u>Arch. Ital. Biol.</u> 96, 361-373.

Cohen, M. I. & Feldman, J. L. (1977). <u>Federation Proc.</u> 36, 2367-2374.

Cohen, M. I. & Feldman, J. L. (1984). <u>J. Neurophysiol.</u> 51, 753-776.

Fedorko, L., Merrill, E. G. & Lipski, J. (1983). <u>Neurosci. Lett.</u> 43, 285-291.

Feldman, J. L. & Gautier, H. (1976). <u>J. Neurophysiol.</u> 39, 31-44.

Kirkwood, P. A. (1979). <u>J. Neurosci. Meth.</u> 1, 107-132.

Lipski, J., Kubin, L. & Jodkowski, J. (1983). <u>Brain Res.</u> 288, 105-118.

Marino, P. L., Davies, R. O. & Pack, A. I. (1981). <u>Brain Res.</u> 219, 289-306.

Merrill, E. G., Lipski, J., Kubin, L. & Fedorko, L. (1983). <u>Brain Res.</u> 263, 43-50.

20

Interactions among Medullary and Pontine Respiratory Neurons

B. G. LINDSEY, L. S. SEGERS and R. SHANNON

Lesioning studies have demonstrated that the respiratory rhythm is generated within the brainstem and that connections between the pons and the medulla must be intact for the generation of eupneic breathing in the decerebrate or anesthetized vagotomized cat. The nature of these connections is not well understood. Interactions between respiratory neurons of the rostral pons (n. parabrachialis medialis, Kölliker-Fuse n.) and the ventral respiratory group (VRG; n. retroambigualis, n. ambiguus, retrofacial n.) were investigated (1) because of neuroanatomical and electrophysiological evidence for such connections (2,3,4; references therein). Preliminary data from a study of interactions among medullary respiratory neurons within the VRG is also reported.

Decerebrate cats were vagotomized, paralyzed and artifically ventilated. Phrenic nerve and pontine and medullary single neuron activity were recorded extracellularly, amplified and displayed using conventional methods. Cross-correlation analysis of simultaneously recorded spike trains was used to detect and evaluate evidence of functional interactions between neurons. Cycle triggered histograms (CTH) and autocorrelograms were generated for each neuron. Anesthetized (Dialurethane) cats were used in experiments in which recordings were confined to the VRG.

<u>Pontine-Medullary Neuron Interactions</u>

Eighteen (7%) of the 255 cell pairs analyzed showed evidence of short time scale correlations indicative of a functional interaction.

The interpretations of the detected correlations suggest that some cell pairs were correlated due to mono- or paucisynaptic interactions while the others were correlated due to the influence of an unobserved shared input. Three of the positive correlations were characterized by a central peak in the cross-correlation histogram (CCH), 8 by a peak or trough located to one side of the origin, and 7 by the occurrence of multiple peaks and troughs in the absence of obvious primary characteristics. Only 1 of 67 pairs of inspiratory neurons exhibited correlated activity that could be interpreted as evidence of a connection between them. Respiratory cells of the rostral pons and the VRG appear to influence one another throughout the entire respiratory cycle. The interpretations for ten of the thirteen cell pairs for which a mono- or paucisynaptic connection may be postulated involved a projection from a tonically active respiratory neuron. Twelve of the eighteen positive correlations involved neurons whose maximum rates of discharge occurred during different parts of the respiratory cycle.

<u>Medullary Neuron Interactions</u>

One hundred twenty-six pairs of ventral medullary respiratory neurons have also been analyzed. Seventeen percent exhibited short time scale correlations:

<u>Pair*</u>	N	# CORRELATED
I-I	80	17 (excluding high frequency oscillation)
E-E	15	1
IE-I	15	3
IE-E	4	1
I-E	12	0

(tonic; with * I = Inspiratory
overlapping activity)   E = Expiratory
IE = I-E Phase Spanning

The figure shows 4 examples (A-D) of the correlations detected. CTHs for reference and target neurons are shown to the left and right, respectively, of the corresponding CCH. Phrenic activity is also shown in each CTH (dotted line). Row A shows the correlation of 2 I neurons. The reference cell A (+4.0AP,3.5R,4.6D; not antidromically activated from the spinal cord, C3) reaches its peak activity during the inspiratory ramp. The firing probability of the target cell B (+0.5AP,3.5R,

3.9D; bulbospinal) follows a time course similar to phrenic activity.
The CCH shows an increase in B's firing rate 1.5 msec. following A's
spikes. Both the time course of the change in firing probability and
(not shown) spike triggered averaging data are consistent with an excitatory projection from A to B. Row B shows data from an IE-I neuron
pair recorded on a single electrode (+4.0AP,3.3R,5.0D). Analysis of
spike waveforms indicates that the sorting artifact cannot account for
the prolonged decrease in the I cell's firing rate (greater than 21
msec. in duration). The data are consistent with inhibition or dis-
facilitation of B by A. Row C shows another correlation suggesting
an inhibitory interaction. Cell A is a propriobulbar decrementing I
neuron (+0.5AP,3.5L,3.4D). Cell B is a propriobulbar IE phase spanning neuron (+4.5AP,2.9L,4.2D). The CCH in row D shows that cell B,
an E neuron (+4.5AP,3.3L,3.8D) has an increased firing probability following action potentials of cell A, an IE phase spanning cell (+4.5AP,
3.8L,3.8D). A periodicity in the autocorrelogram of cell B is mapped
on the right side of the CCH when computed with larger bin widths
(data not shown), suggesting an excitatory effect of A on B.

Our data provide evidence for interactions among various types of
pontine and ventral medullary respiratory neurons (cf. 5,6,7) and have
numerous implications for models of respiratory phase switching.

Supported by BRSG 5/S07/RR05749 and NS 19814 from NIH, and by The
Araya Fund.

References
1. Segers, L.S., Shannon, R., and Lindsey, B.G. (1983). Short time
scale correlations between pontine and medullary respiratory neurons.
Soc. Neurosci. Abst. 9, 1161
2. Bianchi, A.L. and Barillot, J.C. (1982). Respiratory Neurons in
the region of the retrofacial nucleus: Pontile, medullary, spinal and
vagal projections. Neurosci. Lett. 31, 277-282
3. Denavit-Saubie, M. and Riche, D. (1977). Descending input from
the pneumotaxic system to the lateral respiratory nucleus of the medulla, an anatomical study with the horseradish peroxidase technique.
Neurosci. Lett. 6, 121-126
4. St. John, W.M. and Bianchi, A.L. (1983). Comparison of activities
of medullary respiratory neurons in eupnea and apneusis. Respir.
Physiol. 51, 361-377

5. Feldman, J.L. and Speck, D.F.   (1983)   Interactions among inspiratory neurons in dorsal and ventral respiratory groups in cat medulla. J. Neurophysiol. 49, 472-490

6. Long, S.E. and Duffin, J.   (1984).   Cross-correlation of ventrolateral inspiratory neurons in the cat.   Exp. Neurol. 83, 233-253

7. Hilaire, G., Monteau, R. and Bianchi, A.L.   (1984).   A cross-correlation study of interactions among respiratory neurons of dorsal, ventral and retrofacial groups in cat medulla.   Brain Res. 302, 19-31

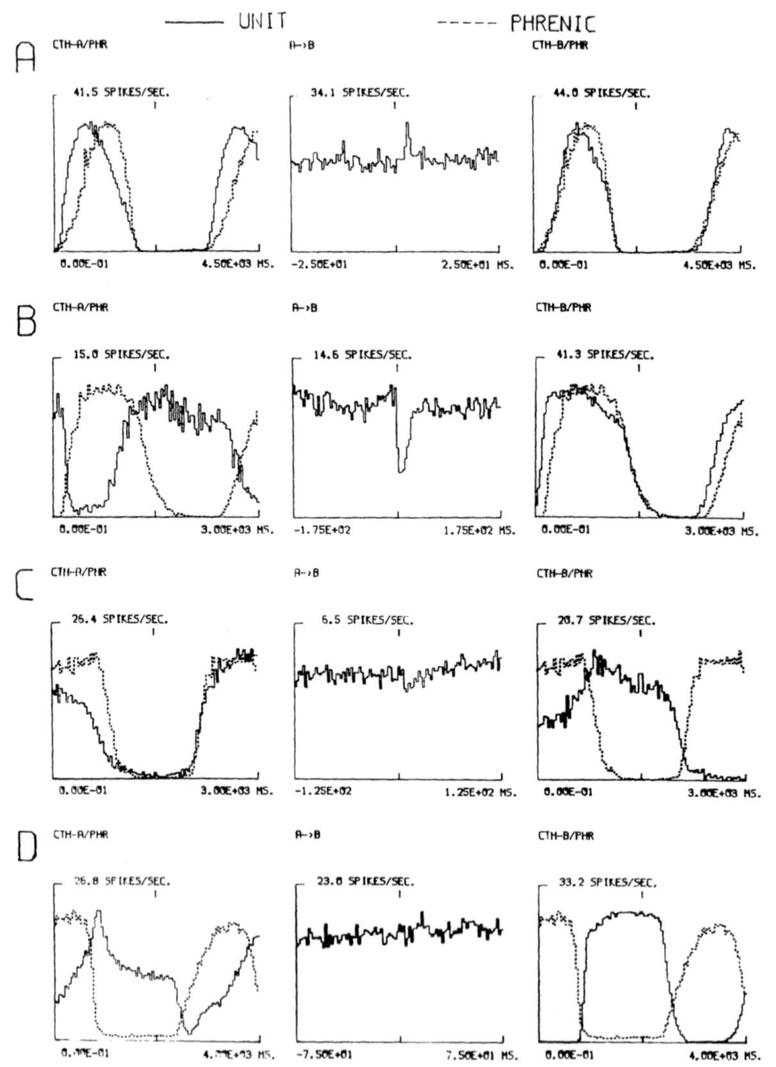

FIGURE: Interactions between VRG neurons.   See text.

21

Interconnections between Central Respiratory Nuclei in the Cat. An Ultrastructural Study

D. RICHE, M. DENAVIT-SAUBIE and J. CHAMPAGNAT

Until now physiological studies have brought to light numerous information about the different neuronal groups involved in respiratory rhythmogenesis, but it is only recently that anatomical and histochemical methods have tried to give more details about their morphological and functional organization. These nuclei are : the parabrachialis medialis (PBM) and Kölliker-Füse nuclei (also named "pneumotaxic center")[2], the ventrolateral nucleus of the solitary tract (or dorsal respiratory group : DRG)[3,9,11], the ambiguus and retroambigualis nuclei (or ventral respiratory group : VRG)[7,11,12,14] and the region of the retrofacial nucleus[12,13].

In 1977 following recordings of respiratory activity, after injections of horseradish peroxidase (HRP) at the level of the VRG, we described[7] a direct projection from the PBM to the VRG, the HRP-reaction product being retrogradely transported. In a more detailed study, using semithin sections we have observed that the labeled cells in the PBM appeared heterogeneous : large pyramidal or multipolar neurons (60/30 um in diameter) as well as small ones (20/15 um) were frequently labeled. Small labeled neurons were more obvious under the electron microscope.

In the present paper after a more documented study, using electron microscopy, we report another connection between the VRG and the PBM, because anterograde transport of HRP, from the VRG to the PBM was also observed. HRP-labeled boutons were seen in the neuropil of the PBM

145

nucleus. Generally these boutons were of medium size (2-3 um) with pleomorphic vesicles and mitochondria. Interestingly a labeled bouton in synaptic contact with a labeled neuron was found (fig. 1A). This observation leads to two questions : 1) Is the labeled terminal the direct projection of an ambiguus neuron projecting to the PBM ? or 2) Is it an axon collateral of the retrogradely labeled PBM neuron ? As HRP is transported anterogradely as well as retrogradely, the two phenomena might have occurred in a same neuron. This might be a recurrent collateral presumably involved in rhythmogenesis[1,5,15,16]. It can be thought that the two possibilities might coexist because spread labeled terminals were more frequent than the monosynaptic pathway. Thus, the anterograde HRP transport demonstrates reciprocal connections between the VRG and PBM and allows to strongly favor a widespread pattern of recurrent collaterals between distant nuclei. In one instance, in the neuropil of the DRG a myelinated axon was also observed dividing at a node of Ranvier to give a collateral branch. Collateral branching has also been observed in the phrenic nucleus[10].

Another study, has been done using kainic (KA) applications and electrolytic lesions in the PBM[8]. New complementary electron microscopic investigations for degenerating boutons in the DRG and the VRG showed that degenerated terminals were less numerous after KA than after electrolytic lesions. According to previous studies[4,17], they were also less numerous in the DRG than in the VRG.

Fig.1. A: An HRP-labeled terminal (arrow) is observed in synaptic contact with a labeled neuronal cell body from the PBM, after HRP injection at the level of the ambiguus nucleus. The dense bodies filled with HRP-reaction product are scattered in the cytoplasm (cyt). Arrow head : another possible labeled terminal.(n = nucleus). B and C : Degenerating terminals (B in the VRG and C in the DRG) showing dark accumulations of lytic material, in a dense matrix. Spherical or pleomorphic synaptic vesicles and synaptic clefts are still visible. (Survival time : 1 week). D : Normal terminal of big size with spherical vesicles and numerous mitochondria, in the DRG, in synaptic contact with a soma (S) and showing "adhesive-like contacts" with other terminals of P-type. E : M-type synaptic bouton with rounded vesicles in the ambiguus nucleus showing "adhesive-like contacts" with   a neuronal soma (S) and one asymmetrical synapse on a somatic spine (arrow). Scale Bar = 1 um.

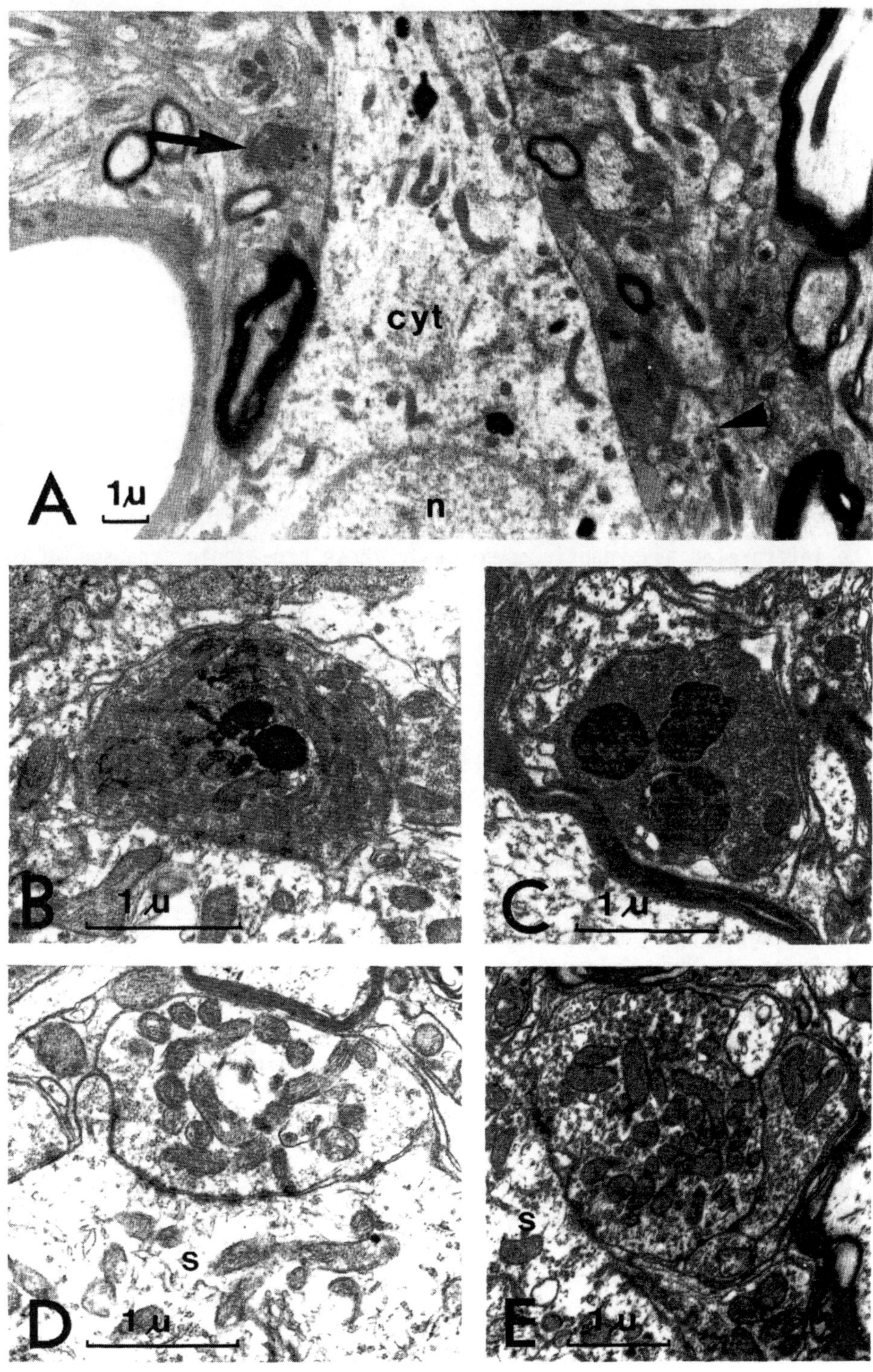
A
1μ
cyt
n
B
1μ
C
1μ
D
s
1μ
E
s
1μ

The characteristics of these different terminals following one week
of survival seem to be close : they are mostly of the dark type, with
an increased density of the axoplasm, rounded or pleomorphic vesicles,
some mitochondria and dark degenerative organelles (fig. 1 B and C).

Until now, in the three different groups of neurons, we have never
observed dendrodendritic appositions, but in the VRG numerous big
normal boutons of the M-type[6,10] containing spherical vesicles and
numerous mitochondria, were found, forming multiple synaptic contacts
(or desmosomes) with the neuronal body and true asymmetrical synapses
on somatic spines (fig. 1E).

In the DRG, similar terminals were observed frequently with juxtaposed P-boutons[6,10], and showing "adhesive-like contacts" between them
(fig. 1D). These contact regions between P- and M-boutons have also
been observed in the phrenic motor nucleus in the rat[10]. They should
be interpreted as synaptic complexes[6]. These axo-axonic synapses in the
DRG, as in the phrenic motor nucleus, may support the theory of presynaptic inhibition.

References

1. Berger, A. J. Federation Proc., 40 : 2378-2383 (1981).
2. Bertrand, F. and Hugelin, A. J. Neurophysiol., 34 : 189-207 (1971).
3. Bianchi, A.L. J. Physiol. (Paris) 63 : 5-40 (1971).
4. Bystrzycka, E.K. Brain Res., 185 : 59-66 (1980).
5. Cohen, M.I. Federation Proc., 40 : 2372-2377 (1981).
6. Conradi, S. Acta Physiol. Scand., suppl. 332 : 5-48 (1969).
7. Denavit-Saubié, M. and Riche, D. Neurosci. Lett., 6 : 121-126 (1977).
8. Denavit-Saubié, M., Riche, D., Champagnat, J. and Velluti, J.C.
   Neuroscience, 5 : 1609-1620 (1980).
9. Euler, C. Von, Hayward, J.N., Marttila, I. and Wyman, R.J. Brain
   Res., 61 : 1-22 (1973).
10. Goshgarian, H.G. and Rafols, J.A. J. Neurocytol., 13 : 85-109(1984).
11. Kalia, M. Federation Proc., 40 : 2365-2371 (1981).
12. Kalia, M., Feldman, J.L. and Cohen, M.I. Brain Res., 171 : 135-141
    (1979).
13. Lipski, J. and Merrill, E.G. Brain Res., 197 : 521-524 (1980).
14. Merrill, E.G. Brain Res., 24 : 11-28 (1970).
15. Merrill, E.G. Federation Proc., 40 : 2389-2394 (1981).
16. Richter, D.W., Camerer, H., Meesmann, M. and Röhrig, N. Pflügers
    Arch., 380 : 245-257 (1979).
17. Saper, C.B. and Loewy, A.D. Brain Res., 197 : 291-317 (1980).

The authors wish to thank H. Hryn for her assistance.
This work was supported by Grants from the C.N.R.S., D.R.E.T. and
Fondation pour la Recherche Médicale.

22

Inhibitory Convergence on Medullary Expiratory Neurones. A Case for 'Common Inhibitors'?

D. BALLANTYNE and D. W. RICHTER

We have previously given evidence that at the level of the output
neurones from the medulla the respiratory cycle may be divided into
a three-phase sequence whose relationship to the rhythmic activity
recorded in the phrenic or internal intercostal nerves enables a
distinction to be made in neural terms between inspiration, stage I
expiration (or post-inspiration) and stage II expiration (1, 2).
The terms 'stage I' and 'stage II' may be taken simply as a means of
highlighting the changing character of synaptic activity as it
develops during the course of expiration, but the central notion
behind this usage is that each stage reflects the sequential
activation of different components of the respiratory network (3).
More recently, it has become apparent that the synaptic relations
between phases are not invariant but are capable of considerable
modification both on a cycle to cycle basis and in the longer term,
a feature which certainly emphasizes the integrative capabilities of
the output neurones as sites of synaptic convergence and which also
bears importantly on the wider issue of how stage I expiration is
controlled and co-ordinated with the rest of the cycle.

We have become increasingly encouraged in the view that this
control depends, in part at least, upon interactions between two
systems of inhibitory neurones with parallel outputs to both
inspiratory and expiratory neurones within the medulla. This view

rests in part on an examination (1,2,3,4) and comparison of the time-intensity profiles of synaptic inhibition in these neurones and in part on a consideration of the properties of two particular classes of neurone: the early inspiratory neurone (5,6) and the post-inspiratory neurone (7) of the ventral respiratory group. In essence, we suggest that at least some fraction of both populations is local (intra-medullary) and inhibitory in action and that the parallel nature of this inhibition preserves phase between inspiration and stage I expiration. The properties of early inspiratory and post-inspiratory neurones which peculiarly suit them to the role of 'common inhibitors' have been dealt with elsewhere (1,2,8). The notion of a control sequence, in the sense that events occurring in inspiration influence those occurring subsequently during stage I expiration follows as a natural consequence of the progressively discovered properties of these neurones and their presumed connectivity, but this control aspect of their relationship has not been explicitly considered previously.

Experimental results which provide a basis for this sequence have been incorporated into Fig. 1 which shows in schematic form what we believe to be the significant relationship between these putative inhibitor-target neurone combinations. It is important to emphasize the distinction between indirect inference and experimental observation in the construction of Fig. 1. The key new elements of this sequence which are based directly on experiment are:
1) Early inspiratory neurones do not discharge in a fixed pattern but are capable of variation in a manner which can be related to the prevailing pattern of phrenic nerve activity.
2) Expiratory bulbospinal neurones of the caudal medulla form a point of convergence for two rhythmic patterns of inhibitory synaptic activity: the first related to the inspiratory component of phrenic nerve activity (9) and the second, present under many but certainly not all experimental conditions, related to the post-inspiratory component of phrenic activity (10).

Before considering Fig. 1 in any detail it may be in order to briefly review some of the evidence for the inferred relationships in this sequence. There are two main inferences. The first is that early inspiratory neurones constitute a shared source of inhibitory

input for both expiratory bulbospinal and post-inspiratory neurones.
Comparison of the time-intensity profiles of the inspiratory-related
inhibition of both kinds of neurones reveals so close a similarity
that it is natural to suppose that they have their origin in a
common input source. Further, the similarity of these profiles -
particularly apparent following i.p.s.p. reversal - to the discharge
pattern of early inspiratory neurones makes the latter a highly
plausible (intra-medullary) candidate for this input source (see also
6). The second inference is that some unspecified fraction of the
population of post-inspiratory neurones is inhibitory to expiratory
bulbospinal neurones. This is based on the close similarity shown by
the time-intensity profile of post-inspiratory i.p.s.p.s - again most
apparent following their reversal - to the abrupt onset, decrementing
pattern of discharge shown by many post-inspiratory neurones. This
is a _necessary_ inference since the existence of neurones showing a
post-inspiratory pattern of activation is demanded by the fact of
post-inspiratory inhibition.

These relationships are depicted by the continuous lines in Fig. 1
which illustrate what might be called a 'control' situation in which
the early inspiratory neurone exhibits a rapid onset, decrementing
pattern of inspiratory depolarization and corresponding pattern of
discharge. In this situation, using the pattern of phrenic discharge
as a reference - smoothly rising with no marked change in slope
during the growth phase or a pronounced late-inspiratory peak - the
corresponding polarization of expiratory and post-inspiratory neurones
implies a rapid onset, decrementing pattern of i.p.s.p. arrival.
Fig. 1 is rather idealized in that i.p.s.p. intensity is shown as
decreasing continuously from the start of inspiration. More usually
there is some evidence of a maintained level of inhibition for at
least a brief period. The shape of the expiratory phase membrane
potential trajectory is dependent, _inter alia_, on several aspects of
phrenic nerve discharge. This has been most extensively explored in
expiratory bulbospinal neurones (10). The situation illustrated is
one in which there is a weak pattern of post-inspiratory i.p.s.p.s,
occupying the same fraction of the expiratory interval as the
post-inspiratory component of phrenic discharge and which, in these

neurones, is the distinguishing feature of stage I expiration.
Correspondingly, the lifting of inspiratory inhibition from post-
inspiratory neurones ('disinhibition') is followed by their activation
along a time course comparable to stage I expiration.

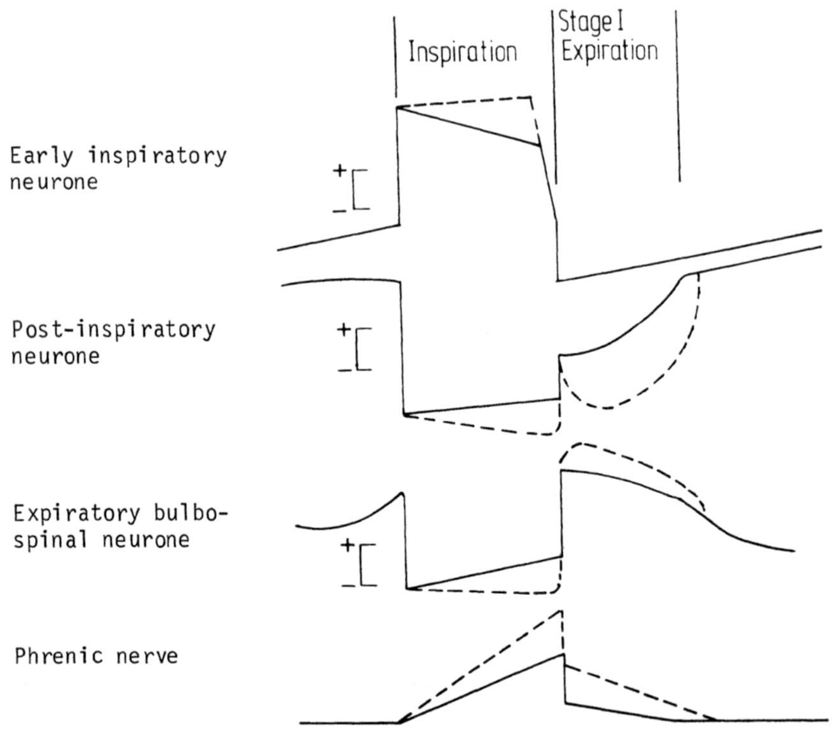

Figure 1   Schematic representation of the changes (continuous versus
dashed lines) in the membrane polarization pattern of an
early inspiratory, post-inspiratory and expiratory bulbospinal neurone in a pentobarbital-anaesthetized, bilaterally
vagotomized, paralyzed and artificially ventilated cat.
This representation is based on results from separate
experiments in which the changes shown were elicited by
electrical stimulation of the superior laryngeal nerve
(low intensity, upto 20 cps for 10-20s. The behaviour of
the post-inspiratory neurone is partly inferential (see
Text) and based on reference 11.

A most important and previously unrecognized fact is that the
discharge behaviour of early inspiratory neurones is not invariant
but modifiable over a fairly wide range with changes in the pattern
of phrenic activity. Now, in order to link this variation to the
behaviour of expiratory and post-inspiratory neurones, some kind of

common reference is required. The most extreme instance of variation
in early inspiratory neurones which we have observed to date is that
elicited by a short period of low intensity stimulation of the
superior laryngeal nerve. The kind of change which occurs is shown
by the dashed line in Fig. 1 and consists in the conversion of the
rapid onset, decrementing pattern of inspiratory depolarization to a
rapid onset, plateau-like or incrementing pattern with a late peak.
This change persists in attenuated form for a variable period beyond
the cessation of stimulation. The critical feature here is that
the pattern of phrenic discharge changes approximately in the manner
shown in Fig. 1. Experimentally, a stimulus regime of this kind
also results in a conversion of the decrementing inspiratory
inhibition of expiratory neurones to a plateau-like or incrementing
pattern. (Doubtless, stimulation of the superior laryngeal nerve
alters the activity of a great many brain-stem neurones but the
relevant point here is the appropriateness of the altered behaviour
of early inspiratory neurones which is exactly of the kind required
to account for the changed profile of inspiratory inhibition of
expiratory neurones). For present purposes, the feature of particular
interest in the response of expiratory neurones to such stimulation
is the large increase in the intensity and usually the time course
(but see below) of their post-inspiratory wave of i.p.s.p.s. This
effect is also seen in response to adequate stimulation of presumed
'irritant' laryngeal receptors (D. Ballantyne, S. Mifflin,S. Backman,
D.W. Richter, unpublished data). Our assumption is that this
intensified and prolonged pattern of post-inspiratory i.p.s.p.s is
the result of a corresponding change in the activation pattern of
post-inspiratory neurones which also results from stimulation of the
superior laryngeal nerve (11). Again, the appropriateness of this
kind of change in the behaviour of post-inspiratory neurones lends
support to the principle assumption in this article that expiratory
bulbospinal neurones represent the common target for post-inspiratory
and early inspiratory neurones.

It is thus possible to envisage a control sequence operating with a
minimum of components in which the trigger event is a modification to
the discharge pattern of early inspiratory neurones. Two of the main
targets for this altered pattern of inhibitory output are also

connected such that increased activity in one target (post-inspiratory
neurone) deepens the inhibition of the other (expiratory neurone).
Since post-inspiratory neurones are themselves inhibited in the later
stage (II) of expiration (1, 11) a functional symmetry is preserved
in the network. This does not of course necessarily imply mutual
inhibition; the projection of Bötzinger expiratory neurones to the
caudal portion of the ventral respiratory group (12) might conceivably
transmit expiratory inhibition to post-inspiratory neurones.

So far as the response to activation of laryngeal afferents is
concerned, it seems not unlikely that the changed activation pattern
of early inspiratory and post-inspiratory neurones is simultaneous
rather than sequential. It should, however, be emphasized that
stimulation of laryngeal afferents simply extends to extremes a range
of variability in behaviour of these neurones which can be encountered
without recourse to such stimulation. What one sees is an entirely
continuous range of variation in intensity and time course of post-
inspiratory neurone activation, post-inspiratory i.p.s.p.s and post-
inspiratory phrenic discharge. Clearly, the most obvious means by
which a modified pattern of inhibitory output from early inspiratory
neurones may exert an influence over subsequent events is by
establishing in post-inspiratory neurones the conditions required
for rebound excitation. Equally clearly, the effect of the latter is
likely to be transient but it takes on at least a potential
importance when it is remembered that the occurrence of post-inspiratory
i.p.s.p.s is by no means restricted to expiratory bulbospinal neurones
but has been described in a number of inspiratory neurones (1,3,4).
Elsewhere (2), we have given reasons for supposing that such
inhibition constitutes the effective element in the irreversible
component of the inspiratory 'off-switch'. Similarly, the inhibitory
output from early inspiratory neurones is probably also not
restricted to the kinds of neurones discussed here. Certainly in
some inspiratory neurones the shape of their depolarizing trajectory
during inspiration is dependent on the occurrence of i.p.s.p.s whose
time-intensity profile, revealed by manipulating the relationship
between i.p.s.p. driving potential and reversal potential, is not too
dissimilar to that seen in expiratory or post-inspiratory neurones
(2,4) at least under the restricted range of conditions in which it

has been recorded. Such inhibition is most readily detected in
neurones of the 'late-inspiratory' type but this may be simply because
they were of this type in the particular circumstances in which they
were recorded. At all events, a shared distribution in the level of
(inhibitory) synaptic noise might reasonably be expected to exert
some synchronizing, as well as a delaying, action on their discharge
but again these kinds of effects might only be detectable in the
sort of situation in which the inhibitory output from early inspiratory
neurones is intensified, i.e. the kind of situation illustrated in
Fig. 1. Whether this is also the kind of situation ensuring a rapid
progression from reversible or transient inspiratory 'off-switching'
(2) to irreversible 'off-switching' is not clear, at least in
synaptic terms.

It will be evident that the sort of relationships discussed here
are based on a set of comparisons which have probably been pushed
about as far as they will go. The encouraging feature of these
comparisons, however, is that the sort of relationships which
emerge, particularly in response to 'stressing' the system with a
powerful afferent input (10, 11), are internally consistent, have
seemed logical and have tempted us to link what, on the surface,
seem to be rather disparate kinds of events. It should, however,
be emphasized that some events do not fit in a simple way. Two in
particular are worth mentioning. First, the pattern of post-
inspiratory i.p.s.p.s may show a simple, nearly proportional
relationship between intensity and duration (the usual situation)
but at times these features may vary inversely. Comparable variations
in the activation behaviour of post-inspiratory neurones have not
yet been observed. Second, while the relationship between post-
inspiratory phrenic discharge and the synaptic correlates of this
period is in general close, there are a sufficiently large number
of occasions to require explanation when these two kinds of events
may be wholly dissociated.

Supported by the DFG

References

1.   Richter, D.W. (1982). J.Exp. Biol. 100, 93-107.

2. Ballantyne, D. and Richter, D.W. (1984). J. Physiol. (Lond.) <u>348</u>, 67-87.

3. Richter, D.W. and Ballantyne, D. (1983). In <u>Central Neurone Environment</u>. eds. Schläfke, M., Koepchen, H.P., See, W.G. pp. 164-174. Springer-Verlag Berlin-Heidelberg.

4. Richter, D.W., Camerer, H., Meesmann, M. and Röhrig, N. (1979) Pflügers Arch. <u>380</u>, 245-257.

5. Merrill, E.G. (1970). Brain Res. <u>24</u>, 11-28.

6. Merrill, E.G. (1974). In <u>Essays on the Nervous System</u>. eds. Bellairs, R., Gray, E.G. pp. 451-486. Oxford: Clarendon Press.

7. Richter, D.W. and Ballantyne, D. (1981). Neurosci. Lett. <u>7</u>, 117.

8. Richter, D.W., Ballantyne, D. and Mifflin, S. (1984). (This volume).

9. Mitchell, R.A. and Herbert, D.A. (1974). Brain Res. <u>75</u>, 345-349.

10. Ballantyne, D. and Richter, D.W. (1984). (Submitted for publication).

11. Remmers, J.E., Richter, D.W., Ballantyne, D., Bainton, C.R. and Klein, J.P. (1984). (This volume).

12. Fedorko, L. and Merrill, E.G. (1984). J.Physiol. (Lond.) <u>350</u>, 487-496.

23

Neuronal Morphology and Synaptic Transmission in the Solitary Complex In Vitro

J. CHAMPAGNAT, K. GRANT, K. F. SHEN
and M. DENAVIT-SAUBIE

<u>INTRODUCTION</u>

The solitary complex (SC) has been subdivided into several subnuclei defined anatomically in coronal sections from rats and cats (9,11,15). The ventral and lateral regions contain the dorsal group of respiration-related neurones. Other subdivisions of the SC are involved rather in cardiovascular or digestive controls. The development of an in vitro preparation of brain stem slices is convenient for both electrophysiological and (see 8) pharmacological studies of SC neurones. However, in these in vitro preparations, the efferent systems are absent and neurones must be identified by morphology or local connectivity. This report summarizes identification criteria based on present data, with a special reference to the ventral and lateral SC. Rat brain stem was selected to differentiate the medial and ventral SC neurones from dorsal vagal neurones identified by antidromic stimulation. Cats were used for study of the respiratory related ventrolateral SC subnucleus.

<u>METHODS</u>

<u>The brain stem coronal slice preparation</u>

Male Wistar rats weighing 100-200 g or 10-20 days old kittens were anaesthetized with halothane or ether, craniotomized and decapitated at the upper cervical spinal cord. During these surgical procedures, cold (6°C) artificial cerebrospinal fluid (ACSF) was dripped on to exposed brain surfaces. The composition of the ACSF (in mM) was : NaCl, 124 ; KCl, 5 ; $MgSO_4(7H_2O)$, 2 ; KH_2PO_4, 125 ; $NaHCO_3$, 26 ; $CaCl_2$, 2 ; glucose, 10. The ACSF was bubbled with 95 % CO_2 in all cases. After transcollicular section, the brain stem was removed, immersed in cold ACSF and separated from the cerebellum. Coronal slices (350-500 um thick) were cut on a tissue chopper, placed in cold ACSF and transfered to warm (31-37°C) ACSF in the recording chamber. The upper surfaces of the slices were first exposed to warm humidified 95 % O_2 5 % CO_2 for 10-45 mn, and after this the slices were completely immersed and continuously superfused at a constant rate (1-2 ml/mn).

Afferent stimulation, intracellular recording and staining procedure

Neurones were recorded in the SC with glass micropipettes filled with K-acetate (40-100 M) for intracellular recording. Concentric bipolar metal electrodes (0.1 mm tip diameter) were used for electrical stimulation. Gross anatomical structures could be identified through a dissecting microscope. Standard recording techniques were used (see 6). Potential transients were averaged using a PDP-8 calculator on-line. Intracellular iontophoresis of horseradish peroxydase (HRP, Sigma, type VI) was performed using depolarizing pulses of 0.6-2 nA, 0.7 s, at 1 Hz for 10-30 min. At least one hour after injection, the slice was fixed for 30 min-1 hour in glutaraldehyde (pH 7.2) and then wahsed in phosphate buffer (pH 7.4) overnight. The HRP was reacted according to the Hanker-Yates method (modified by Roper, see 1).

RESULTS

1. Anatomical location of stimulation and recording sites

In all cases a stimulating electrode was located in the dorsal part of the tractus solitarius (TS). Preliminary experiments using extracellular recording and various stimulation sites have indicated that this location gives rise to orthodromic responses in all SC subnuclei while stimulations at other sites are generally ineffective (fig. 1 left).

A second stimulating electrode was used for antidromic identification of vagal preganglionic neurones. These experiments were performed only in rats, because in this species, axons arising from dorsal vagal preganglionic neurones are clearly separated from vagal afferent fibers

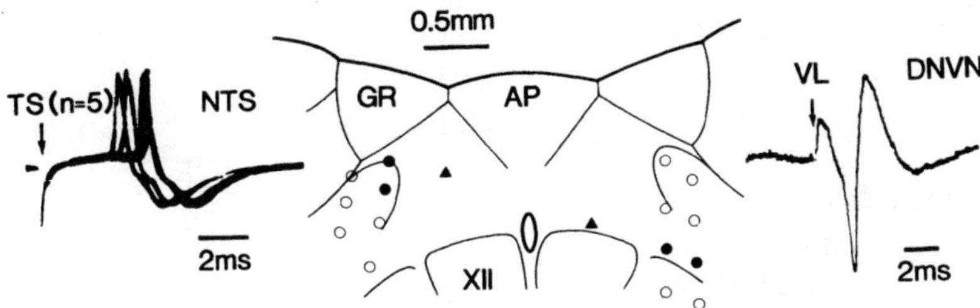

Figure 1 : Exploration of the area of the rat solitary tract with a
bipolar stimulating electrode (circles) while recording extracellularly
(triangles) from the media nucleus tractus solitarius (NTS, left) and
from the dorsal nucleus of the vagus nerve (DNVN, right). Black circles:
position of the stimulating electrode giving rise to a spike response
(illustrated). White circles : positions where the stimulation was not
effective. AP : Area postrema, GR : Gracilis nucleus ; XII : Hypoglossus nucleus ; TS,VL : see text.

and form an intracerebral bundle ventral and ventrolateral to the
solitary tract (3,9). The stimulating electrode was therefore positioned in this ventrolateral position (VL, fig. 1, right).

Recordings were made in all SC subnuclei of the rat. Cats were used
exclusively for recording of ventrolateral SC neurones since electrophysiological studies demonstrating that this nucleus contains the
dorsal respiratory group have mainly been performed in this animal.

2. Electrophysiological responses to stimulations

Preliminary experiments indicated that a large proportion of
neurones in the rat or cat ventrolateral SC exhibit a rhythmic repetitive discharge which can be recorded in stable extracellular recording conditions (5). Some of these were afterwards recorded intracellularly and showed a large after-hyperpolarisation following the
fast component of the action potential.

A. Responses in these following TS stimulation

Coronal brain stem slices maintain excitatory responses to afferent
stimulation (see 2,4,7,13,16). TS stimulation led to an initial
response consisting of a single spike occurring 2-7 ms after the
stimulus (fig. 2). This is a postsynaptic response for the following

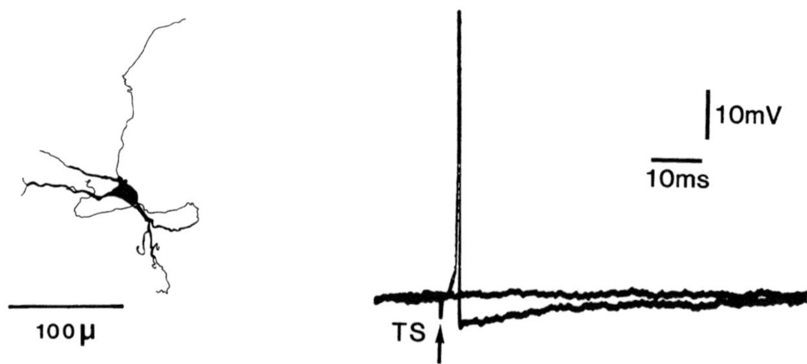

Figure 2 : Left : Morphology of a neurone located in the ventral NTS of the rat. The localization of this neurone with respect to the solitary tract is illustrated in figure 3 (neurone 3). Right : Postsynaptic response following TS stimulation consisting of an EPSP giving rise to a full spike followed by an after-hyperpolarization. The superimposed horizontal trace was recorded without stimulation and gives the resting level of membrane potential (-52 mV).

reasons. 1) The latency of the evoked spike varied by several ms upon slow repetition (less than 0.5 Hz) of the stimulus. 2) The spike was triggered on top of an EPSP-like transient starting 1-3 ms after the stimulus. 3) The EPSP was reduced or abolished after introduction of Mg^{++} (10 mM) in the bathing medium. 4) The EPSP summated with repetitive TS stimulation at 30-50 Hz. 5) The EPSP exhibited an apparent reversal potential of -10 to -30 mV. Membrane depolarisation revealed an IPSP-like transient starting 2-5 ms after the onset of the EPSP.

B. Responses of VPN following VL stimulation

Antidromic action potentials were evoked in VPN after VL stimulation and characterized by different properties. 1) They were able to follow stimulation rates of 500 Hz. 2) The IS and SD components could be distinguished but no subthreshold graded EPSP was found. 3) They collided with a previous orthodromic spike elicited by intracellular depolarisation. 4) They were maintained in the presence of 10 mM Mg^{++} in the bathing medium. 5) After intracellular staining with HRP, the axon could be followed in the direction of the VL area. In one case, an axonal projection was observed close to the lateral border of the hypoglossal nucleus (see fig. 3, neurone n° 1).

3. Different neuronal types in the SC

Variations in the properties of SC neurones were noted from studies
of delayed changes of excitability following TS stimulation and morphological reconstructions were compared.

A) Delayed synaptic responses

Neurones recorded in SC subnuclei could be differentiated according
to their increased or decreased excitability during several hundred
milliseconds following TS stimulation. In only a few cases compound
responses were obtained consisting of an initial inhibition followed
by an excitation.

The mechanisms underlying prolonged "inhibitions" in a first type
of SC neurone have been reviewed previously. These include both the
voltage dependent after hyperpolarization which follows action potentials, and subthreshold mechanisms leading to a decrease in the amplitude of EPSPs when the TS stimulus was repeated 50-200 ms after an
initial conditioning stimulus. Remote dendritic or presynaptic
mechanisms are probably involved in the latter phenomenon (6).

In the second type of neurone, prolonged excitation was related to
both an increase of the synaptic bombardment and slow transient waves
of depolarizations. 1) The synaptic background activity was increased
for several hundreds ms by TS stimulation, giving rise to ripples on
the potential recording. These ripples were recorded with K-acetate
electrodes ; they are EPSPs different from previously described
reversed IPSP observed with KCl filled microelectrodes (2). In
addition, the initial EPSP was followed in these neurones by a depolarization of 1-4 mV and up to 800 ms. This depolarizing wave appeared
as a prolongation of the initial EPSP but could be clearly distinguished from it by its slower time course.

B) Neuronal morphology

SC neurones showed large morphological variations. In all cases,
soma dimensions were smaller than those of dorsal vagal neurones. Larger
neurones, on the one hand, had a medium sized soma that could be spherical or elongated (range of diameters : 15-40 u). In these neurones the
soma gave rise to 2-5 large dendritic trunks. The total extension of
the dendritic tree ranged between 80 and 300 u from the soma (figure 3,

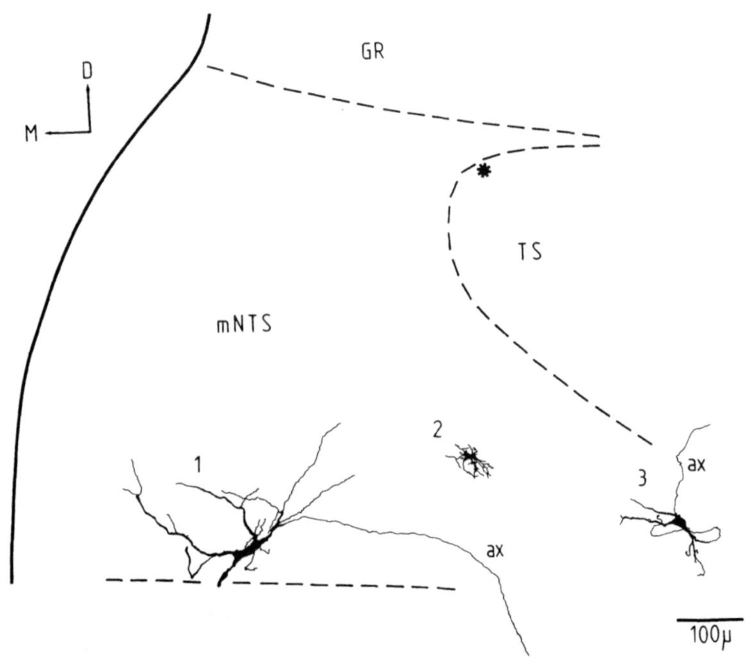

Figure 3 : Semi-schematic illustration of morphology and location of types of neurones identified in the coronal brain stem slice. 1 : Neurone from the dorsal vagal motor nucleus. In this particular case, the axon could be followed along the medial border of the hypoglossus nucleus. 2 : Small neurone from the ventral NTS. 3 : Medium-sized neurone of the ventral SC (illustrated in more details in figure 2). These three neurones were found in different slices. ax : axone. GR : gracilis nucleus. mNTS : medial NTS. TS : tractus solitarius. D : dorsal. M : medial.

neurone n°2 ; figure 2). In these neurones we have observed early EPSPs and delayed "innibitions" in response to TS stimulation. Smaller neurones on the other hand exhibited prolonged excitations. They were multipolar somata with a dense arborization of thin dendrites and no large dendritic trunks. The extension of the dendritic tree was less than 100 u from the soma (figure 3, neurone n°2).

DISCUSSION

These results raise new aspects of synaptic transmission in SC sub-nuclei including the dorsal respiratory group. 1) Of the two classes of neurones found in the SC, the dendrites of the larger neurones may extend over a wide area, covering several subnuclei. Synaptic contacts

with sensory afferents which spread over different parts of the SC
(2,10,14) are therefore probable. In addition, the same SC neurone
may be a site for the integration of afferent inputs of several
different sensory modalities. The latter is a possibility which could
be explored in   in-vivo preparations. 2) A new class of SC neurones
have been identified which do not seem subject to the rate-limiting
controls described in other cells (6,12) and in which sustained post-
synaptic activity and prolonged synaptic potentials are characteristic. This is relevant to the understanding of interaction of reflex
activation with long term or tonic central activities   such as, for
instance the interaction of vagal afferents with the respiratory
rhythm (see 8).

ACKNOWLEDGEMENTS

Supports from Fondation pour la Recherche Médicale and INSERM are
gratefully acknowlege as well as grants from CNRS and DGRST.

REFERENCES

1. Bell, C.C., Finger, T.F. and Russell, C.J. (1981) Exp. Brain Res.
   49 : 9-22.

2. Berger, A.J. and Averill, D.B. (1983) J. Neurophysiol. 49 : 819-830.

3. Cajal, B.R. (1909) Histologie du système nerveux de l'homme et des
   vertébrés, Maloine, Paris.

4. Camerer, R.H., Richter, D.W., Röhrig, N. and Meesmann, M. (1978)
   In : C. Von Euler and H. Lagernientz (eds) Central Nervous Control
   Mechanisms in Breathing, Vol. 32, pp. 261-266 (Oxford : Pergamon).

5. Champagnat, J., Denavit-Saubié, M. and Siggins, G.R. (1983) Brain
   Res. 280 : 155-159.

6. Champagnat, J., Siggins, G.R., Koda, L.Y. and Denavit-Saubié, M.
   (1984) Brain Res., in the press.

7. Cohen, M.I. (1979) Physiol. Rev. 59.

8. Denavit-Saubié, M., Champagnat, J. and Morin, M.P. (1984) this
   volume.

9. Kalia, M. and Sullivan, J.M. (1982) J. Comp. Neurol. 211 : 248-264.

10. Kubin, L. and Davies, R.D. (1984) this volume.

11. Loewy, A.D. and Burton, H.J. (1978). J. Comp. Neurol. 181 : 421-264.

12. Miles, R. (1984) this volume.

13. Pack, A.T. (1981) Ann. Rev. Physiol. 43 : 73.

14. Spyer, K.M. (1982) J. Exp. Biol. 100 : 109-128.

15. Van der Kooy, D., Koda, L.Y., Mc Ginty, J.F., Gerfen, C.R. and
    Bloom, F.E. (1984) <u>J. Comp. Neurol.</u> <u>224</u> : 1-24.

16. Von Euler, C. (1980) <u>TINS</u>, <u>3</u> : 275-277.

24

Synaptic Integration in the Nucleus of the Solitary Tract Studied In Vitro

R. MILES

Afferent fibres from respiratory and cardiovascular receptors synapse with neurons in the nucleus of the solitary tract. A longitudinal medullary slice has been developed to examine synaptic responses of cells in this area to selective stimulation of solitary tract fibres. The resulting intracellular synaptic potentials are profoundly depressed at stimulus frequencies as low as 5 Hz. This effect is not due to changes such as a build up of inhibition, in local circuits within the nucleus, but is instead located at the synapses made by afferent fibres.

Experiments were performed using 400 μm thick tissue slices cut from the guinea pig medulla to contain the longitudinal extent of the solitary tract and its nucleus [1]. In vitro the tract was seen to begin rostral to the obex at the junction of medially directed fascicles from the cranial nerve roots. Bipolar tungsten electrodes were placed at this point to activate solitary tract axons with stimuli of duration 0.1 ms and amplitude up to 10 V. The tract then travels caudally with fibres leaving it to terminate on neurones in the surrounding nucleus. Recordings were made from these cells at levels within 500 μm rostrally from the obex and up to 200 μm medially and laterally from the tract, with electrodes filled with 2 M KAc or CsAc and bevelled to a resistance of 80-120 MΩ.

Recordings were made from more than 50 neurones with resting potential greater than 60 mV and overshooting action potentials.

Membrane time constants were in the range 8-15 ms and input resistances varied between 30 and 80 MΩ. Action potentials either occurred spontaneously at low frequencies or could be elicited by intracellular current injection and were of duration 1.0-1.5 ms. They were invariably followed by an after-hyperpolarization of amplitude up to 20 mV and duration up to 30 ms.

Synaptic responses to solitary tract stimulation were more variable. An epsp was elicited in virtually all cells and when maximally evoked usually led to neuronal firing. It reversed at about 0 mV, following intracellular Cs injection, which by reducing K currents [2] allows large membrane depolarizations to be applied. Epsp latency was 1-6 ms for conduction distances of 1.5-2.5 mm. Slowing of axonal conduction within the medulla [3] may account for these comparatively long conduction times. In some neurones a second epsp occurred with a longer, relatively constant latency suggesting that slower conducting fibres were also activated. In other cells longer latency epsp's fluctuated widely in onset and were probably due to activity elicited in multi-synaptic circuits. Ipsp's were observed at resting potential in a minority of cells. However in other neurones both spontaneous and evoked ipsp's were only apparent at depolarized levels (less than -55 to -65 mV) following Cs injection.

In all neurones, both medial and lateral to the tract, synaptic potentials were depressed as stimulus frequency was increased (Fig. 1). Cells were held hyperpolarized to avoid action potential generation and a maximal stimulus was applied at 0.5 Hz. On increasing the frequency to 1-50 Hz for period of 30 s the synaptic potential was always depressed. Both the time taken to reach a steady state depression and its magnitude (Fig. 1b) were dependent on stimulus frequency. At 20 Hz, epsp amplitude was reduced by 80% (mean of 18 cells) and steady state depression was typically reached in about 5 s. Membrane potential in the post-synaptic cell changed by less than 5 mV and conductance by less than 10% during the depression.

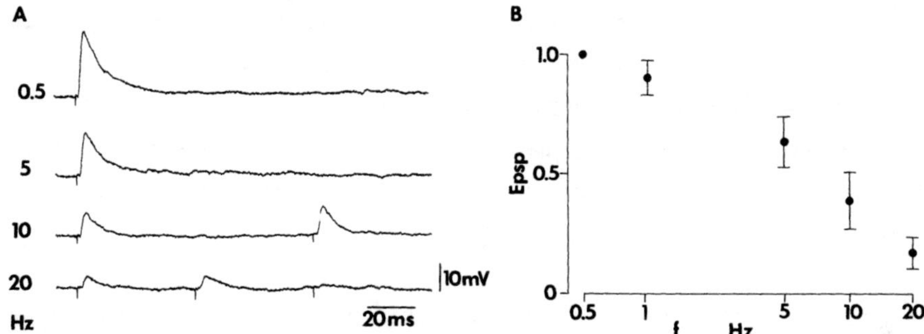

Figure 1. a. Frequency dependence of synaptic responses to solitary tract stimulation for a cell medial to the tract. Control stimulus frequency was 0.5 Hz and other responses were obtained 5 s after starting stimulation at each frequency. b. Mean steady state epsp amplitude plotted against frequency (log scale) for 13 cells. Amplitudes normalized to 1 at 0.5 Hz.

Could changes in synaptic efficacy in local circuits within the nucleus account for the depression of synaptic potentials? Ipsp facilitation at higher frequencies might explain the results. This was examined by stimulating at 20 Hz at membrane potentials held close to the epsp reversal after Cs injection. The resulting inhibitory potential was also depressed showing that epsp amplitude does not fall due to a potentiation of post-synaptic inhibition. Alternatively, a frequency dependent depression of poly-synaptic epsp's might seem to reduce the maximal synaptic potential. This was tested by applying low intensity stimulation to elicit a minimal response consisting of epsp's with similar amplitude to spontaneous synaptic events together with some transmission failures. When the stimulus frequency was increased to 20 Hz these responses were depressed in a similar way to maximal responses recorded in the same cell. Thus, low frequency depression is located at synapses made by afferent fibres rather than in local synaptic circuits within the nucleus.

Significant neuronal processing - a low pass filtering of synaptic transmission - has been demonstrated at synapses made by afferent fibres in the nucleus of the solitary tract in vitro. Does it occur in the intact animal? A similar frequency dependence of solitary tract field potentials evoked by carotid sinus nerve stimulation in the cat was described by Seller and Illert [4]. However since low frequency depression of synaptic transmission was seen in all neurones tested it is possible that synapses made by afferents of other modalities onto second order cells in the nucleus also share this property.

This work was supported by the American Heart Association, Texas Affiliate.

REFERENCES

1. Allen, W.F. (1923). Origin and distribution of nucleus tractus solitarius in the guinea pig. J. Comp. Neurol., 35, 171-204.

2. Tillotson, D. (1979). Inactivation of a Ca-conductance dependent on entry of Ca ions in molluscan neurons. Proc. Nat. Acad. Sci. 76, 1497-1500.

3. Donoghue, S., Felder, R.B., Jordan, D. and Spyer, K.M. (1984). The central projections of carotid baroreceptors and chemoreceptors in the cat: a neurophysiological study. J. Physiol., 347, 397-409.

4. Seller, H. and Illert, M. (1969). The localization of the first synapse in the carotid sinus baroreceptor reflex and its alteration of the afferent input. Pflug. Arch. 306, 1-19.

25
Characterization of NTS Neurons in Brain Stem Slices from the Guinea Pig

M. S. DEKIN, P. A. GETTING and G. RICHERSON

We have developed a brain stem slice preparation for the guinea pig which permits intracellular recording and dye injection for the characterization of neurons by electrophysiological and morphological criteria. We have used this preparation to study the properties of neurons in the ventral and ventro-lateral regions of the nucleus tractus solitarius (NTS). Our approach has been two-fold.

First we have mapped and characterized the firing patterns of neurons which burst in relation to the phrenic nerve discharge during respiration. For this purpose we have used stereotaxic, extracellular recording from an anesthetized (penthrane and nitrous oxide), paralyzed animal artifically respirated by a cycle-triggered pump. Respiratory related units (RRU) were found in locally concentrated groups corresponding to a dorsal respiratory group (DRG) and a ventral respiratory group (VRG). The DRG corresponded to the ventral and ventro-lateral regions of the NTS extending from the obex to approximately 2 mm rostral. The VRG corresponded to the general region of the nucleus ambiguus. Within these regions we encountered four major types of RRU: inspiratory neurons, expiratory neurons, pump-excited neurons, and pump-inhibited neurons. All four types were observed in both the DRG and VRG although the relative proportion of inspiratory to expiratory neurons was higher in the DRG than the VRG. The distribution of RRU in the guinea pig closely parallels that of the cat.

An _in vitro_, brain stem slice preparation was used to characterize the properties of neurons within the ventral and ventro-lateral regions of the NTS. Slices were cut on a vibrotome at 300-500 micron thickness and subfused with oxygenated (95% CO_2 - 5% O_2) Ringer's solution at 35-37°C. Neurons were penetrated with glass microelectrodes filled with either 3M KCl or 4% Lucifer yellow in 100 mM. The NTS was localized by visual criteria. All "healthy" neurons had resting potentials between -50 to -70 mV and had input resistances in the range of 100 to 150 megohms. On the basis of their spontaneous firing properties and responses to injected current pulses, three classes of neurons were observed in the ventral and ventro-lateral NTS. While all three cell types fired action potentials at an average rate of 1-2 spikes/sec at resting potential, none showed spontaneous bursting characteristic of the respiratory rhythm recorded _in vivo_. The three cell classes have been termed Type A, B, and C.

Type A neurons fired repetitively in response to depolarizing current pulses with a maximum firing rate of 100-200 spikes/sec. With prolonged depolarizations, spike frequency declined indicative of adaptation. The most distinctive characteristic of Type A neurons was a delay between the onset of the depolarizing current pulse and the first spike. This delay (termed "delayed excitation") could be as long as 500-600 msec and depended upon the membrane potential immediately preceeding the depolarizing stimulus pulse. In addition, Type A cells showed no endogenous pacemaker activity. Type C neurons were autoactive, pacemaker neurons and fired at about 1 spike/sec superimposed upon an endogenous membrane potential oscillation. Hyperpolarization abolished both the oscillations and the pacemaker firing. In response to rapid depolarizing pulses, these cells fired short, self-terminating bursts with little or no delay. Type B neurons shared some features in common with both the Type A and C cells including endogenous pacemaker oscillations (like Type C) and delayed excitation (like Type A).

We believe these properties represent different cell types because they were stable in time, did not depend upon the orientation of the slice within the brain stem, and were non-uniformly distributed. Type A neurons were found most commonly in the ventro-lateral regions of the NTS, 0.5-2.2 mm rostral to the obex. Type B neurons were the most numerous and were found in the same general rostral/caudal axis but were concentrated more medially in the ventral NTS. Type C neurons were found almost exclusively from the obex to 0.5 mm rostral and from the midline to 0.5 mm lateral. These results suggest that the variety of firing patterns displayed by neurons in these regions may reflect differences in their intrinsic properties as well as synaptic drive. It is interesting to note that the firing pattern and the magnitude of the delay of Type A neurons was heavily dependent upon the membrane potential immediately preceeding a depolarizing stimulus. It is possible that this one class of neurons may account for several different firing patterns depending upon the magnitude of the synaptic inhibition preceeding excitation.

Section 3
Spinal and Brainstem Neurons Involved in Respiratory Movements

26

Interaction between Postsynaptic Activities and Membrane Properties in Medullary Respiratory Neurones

D. RICHTER, D. BALLANTYNE and S. MIFFLIN

Recently we described a three phase organization of the respiratory rhythm, and based on their patterns of synaptic activity we predicted various sorts of connections between different types of medullary respiratory neurones. We indicated how some of these connections could provide a mechanism for the "reversible" and "irreversible" components of the inspiratory "off-switch" and the delayed onset of the next inspiration (3, 6, 7).

However, a mechanism based on synaptic interactions alone is probably not enough to explain the rhythmic character of respiration, and it seems likely that additional mechanisms based on intrinsic membrane properties may be involved in organizing "trigger","delay" and "fatigue" functions.

There is abundant evidence from other neuronal systems of invertebrates and vertebrates that a combination of synaptic activity and intrinsic membrane properties, i.e. calcium (I_{Ca}) and potassium conductances (I_C, I_A, I_M) cooperate to determine neuronal excitability (1). The patterns of postsynaptic response of bulbar respiratory neurones provide several clues as to a similar sort of cooperation between synaptic connectivities and membrane properties. We, therefore, investigated medullary respiratory neurones for these membrane conductances.

The experiments were performed on cats which were anaesthetised with sodium pentobarbital, thoracothomized, paralyzed and artificially ventilated.

A. The existence of a potassium conductance which is activated by intracellular calcium, i.e. I_C is indicated by the following findings:

1. Two classes of neurones - early-inspiratory and postinspiratory neurones - reveal reciprocal features to a remarkable extent, though they do not seem to be reciprocally inhibitory, i.e. early-inspiratory neurones stop firing towards the end of inspiration before post-inspiratory neurones start to discharge action potentials. The amplitude of membrane depolarization declines continuously throughout inspiration but at an increasing rate during the later part of inspiration; this is accompanied by a reduction in the intensity of synaptic noise and neurone input resistance. At the point of transition to postinspiration there occurs a low noise, rapid phase of membrane hyperpolarization to a maximum potential and then the membrane depolarizes slowly at a fairly uniform rate throughout expiration. This behaviour is probably not simply the result of synaptic interactions but is suggestive of a nonsynaptic, i.e. intrinsic, membrane conductance. The decay of membrane depolarization during inspiration in particular and the subsequent membrane hyperpolarization are highly suggestive of a potassium current which is activated by a calcium influx during early inspiration, i.e. I_C.

2. In the extracellular vicinity of inspiratory and expiratory neurones calcium activity decreases during the periods of activation (see: Acker and Richter, this meeting).

3. Intracellular analyses reveal that this decrease in extracellular calcium at least partly results from an I_{Ca} into respiratory neurones and that I_C can be blocked by intracellular EGTA injection (5). The consequence of such an EGTA injection into an inspiratory neurone (Fig. 1 B) is a decrease in neurone input conductance, a steepening of membrane depolarization in response to inspiratory e.p.s.p.s and the disappearance of a transient membrane hyperpolarization which followed inspiratory depolarization during control (Fig. 1 A).

The action of I_C would explain Merrill's (4) description of temporal changes in the amplitude of afterhyperpolarizations of action potentials during a burst discharge. It suggests that an I_C is functionally most important in early-inspiratory and possibly postinspiratory neurones.

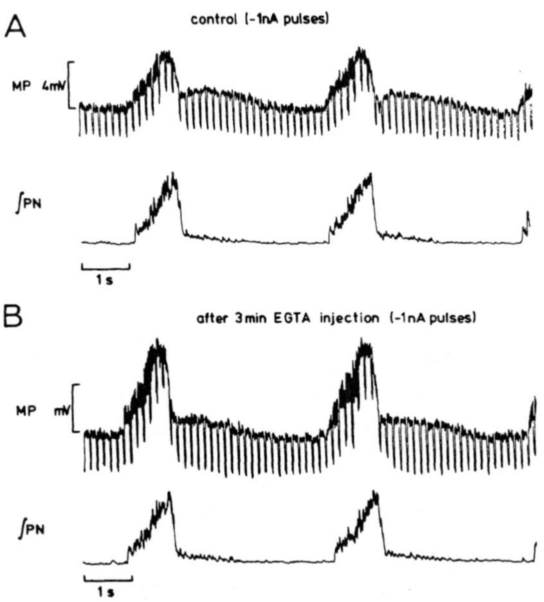

FIG. 1: Intracellular EGTA injection into an
inspiratory neurone. The neurone input resistance
was measured by 1 nA negative current pulses.
The maximal membrane potential (MP) was −50 mV
under control conditions (A) and depolarized by
3 mV after EGTA injection (B). PN: phrenic nerve
activity

The I_C, activated during the early part of inspiration, may
produce the slow repolarization of the membrane potential
during the later periods of inspiration and "accommodation" of
the discharge (4) although the excitatory drive of the cells may
remain constant (Fig. 2).

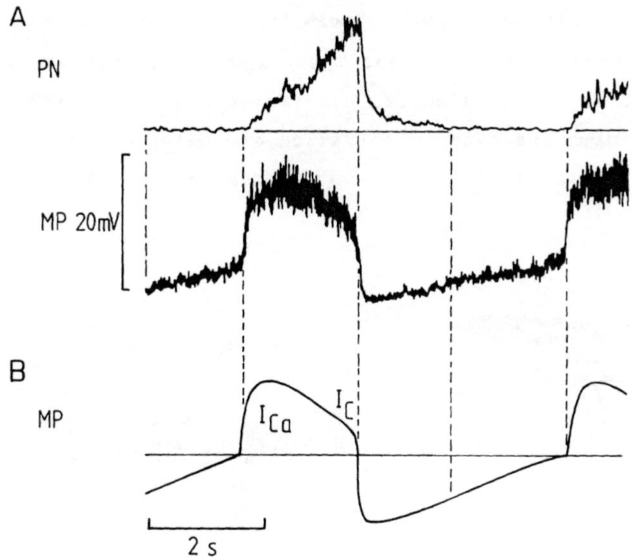

FIG. 2: Original (A) and schematic (B) pattern
of membrane potential(MP) in an early-inspiratory
neurone. PN: phrenic nerve activity; I_{Ca} :
calcium inward current; I_C: calcium dependent
potassium outward current;
For further explanation see text.

B. There is also evidence for the existence of two
voltage-dependent potassium conductances, i.e. I_A and I_M:
1. The effect of I_A is seen when respiratory neurones were steeply
depolarized after preceding long-lasting hyperpolarization. When
neurones were depolarized by a rapid onset of e.p.s.p.s following
such a period of hyperpolarization or simply by release from
strong synaptic inhibition, this I_A is activated (Fig.3 A). The
activation of I_A takes place with some delay, thus allowing a
transient rebound depolarization to occur (Fig. 3 C), but when
activated reduces the steepness of membrane depolarization (Fig.
3 A, B and D).
A phenomenon resembling rebound excitation followed by some
decay of activity can even be seen in phrenic nerve discharge
when the inspiratory drive and probably the expiratory activity
is large which indicates its potential importance for triggering

inspiration and following slow augmentation of inspiratory activity. At the neuronal level it was typically seen in ramp inspiratory neurones (Fig. 3 C and D) and postinspiratory neurones (Fig. 3 B) which start to depolarize fairly quickly from a level of membrane hyperpolarization and often reveal an overshoot during the initial period of depolarization.

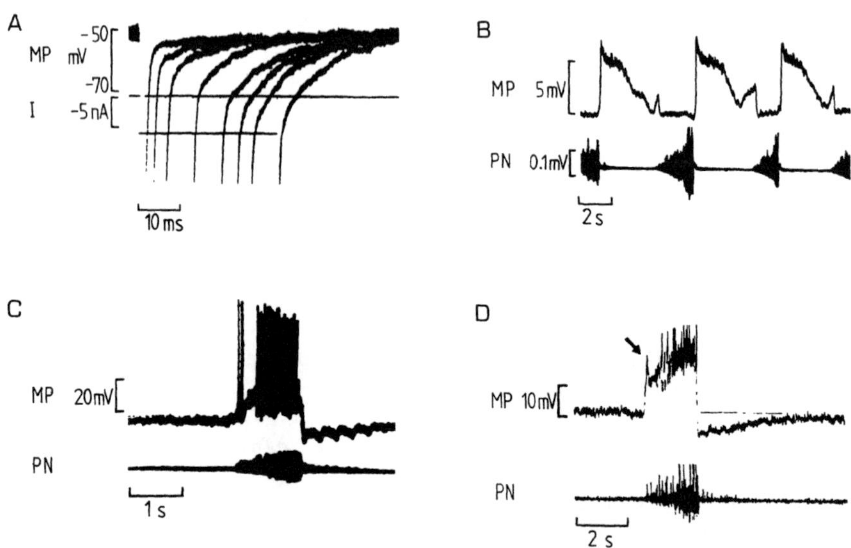

FIG. 3: Membrane potential patterns (MP) indicating the presence of a voltage-dependent transient outward current I_A in inspiratory (A, C and D) and postinspiratory (B) neurones. Inactivation of I_A is removed by membrane hyperpolarization to more than 60 mV which lasts for longer than 25 ms (A). After preceding hyperpolarization I_A is activated by steep depolarizations as can be seen in the time course of the electrotonic potentials. Effects of I_A on spontaneous membrane potential patterns can be seen in B, C and D. For further explanation see text.
PN: phrenic nerve activity

2. Some neurones reveal rectification in the steady state current-voltage relationship when they are depolarized (Fig.4) and preliminary results obtained from single electrode voltage-clamp measurements indicate that an outward current is activated when neurones are depolarized to a level of −40 mV. Such a behaviour is consistent with the possible existence of an I_M (2) in addition to the I_C.

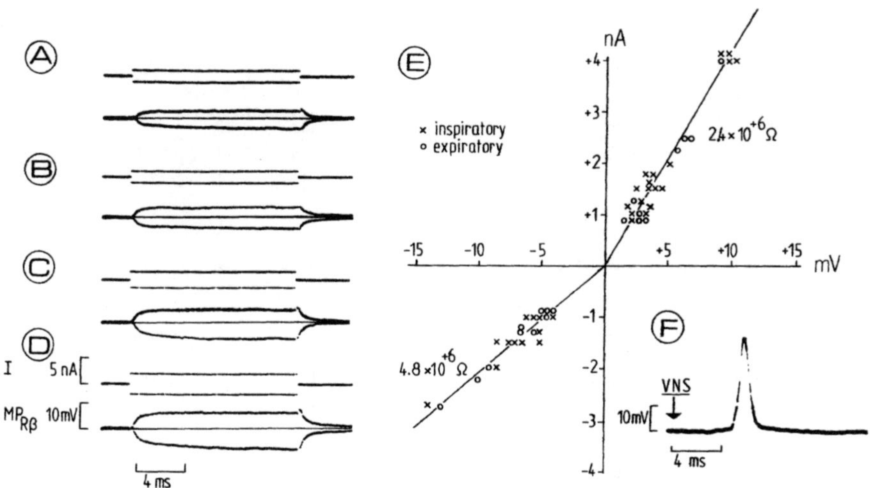

FIG. 4: Current-voltage relation of an inspiratory beta neurone. The cell was identified by vagus nerve stimulation (VNS) shown in F. Electrotonic potentials (MP) were evoked by constant current pulses (I) of varying intensities (A-D). The current-voltage relation reveals a decrease in resistance when the membrane was depolarized from −45 mV (no current injection) to −35mV. This rectification was seen during inspiration and expiration (E).

The findings imply that different membrane conductances may well modulate the excitability of medullary respiratory neurones depending not only on the ongoing, but also on the preceding synaptic activity (Fig. 5). They seem to be involved in the mechanisms responsible for "trigger", "delay" and "fatigue" functions which are important for the generation of the central respiratory rhythm.

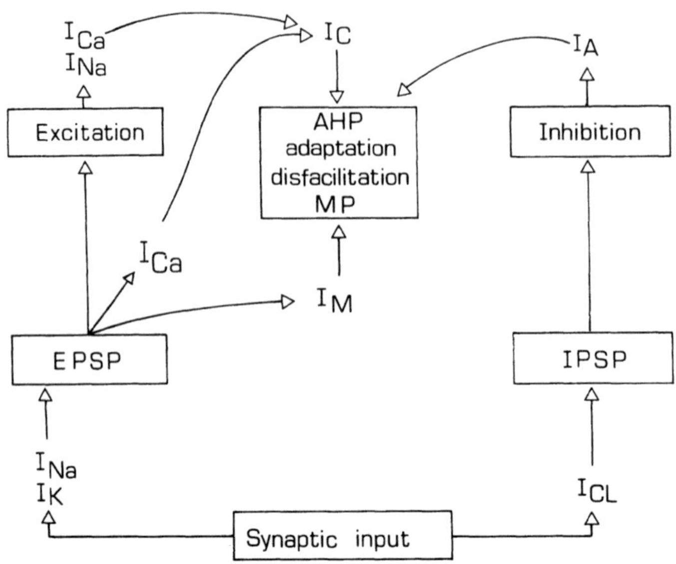

FIG. 5: Scheme   showing how synaptic activities
and membrane conductances may   interact.

(This work was   supported by the DFG.)

References

1. Adams, P. R. (1982): Trends in Neuroscience 5, 116-119
2. Adams, P. R., Brown, D. A., and Constanti, A. (1982):
            J. Physiol. (Lond.) 330, 537-572
3. Ballantyne, D., and Richter, D. W. (1984):
            J. Physiol. (Lond.) 348, 67-87 (1984)
4. Merrill, E. G. (1974): in "Essays on the Nervous System",
            Bellairs, R. and Gray, E. G., Eds,
            Clarendon Press, Oxford, pp.451-486
5. Mifflin, S. W., Ballantyne, D., Backman, S. B., and
   Richter, D. W. (1984): Pflügers Arch. 400, R48
6. Richter, D. W. (1982): J. exp. Biol. 100, 93-107
7. Richter, D. W., and Ballantyne, D. (1983):
            in "Central Neurone Environment" (M. E. Schläfke,
            H.-P. Koepchen, W. R. See, Eds.), Springer
            Verl., Berlin, Heidelberg, pp. 164-174 (1983)

27

Evidence for a Calcium Activated Potassium Conductance in Medullary Respiratory Neurones

S. MIFFLIN, D. BALLANTYNE, S. BACKMAN and D. RICHTER

Currently accepted models of respiratory control are based solely upon consideration of the synaptic interconnections between the various neurones within the rhythm generating network. However, studies in vertebrate and invertebrate neurones indicate that a combination of both synaptic inputs and intrinsic membrane properties ultimately determine discharge (1,2,3). An investigation of the membrane properties of medullary respiratory neurones is warranted to determine to what extent these properties influence discharge.

Calcium has been shown to modulate excitability in a variety of cells (4). To investigate the effects of intracellular calcium on medullary respiratory neurones, the calcium chelator EGTA was injected into these cells. Intracellular recordings were made from inspiratory and expiratory neurones in the retroambigual region of pentobarbital anaesthetized cats. Glass microelectrodes were filled with a solution of either 3 M KCl or 4 M KAc and 0.5 M K-EGTA (pH = 7.1). EGTA was injected by passing 1-8 nA negative d.c. current for up to 10 minutes. There was no apparent difference in the changes observed following EGTA injection in inspiratory compared to expiratory cells, therefore the data were pooled and the ranges given in this report.

Measurements of input resistance made after EGTA injection showed the normal, relative rhythmic pattern of variation within the respiratory cycle (5), however, input resistance at any given time in the cycle was increased by 22-43% (n=10).

Under control conditions action potentials (APs) were followed by two afterhyperpolarizations (AHPs): an early AHP lasting less than 2 msec and a late AHP lasting 12-32 msec (n=22). The relative magnitude of early AHPs (range: 5.8-9.1 mV) compared to late AHPs (range: 3.2-10.8 mV) varied from cell to cell, nonetheless the amplitude of both AHPs decreased as a cell was hyperpolarized. The AHPs approached an equilibrium potential which ranged from -75 mV to -95 mV (n=6). In cells where i.p.s.p.s were reversed, using electrodes filled with 3 M KCl, the AHPs persisted and the equilibrium potential of the AHPs remained at a hyperpolarized level. These findings suggest that the AHPs result from an increase in potassium conductance(s). After injection of EGTA there was a reduction of the early AHP (19-41%) and the late AHP (48-100%) following spontaneous and antidromic APs (n= 22) (Figure 1). EGTA completely abolished the late AHP in 14 cells but never abolished the early AHP. The late AHP appears particularly sensitive to EGTA and is likely the result of an increase in potassium conductance following an increase in intracellular calcium activity.

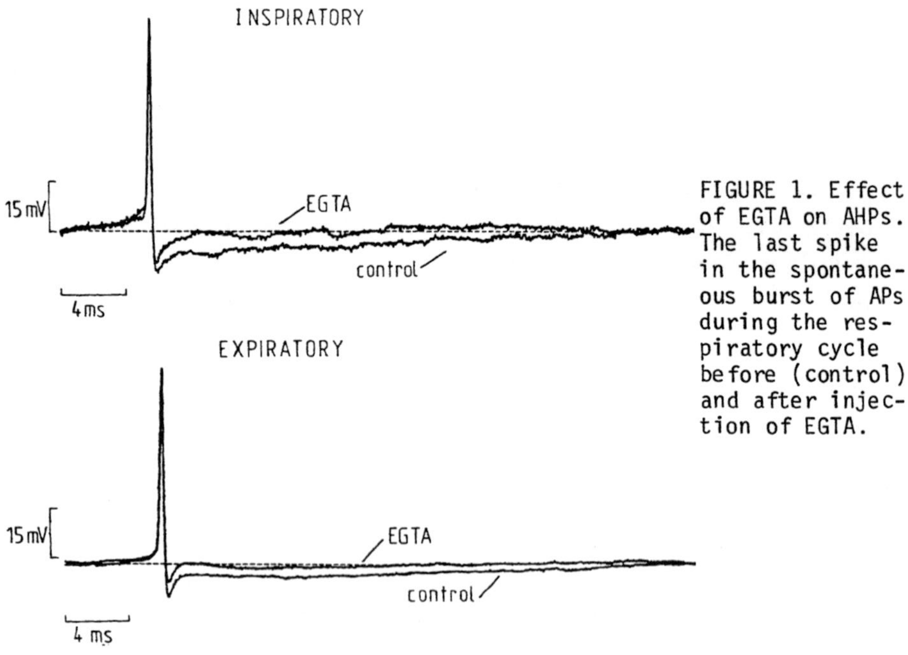

FIGURE 1. Effect of EGTA on AHPs. The last spike in the spontaneous burst of APs during the respiratory cycle before (control) and after injection of EGTA.

A hyperpolarization also followed a burst of APs elicited by a depolarizing current pulse. Such "post-tetanic hyperpolarizations" (PTHs) increased in amplitude and duration as the intensity or the duration of a depolarizing current pulse was increased. As a cell was hyperpolarized by d.c. current injection the PTH approached an equilibrium potential similar to that of the AHPs. Following injection of EGTA, the PTH decreased in amplitude an average of 65% (range: 47-79%;n=4) (Figure 2). During a PTH input resistance decreased by 22-46% as measured using constant current pulses (n=3). Following EGTA injection input resistance during the diminished PTH decreased by only 12-33%. A PTH elicited during the burst of APs associated with the respiratory cycle abolished the spontaneous AP discharge. Under conditions where the PTH was maximal, discharge was suppressed for up to 200 msec after cessation of the depolarizing pulse. This suppression of discharge was decreased after EGTA injection (from 42-84 msec to 23-40 msec).

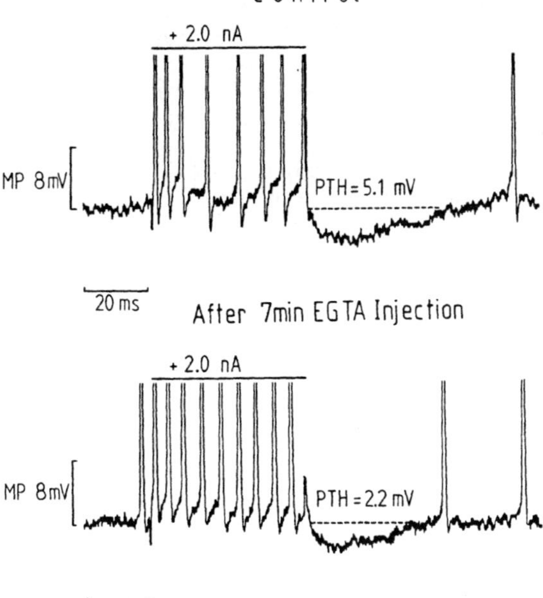

FIGURE 2. Effect of EGTA on PTHs. The hyperpolarization which followed a depolarizing current pulse was reduced after injection of EGTA. The AHPs after spontaneous and current elicited APs were reduced and discharge frequency increased after EGTA injection. The tops of the APs have been clipped.

Therefore, in conclusion:

1) Injection of EGTA increases input resistance throughout the respiratory cycle.

2) The late AHP following spontaneous and antidromic APs results from a potassium conductance activated by an increase in intracellular calcium activity.

3) The PTH observed after a burst of APs results, at least in part, from a calcium activated potassium conductance.

This calcium activated potassium conductance might play a role in regulating the discharge behaviour of medullary respiratory neurones. Synaptic modulation of this conductance provides a mechanism by which the response of a cell to a given level of synaptic input could be altered (6,7,8)

This work supported by the DFG and the Alexander von Humboldt Stiftung

References

1. Wilson, W.A. and H. Wachtel. Science 202; 772-775, 1978.
2. Adams, P.R. Trends in Neuroscience 5; 116-119, 1982.
3. Jahnsen, H. and R. Llinas. Jr. Physiology 349; 227-247, 1984.
4. Meech, R.W. Ann. Rev. Biophysics and Bioengineering 7; 1-18, 1978.
5. Richter, D.W., F. Heyde and M. Gabriel. Jr. Neurophysiology 38; 1162-1171, 1975.
6. Horn, J.P. and D.A. McAfee. Jr. Physiology 301; 191-204, 1980.
7. Dunlap, K. and G.F. Fischbach. Jr. Physiology 317; 519-535, 1981.
8. North, R.A. and T. Tokimasa. Jr. Physiology 342; 253-266, 1983.

28

Changes in Potassium Activity, Calcium Activity, and Oxygen Tension in the Extracellular Space of Inspiratory Neurones within the NTS of Cats

H. ACKER and D. RICHTER

Studies in various species have shown that changes in the ionic environment of neurones influence their intrinsic membrane properties and their response to synaptic inputs. These control mechanisms act not only by influencing the membrane potential of neurones but by modulating different sorts of membrane conductances (e.g. 1, 2, 3). Potassium and calcium ions are potent canditates for such a control mechanism of neurone excitability. To see whether this might be involved in modulating the responsiveness of respiratory neurones, we measured these ion activities in combination with the oxygen tension in the vicinity of respiratory neurones.

The experiments were performed in pentobarbital-anaesthetized cats which were paralyzed, thoracotomized and artifically ventilated. Phrenic nerve activity was recorded for monitoring the central respiratory rhythm. The spinal cord (at C-3) was stimulated for antidromic identification and the vagus nerve for synaptic activation of neurones. Three-barrelled microelectrodes were used for recording potassium activity and calcium activity, or calcium activity and oxygen tension in addition to the d.c. potential and unitary spike discharge of respiratory neurones. Typically, multi-unit inspiratory spike discharges were recorded within the ventrolateral region of the tractus solitarius where vagus nerve stimulation evoked focal potentials reflecting the synaptic response of inspiratory beta neurones. Unitary spikes were recorded from

single inspiratory beta neurones in only some cases.
Extracellular potassium activity and calcium activity varied in synchrony with the discharge of inspiratory beta neurones. Potassium activity increased to 3.6 mM during inspiration and decreased to 3.1 mM during stage-II expiration. When correction was made for the response time of calcium electrodes, calcium activity was sometimes seen to decrease to 1.2 mM during inspiration and to increase to 1.6 mM during stage-II expiration (Fig.1). When animals were ventilated with normal air, oxygen tension sometimes varied within the respiratory cycle, falling during inspiration and rising during stage II expiration. However, oxygen tension remained fairly constant at about 60-80 mmHg when oxygen was added to the inspired air.

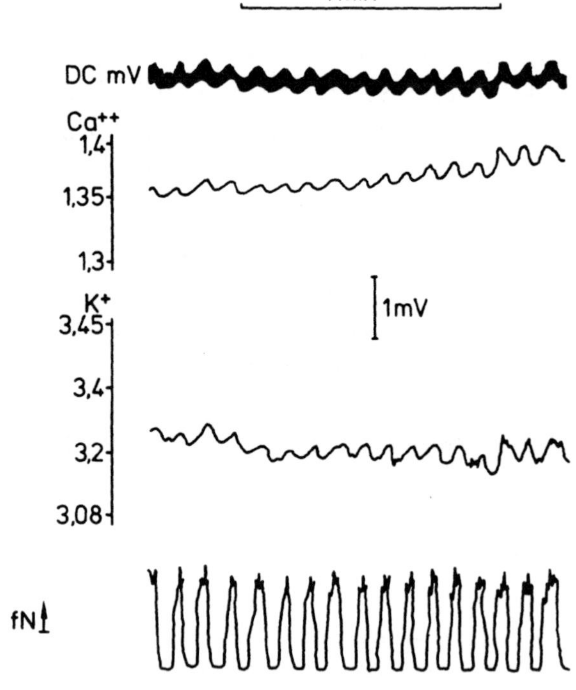

FIG. 1   Respiratory related fluctuations in extracellular calcium and potassium activities. Extracellular calcium (Ca++) , potassium (K+) activities (values in mM) and the d.c. potential (DC, see mV scale) were measured in parallel with the discharge of the phrenic nerve which is shown in a digital ratemeter record (fN).

Enhanced central respiratory drive to inspiratory neurones, provoked by hypercapnia, led to an increase in the mean value and the amplitudes of periodic fluctuations of extracellular potassium activity without major changes in oxygen tension. Hypoxia   or asphyxia, however, evoked a fall in oxygen tension to zero level within 60 s. At the same time, potassium activity increased by 1-1.5 mM and calcium activity decreased by 0.3 -0.5 mM. The increase in extracellular potassium activity during hypoxia or asphyxia   often produced a prolongation of the phrenic and unit discharge and sometimes this was followed by block of both phrenic and unit activities.

After asphyxia ended, potassium activity, calcium activity and oxygen tension recovered slowly within 8-10 min. and normally revealed an overshoot.   During and after asphyxia, respiratory-related fluctuations of potassium activity and calcium activity more than doubled in amplitude while the frequency of unitary spike discharge and the strength of phrenic nerve activity increased.

Repetitive vagus nerve stimulation (100Hz; 0.3 mA; 0.1ms) over a period of 30-40 s evoked a Hering-Breuer response in the phrenic nerve activity which was accompanied by a fall in oxygen tension to about 50 per cent of control values, an increase in potassium activity by 1.0 mM and a decrease in calcium activity by 0.4 mM.

Spinal cord stimulation also produced a decrease in extracellular calcium activity and a fall in oxygen tension.

The rhythmic changes in the ionic environment of respiratory cells in parallel with their synaptic activation are indicative of various membrane conductances and transport systems which are involved in generating the pattern of postsynaptic responses in respiratory neurones.

(This work was supported by the DFG)

References:

1. Adams, P. R., Brown, D. A., and Constanti, A.:
        J. Physiol. (Lond.) 330, 537-572 (1982)
2. Neher, E., and Lux, H. D.:
        J. gen. Physiol. 61, 385-399 (1973)
3. Richter, D. W., Camerer, H., and Sonnhof, U.:
        Pflügers Arch. 376, 139-149 (1978)

29

Post-Synaptic Potentials of Bulbar Respiratory Neurons of the Turtle

R. TAKEDA, J. E. REMMERS, J. P. BAKER, K. P. MADDEN
and J. P. FARBER

The mechanics of breathing is similar in reptiles and mammals, in
that both employ an aspiration mechanism. Despite this similarity,
the respiratory cycle in these two classes differs qualitatively. The
reptilian mechanical cycle consists of three distinct phases: inspiration, a long breath-hold, and expiration (1); the mammalian cycle
generally has two phases: inspiration and expiration. However,
during the early part of mammalian expiration, airflow is retarded by
contraction of inspiratory muscles and by glottal narrowing. Later in
expiration, airflow may be facilitated by contraction of expiratory
muscles and glottal dilation (2). This initial phase of mammalian
expiration may be analogous to the reptilian breath-hold, and the
later phase appears to correspond to the phase of expiration in
reptiles. A possibly important corollary of these global features of
the motor act of breathing is the finding, derived from the timing of
post-sympatic potentials in respiratory neurons of anesthetized cats,
that the mammalian neural respiratory cycle consists of three phases:
inspiration, post-inspiration, and neural expiration (3). This 3-
phase mammalian neural cycle would appear to correspond to the three-
phase mechanical cycle of reptiles, i.e., mammalian post-inspiration
might correspond to the reptilian breath-hold. In the present paper,
we refer to the interval between inspiration and expiration as the
post-inspiratory phase.

In order to evaluate this proposed reptilian-mammalian correspon-

dence, we recorded membrane potential (MP) from bulbar respiratory
neurons of decerebrate, paralyzed turtles (Pseudemys scripta). Inspiratory (serratus) and expiratory (pectoralis) efferent neurograms were
recorded, and used to drive a servo-respirator. In this preparation
the respiratory cycle consists of an expiratory burst that deflates
the lungs followed by augmenting inspiratory (serratus) activity
which inflates the lungs followed by a post-inspiratory phase where
the lungs are held inflated at +3 to +5 cm H_2O. The animals breathed
rhythmically for 48-72 hours. Body temperature was maintained at 30-
33 degrees C and the fractional concentration of CO_2 in the end-tidal
gas was 0.06-0.08.

Stable recordings were obtained from 43 neurons none of which
could be antidromically activated by vagal stimulation. Spinal projections were not assessed. Based upon the phase(s) in which a neuron
displayed action potentials (AP), we identified the following 5 groups
of respiratory neurons; inspiratory (1), post-inspiratory (PI),
expiratory (E), post-inspiratory/expiratory (pi-E) and expiratory/
inspiratory (E-i). The last two displayed APs in two respiratory
phases and depolarization was greatest during expiration in each
groups. The appearance of inhibitory post-synaptic potentials
(IPSPs) was identified by reversal of MP deflection after iontophoresis
of chloride ions and by a decrease in "input" resistance.

Table 1 lists the time of depolarization and the time of occurrence
of IPSP's in the 5 groups of respiratory neurons observed in this
study. Also listed is the respiratory muscle of the turtle that
corresponds in its temporal firing pattern to each neuron.

The behavior of post-synaptic potentials in most turtle neurons
resembled that described for anesthetized cat (3) and, in particular,
PI neurons of the turtle behaved like those of the cat. This supports
the notion that the mammalian 3 phase neural respiratory cycle corresponds to the reptilian respiratory cycle and that the breath-hold
phase represents a prolonged post-inspiratory phase. The slow
respiratory rhythm and prolonged post-inspiratory phase of the reptile
may provide insights into respiratory rhythmogenesis. For instance,
events relating to the onset of expiration are of considerable
interest. During the post-inspiratory phase, PI neurons remain depolarized. Termination of post-inspiration is heralded by the

appearance of EPSPs in E neurons 5-8 sec before occurrence of APs in these neurons. The source of this EPSP input to E neurons remains to be elucidated.

Table 1. Respiratory phases of the turtle

Neuron Type	Expiration	Inspiration	Post-Inspiration	Corresponding Respiratory Muscle
E	Ramp depol.	?	Pre-E depol.	Pectoralis
pi-E	Ramp depol.	IPSPs	Steady depol.	Transversus
E-i	Depol.	Depol.	?	Laryngeal Abductor
I	IPSPs	Depol.	?	Serratus & Oblique
PI	IPSPs	IPSPs	Abrupt depol.	Laryngeal Adductor

* Depol: depolarization

REFERENCES

(1) Cans, C. and Hughes, G.M. (1967). J. Exp. Biol., 47: 1.

(2) Gautier, H., Remmers, J.E. and Bartlett, Jr., D. (1973). Resp. Physiol., 18: 205

(3) Richter, D.W. (1982). J. Exp. Biol., 100: 93.

30

High-Frequency Components of Inspiratory Discharges

M. I. COHEN, J. L. FELDMAN, D. F. DONNELLY and A. L. SICA

In addition to being synchronized on the time scale of the respiratory
cycle, discharges of inspiratory (I) neurons are synchronized on a
shorter time scale, in the form of high-frequency oscillations (HFO)
in the frequency range 30-100 Hz. These oscillations have been found
in I motor nerves and in medullary neuron and population activity.
The phenomenon has been studied by power spectral analysis (Richardson
& Mitchell, 1982; Bruce & Goldman, 1983; Ackerson & Bruce, 1983) and
by crosscorrelation analysis (Cohen, 1973, 1976; Cohen & Feldman,
1984).

It was found by Richardson & Mitchell (1982) that the distributions
of HFO frequencies were different in the phrenic and in the recurrent
laryngeal nerves. Both nerves had prominent peaks at 88 Hz; in addition phrenic activity had a peak at 37 Hz, whereas laryngeal activity
had a peak at 55 Hz. In this laboratory, we have also analyzed efferent hypoglossal activity, and we have found that the HFO frequencies
in hypoglossal and recurrent laryngeal activity resemble each other.
This is of interest because both activities are involved in control of
airway resistance and both are more responsive to pulmonary afferent
inputs than is phrenic discharge.

In a study of DRG I neurons (Cohen & Feldman, 1984), we computed
crosscorrelation histograms (CCH) between unit and phrenic activity.
The neurons were classified according to their responses to the withholding inflation test. Although for the majority of augmenting I
neurons peak unit activity led peak phrenic activity (range 2-4 msec),
there were significant average differences of HFO phase between the
groups as classified by inflation response.

In contrast, the HFO correlation patterns of the decrementing I
neurons were quite different: peak unit activity lagged peak phrenic
activity by 1-3 msec. This pattern is similar to that found in
phrenic-laryngeal CCHs.

The observed diversity of HFO correlation phases presumably reflects differences in the connectivity of different neuron populations. In particular, these phase differences suggest the existence of excitatory and inhibitory loops between different types of I neuron.

As a preliminary approach to study of these interactions, phase-resetting experiments have recently been done. The experimental method was the application of an I-inhibitory input, single-shock stimulation to the afferent superior laryngeal nerve, at different phases of the HFO cycle. The phase shift of the HFO in phrenic activity that was produced by the stimulus depended markedly on the time of stimulus delivery. We assumed that phrenic activity changes are produced by changes in oscillatory activity of DRG I neurons (which drive phrenic motoneurons), and corrected the observed phases for the known conduction times in the afferent and efferent pathways. We found that when a stimulus arrives during the inactive phase of the oscillation in DRG neurons, there is a prolongation of the inactive phase and a phase shift of subsequent oscillations; whereas when a stimulus arrives during the active phase, although there is an inhibition of ongoing activity, there is no phase shift of the subsequent cycle.

These results can be interpreted as indicating that the afferent volley is transmitted to different points of the neural circuit(s) generating the oscillation, and that the response depends on the activity phase of the affected neurons. Obviously, the analysis of the HFO is still at a primitive stage, but it is hoped that extension of the phase-resetting protocol will lead to more information about the interactions that generate the HFO.

References

Ackerson, L. M. & Bruce, E. N. (1983). Brain Res. 271, 346-348.
Bruce, E. N. & Goldman, M. D. (1983). Brain Res. 269, 259-265.
Cohen, M. I. (1973). Acta Neurobiol. Exp. 33, 189-218.
Cohen, M. I. (1976). In Respiratory Centres and Afferent Systems, ed. Duron, B., pp. 19-29. Paris, Colloq. Inst. Natl. Sante Rech. Med., vol. 59.
Cohen, M. I. & Feldman, J. L. (1984). J. Neurophysiol. 51, 753-776.
Richardson, C. A. & Mitchell, R. A. (1982). Brain Res. 233, 317-336.

31

Spinal Motoneurons Related to Respiratory Movements: Their Relationships with Supraspinal Levels

R. MONTEAU, M. KHATIB and G. HILAIRE

The aim of the studies reported herein was to analyse the respiratory drive on the phrenic motoneurones (PM). The efferent projections from brainstem respiratory neurons to spinal motoneurones have been studied by several computer based techniques :

- cross-correlation histogram (CCH) (7) between medullary neuron activities and phrenic nerve discharge,

- analysis of the "short term synchrony" between pairs of unitary discharges of phrenic motoneurons (8, 9),

- spike triggering average (STA) between medullary neurons and intracellular recording of phrenic motoneurons.

Antidromic stimulation experiments (2, 11) have previously shown that two groups of medullary neurons send their axon to the spinal cord ; these inspiratory bulbospinal neurons (IBSN) are located in the dorsal (DRN) or in the ventral (VRN) respiratory nuclei. Thus the respiratory drive on phrenic motoneurons could originate from DRN or VRN inspiratory bulbospinal neurons ; axons of the former generally do not descend below C7 (2), and many axons of the later arborize in the phrenic nucleus (11). CCH suggests that both DRN and VRN neurons (3, 7) monosynaptically drive contralateral PM. The fig.1 shown two examples of analyses obtained for DRN and VRN neurons. Similar results were obtained for 80 % of DRN neurons and 60 % of VRN neurons. In both cases, the waves on the CCH have short latencies

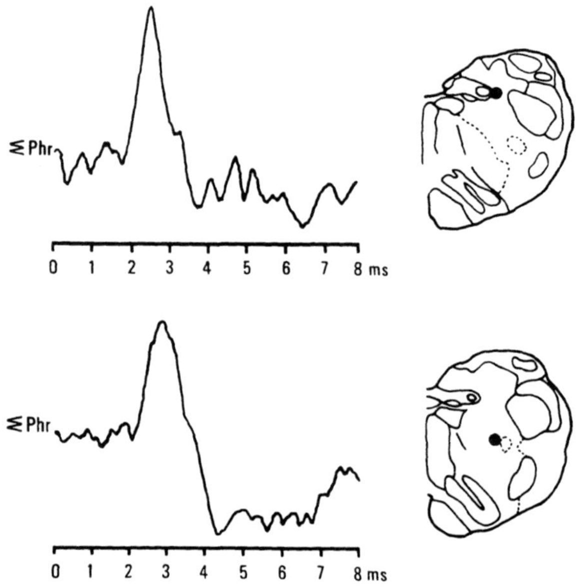

FIGURE 1 : Time relation between spike activity of medullary neurons
and contralateral phrenic efferent discharge. On the right
: localization of the trigger neuron in the DRN (top) on
the VRN (bottom). On the left : summed phrenic nerve
activity. In both cases short latency, fast rise time
waves appear, indicating a monosynaptic drive on P.M.

(mean values : 1.6 +- 0.1 ms for DRN ; 1.9 +- 0.2 ms for VRN) and
similar fast rise time (0.9 +- 0.3 ms and 0.8 +- 0.3 ms respec-
tivelly). Thus it is assumed that both DRN and VRN neurons monosynaptically drive the contralateral PM.

For DRN neurons this hypothesis is fully confirmed by STA
experiments performed by us (Fig 2) or by others (4, 9). These experiments clearly demonstrate the existence of monosynaptic EPSPs.

For VRN neurons STA experiments do not confirm CCH
results since only few monosynaptic connections are found (4). Thus,

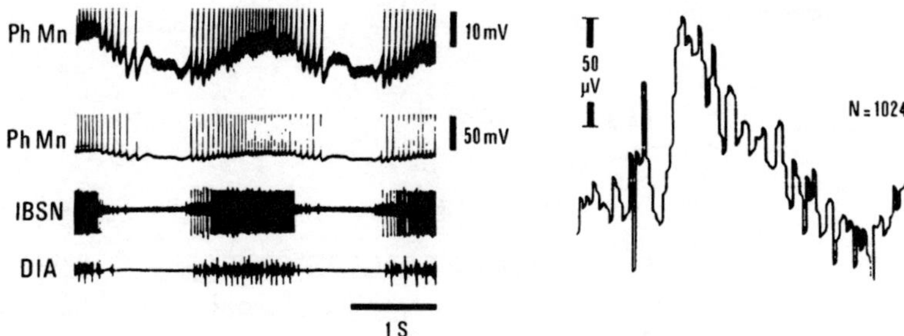

FIGURE 2 : Spike triggering average of P.M. Left, from top to bottom
: intracellular record at high and low gain of a P.M.,
extracellular record of an inspiratory bulbospinal neuron
of the DRN, diaphragm EMG. Right : using each IBSN spike
to trigger the averaging of the PM membrane potential
(after supressing the spikes by a moderate hyperpolarization) reveal an unitary EPSP.

if monosynaptic connections between DRN neurons and PM are well
documented, some discrepencies exist concerning the nature (monosynaptic or not) of the connections between VRN neurons and PM.

For lumbosacral motoneurons HENNEMAN and co-workers (5)
have shown that cell size is the determining factor of the recruitment sequence, the smaller motoneurons being recruited before the
larger ones. Phrenic motoneurons as other motoneurons are not recruited simultaneously and have been classified (6) as early (E) or
late (L). These differences of recruitement order could not be explained only by size differences and a central determination of
recruitment was assumed on the basis of indirect arguments (8).

Cross-correlation

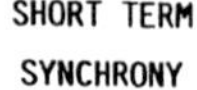

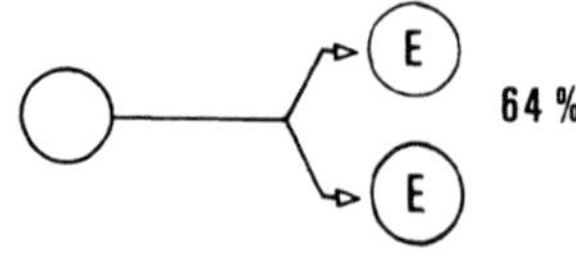

FIGURE 3

Short term synchronization be -
tween phrenic motoneurons. 85 pairs were
tested ; the short term synchrony
appears mainly in homogeneous pairs (EE
or LL) and is scarce for heterogeneous
pairs (EL).

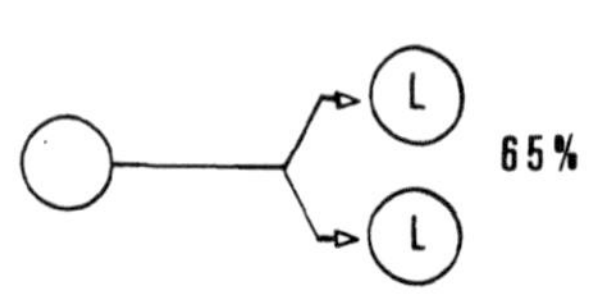

analysis performed between PM pairs revealed a short term synchrony
due to common inputs from the inspiratory neurons. However this synchrony appears mainly (65%) between motoneurons with similar
recruitment (i.e both E or both L) and is unfrequent (23 %) among
mixed pairs including E and L PM (9) (Fig. 3). As two groups (E and
L) could be distinguished among the inspiratory bulbospinal neurons,
STA experiments were performed in order to examine if monosynaptic
EPSPs were randomly distributed irrespective of the recruitments of
neurons and motoneurons. Eighty pairs were analysed and classified
into four groups (FIg. 4) according to their recruitments (EE, EL,
LE, LL). As expected the occurence of EPSPs is not randomly distributed ; they are frequently observed between neurons and motoneurons
with a similar recruitment (both E or L) and were unfrequent (23 %)
between neurons and motoneurons with different recruitment (one E,one
L) (Fig 5).

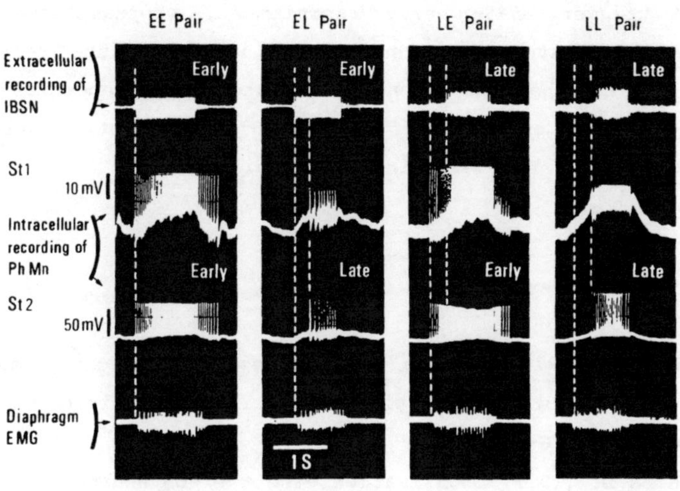

FIGURE 4 : Differents groups of recorded pairs of neurons and motoneurons. From top to bottom : IBSN activity, PM activity, (high and low gain DC records), diaphragm EMG used to define the onset of inspiration. According to the delay of recruitment four groups were distinguished.

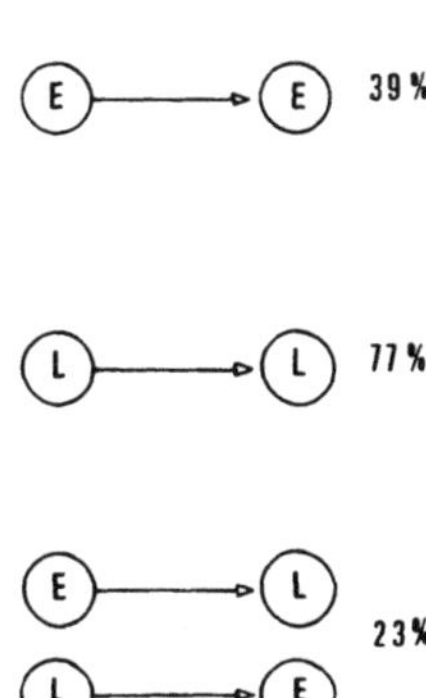

FIGURE 5

Schematical organization of the connections between inspiratory bulbospinal neurons of the DRN and phrenic motoneurons. The percent of revealed EPSPs is indicated for the differents groups. 80 pairs were tested.

These results clearly confirm that early recruited inspiratory bulbospinal neurones preferentially drive some motoneurons and trigger their early recruitment, whereas late neurons preferentially drive other PM determining their late recruitment. Although other mechanisms depending on motoneuron properties could be involved (1), the existence of an inhomogeneous central drive is an important factor for determining PM recruitment

References

1 BERGER A.J. (1979) Phrenic motoneurones in the cat : subpopulations and nature of respiratory drive potentials. J. Neurophysiol., 42, 76-90.

2 BIANCHI A.L. (1971) Localisation et étude des neurones respiratoires bulbaires. J. Physiol. (Paris), 63, 5-40.

3 COHEN M.I., PIERCEY M.F., GOOTMAN P.M. and WOLOTSKY P. (1974). Synaptic connections between medullary inspiratory neurons and phrenic motoneurons as revealed by cross-correlations. Brain Res., 81, 319-324.

4 FEDORKO L., MERRILL E.G., LIPSKI J. (1983). Two descending medullary inspiratory pathways to phrenic motoneurones. Neurosci. Lett., 43, 285-291.

5 HENNEMAN E., SOMJEN G., CARPENTER D. (1965). Functional significance of cell size in spinal motoneurons. J. Neurophysiol, 28, 560-580.

6 HILAIRE G., MONTEAU R., DUSSARDIER M. (1972). Modalités du recrutement des motoneurones phréniques. J. Physiol. (Paris), 64, 457-478.

7 HILAIRE G., MONTEAU R. (1976). Connexions entre les neurones inspiratoires bulbaires et les motoneurones phréniques et intercostaux. J. Physiol. (Paris), 72, 987-1000.

8 HILAIRE G., MONTEAU R. (1979). Facteurs déterminant l'ordre de recrutement des motoneurones phréniques. J. Physiol. (Paris), 75, 765-781.

9 HILAIRE G., GAUTHIER P., MONTEAU R. (1983). Central respiratory drive and recruitment order of phrenic and inspiratory laryngeal motoneurones. Resp. Physiol., 51, 341-359.

10 LIPSKI J., KUBIN L., JODKOWSKI J. (1983). Synaptic action of Rβ neurons on phrenic motoneurons studied with spike triggered averaging. Brain Res., 288, 105-118.

11 MERRILL E.G. (1974). Finding a respiratory function for the medullary respiratory neurons. Essays on the nervous system Bellairs & Gray Ed.

32

Recurrent Inhibition in the Control of Respiratory Motoneurons

J. LIPSKI, R. E. W. FYFFE and J. JODKOWSKI

The presence of recurrent inhibition in spinal motoneurons supplying respiratory muscles has been a matter of controversy. In two
early intracellular studies of phrenic and intercostal motoneurons
(3,9) no significant degree of recurrent inhibition was detected and
subsequently these motoneurons have been frequently cited as examples
which lack this type of inhibition. More recently, however, Kirkwood
et al.(5) reinvestigated the problem of the presence of recurrent
inhibition in intercostal motoneurons. They concluded that the
inhibition is present in these motoneurons and that the strength of
the overall inhibitory effect was comparable with that seen in some
hind limb motor nuclei.

The results of more recent studies on phrenic motoneurons are
equivocal. Cameron et al. (1) found no initial axon collaterals in
eleven cat phrenic motoneurons that were intracellularly labelled with
horseradish peroxidase (HRP), while Webber and Pleschka (10) reported
a motor axon collateral in one out of four motoneurons (stained with
Procion Yellow). The electrophysiological studies also resulted in
different conclusions. In contrast to the results obtained by Gill
and Kuno (3), Hilaire et al.(4) have recently recorded recurrent
hyperpolarizing potentials in a few phrenic motoneurons as well as
Renshaw-like units in the phrenic nucleus. The present experiments
were undertaken to reinvestigate the problem of whether recurrent
inhibition and initial axon collaterals are present in cat phrenic

motoneurons. This brief account is based upon recent preliminary
reports (6,7); a more detailed description of the experimental methods
and results will be presented elsewhere (8).

Four results strongly support the presence of a recurrent inhibitory
pathway controlling the activity of cat phrenic motoneurons:

(I). In cats anaesthetized with chloralose and urethane (and with
the C_4 to C_7 dorsal roots cut) the firing probability of the C_5 or
C_6 phrenic root was measured following antidromic activation of phrenic
motoneurons in the adjacent segment. Stimulation of the C_5 phrenic
root clearly reduced the firing probability in the C_6 root. On the
other hand, stimulation of the C_6 phrenic root resulted only in a
small inhibitory response observed in the C_5 root   (see Fig. 1).

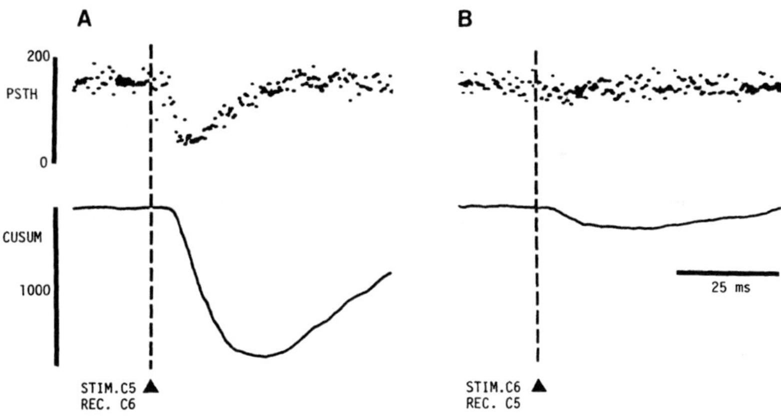

FIGURE 1.   Peri-stimulus time histograms (PSTH) and their cumulative
            sums (CUSUM) of the efferent activity recorded from the
            C_6 phrenic root following stimulation of the C_5 root (A)
            and from the C_5 root following stimulation of the C_6
            root (B). Dorsal roots (C_4-C_7) were cut. Bin width
            0.5 ms. Stimuli marked by filled triangles and vertical
            dashed lines.

This pronounced assymetry of the inhibitory responses of the C_5 and
C_6 roots was observed in all (n = 4) animals studied. These results
indicated the presence of recurrent inhibition in the phrenic nucleus,
at least between phrenic motoneurons located in adjacent spinal
segments, and particularly from motoneurons in C_5 segment onto motoneurons located in C_6 segment.

   (II). In decerebrate spinal cats (C_1 transection of the cord)
intracellular recording and averaging techniques were used in an
attempt to record recurrent IPSPs. Again, dorsal roots C_4-C_7 were
cut. Following suprathreshold stimulation of the appropriate phrenic
root , antidromic action potentials that were followed by a longlast-
ing afterhyperpolarization could be recorded. When the phrenic
root was stimulated with an intensity just subthreshold for the axon
of the impaled motoneuron, intracellular recording and averaging
revealed hyperpolarizing potentials similar (though usually smaller)
to the one shown in Fig. 2.

FIGURE 2.   Potentials recorded intracellularly from a C_5 phrenic
motoneuron following stimulation of the corresponding
phrenic root. In the upper record, which is the average
of 16 sweeps, the stimulus was just suprathreshold for
the axon of the tested motoneuron. In the lower record,
which is the average of 256 responses, the stimulus was
just subthreshold for the axon of this motoneuron.
Stimuli marked by filled triangle. Spinalized cat
with dorsal roots (C_4-C_7) cut.

This type of hyperpolarizing potential was observed in 11 out of 28 motoneurons tested. The amplitudes of these hyperpolarizations ranged from 45 to 260 μV and their durations ranged from 12 to 25 ms. Since the dorsal roots were cut, they were presumed to be recurrent IPSPs.

(III). Extracellular recordings in the vicinity of phrenic motoneurons occasionally revealed units which fired high frequency bursts of spikes, at short latency (mean 2.4 ms, n = 4), following stimulation of the phrenic roots. (Fig. 3). The maximal duration of the bursts (8-21 ms) corresponded to both the duration of the recurrent IPSPs and to the reduced firing probability of the adjacent phrenic root. These units, which exhibited similar characteristics to the neurons recently reported by Hilaire <u>et al.</u> (4), were presumed to be Renshaw cells.

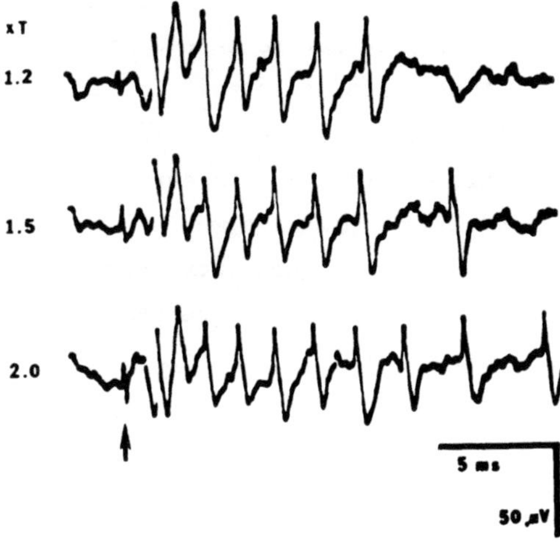

FIGURE 3.   Renshaw cell-like discharge recorded in the C_5 ventral horn following stimulation of the C_5 phrenic root . Stimuli (arrow) at a strength given in multiples of threshold. Spinalized cat with dorsal roots (C_4-C_7) cut.

(IV). Intracellular staining of single phrenic motoneurons was
performed using microelectrodes filled with a solution of horseradish
peroxidase. Forty-nine phrenic motoneuron axons were identified
following iontophoretic injection of HRP and were examined under oil
immersion at a total magnification of 1000 times to detect the origin
and trajectory of initial collaterals. Light microscopic analysis
revealed recurrent axon collaterals in 6 out of the 49 phrenic axons
examined (approx. 12%). Of these 6 axons, five issued single collaterals and one had two collaterals (Fig. 4). All collaterals arose
within the grey matter of the ventral horn, at distances up to 450 μm
from the parent cell body. Because of the limited extent and staining
intensity within these fine collateral branches it is not possible to
comment definitely on their destination. It seemed, however, that
they ramify close to, or in, the phrenic nucleus where they most
likely terminated on Renshaw cells.

In conclusion, these electrophysiological and anatomical data,
are interpreted as evidence that recurrent inhibition is also present
in phrenic motoneurons. Axon collaterals present only in relatively
few phrenic motor axons seem to subserve recurrent inhibitory action
in a larger percentage of phrenic motoneurons located in the same,
as well as in the adjacent, spinal segment. This can be explained by
the divergence of Renshaw cell axons with terminations on a larger
number of motoneurons. It is also suggested that the asymmetry
observed in the recurrent inhibitory action on C_5 and C_6 motoneurons
has a functional role. The recurrent pathway may limit the activity
of phrenic motoneurons which are known to supply the dorsal part
of the diaphragm through the C_6 root (2) when the more ventral part
of the diaphragm is strongly activated.

FIGURE 4.   A. Reconstruction from serial transverse sections of the
            spinal cord of a phrenic motoneuron in C_5 segment. Note
            that many of the dorsomedially directed dendrites cross
            the mid line ventral to the central canal (CC) and that
            the dorsolaterally directed dendrites penetrate for
            some distance into the lateral white matter (the
            boundary between the grey matter of the ventral horn
            and the white matter is represented by the dashed line.)

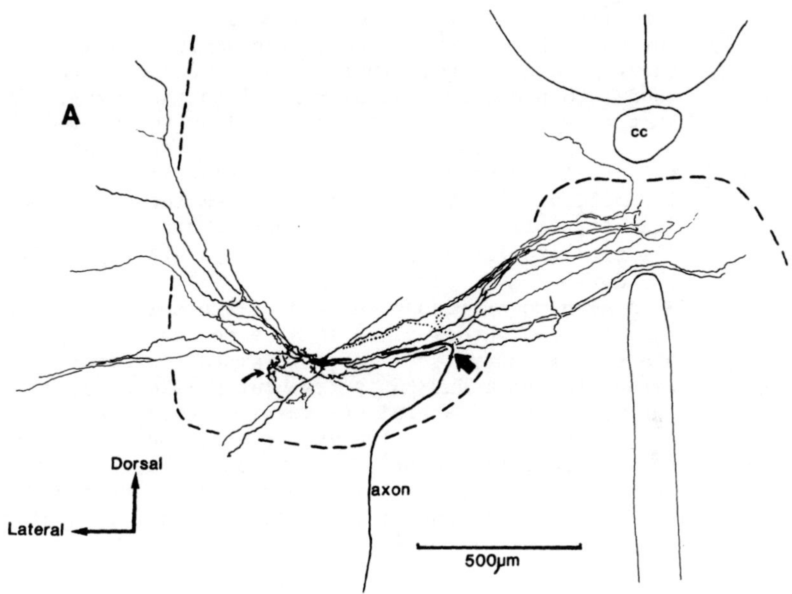

A
cc
Dorsal
Lateral
axon
500µm

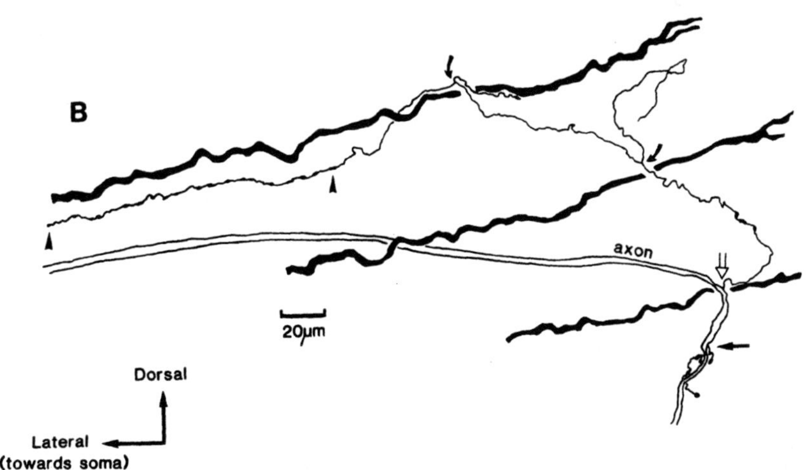

B
axon
20µm
Dorsal
Lateral
(towards soma)

There are no dendrites directed dorsally in the grey
matter, and sparse ventral projections. The group of
dendrites indicated by the curved arrow are oriented
perpendicular to the plane of reconstruction and actually
cover a total longitudinal extent of 3,300 µm (2,400 µm
caudal to the soma, 900 µm rostral to the soma). The axon
initially courses medially then, close to the medial
border of the ventral horn, it curves ventrolaterally to
exit in the ventral root. At the position indicated by
the bold arrow (about 380-400 µm from the soma) two fine
diameter collateral branches emerged from the axon.
The trajectories of these collaterals are represented by
the dots.
B. A high power reconstruction of the collaterals arising
from the neuron described in A; some dendritic profiles
are also illustrated. The more distal collateral (straight
arrow) is short and has a few bouton-like enlargements.
The proximal collateral however is longer, has at least
two daughter branches (curved arrows), one of which
carried several (about 18) en passant boutons along its
terminal 150 µm or so (arrowheads). This terminal branch
runs parallel to one of the dendrites and terminates
within 100 µm of the motoneuron soma.

REFERENCES

1. Cameron, W.E., Averill, D.B. and Berger, A.J. (1983). Morphology
of cat phrenic motoneurons as revealed by intracellular injection of
horseradish peroxidase. J. Comp. Neurol., 219, 70-80.

2. Duron, B., Marlot, N., Larnicol, M.C., Jung-Caiollol, M.C.
and Macron, J.M. (1979). Somatotopy in the phrenic motor nucleus of
the cat revealed by retrograde transport of horseradish peroxidase.
Neurosci. Lett., 14, 159-163.

3. Gill, P., and Kuno, M. (1963). Properties of phrenic motoneurons.
J. Physiol., 168, 258-273.

4. Hilaire, G., Khatib, M. and Monteau, R. (1983). Spontaneous
respiratory activity of phrenic and intercostal Renshaw cells.
Neurosci. Lett., 43, 97-101.

5. Kirkwood, P.A., Sears, T.A. and Westgaard, R.H. (1981). Recurrent
inhibition of intercostal motoneurones in the cat. J. Physiol., 319,
111-130.

6. Lipski, J. and Fyffe, R.E.W. (1984). Evidence for recurrent
axon collaterals in cat phrenic motoneurons. Proc. Austral. Physiol.
Pharmacol. Soc., 15, 163p.

7. Lipski, J. and Jodkowski, J. (1984). Evidence for recurrent inhibition of cat phrenic motoneurons. Neurosci. Lett., Suppl. 15, S 45.

8. Lipski, J., Fyffe, R.E.W. and Jodkowski, J. (1985). Recurrent inhibition of cat phrenic motoneurons. J. Neurosci., submitted.

9. Sears, T.A. (1964). Some properties and reflex connections of respiratory motoneurons of the cat's thoracic spinal cord. J. Physiol., 175, 386-403.

10. Webber, C.L. Jr. and Pleschka, K. (1976). Structural and functional characteristics of individual phrenic motoneurons. Pflügers Arch. 364, 113-121.

33

Direct Excitatory Interactions between Motoneurons of a given Pool: A new Linkage Mechanism

G. HILAIRE, M. KHATIB and R. MONTEAU

Since original data from RENSHAW (5) several experimental results have suggested the existence of direct excitatory relationships between motoneurons (2,4). It has been shown that stimulation of motor axons evokes, after the direct orthodromic spike, a second recurrent response (1,3). The aim of the present work was to look for the occurrence of such a response on phrenic motoneurons.

Experiments were performed on cats anaesthetized by I.P injections of Nembutal, paralyzed and artificially ventilated. One phrenic nerve was dissected in the neck, a stimulating electrode was placed on the central part of the nerve whereas the distal part was dissected into thin filaments for fiber recording. A cervical laminectomy allowed the dorsal roots to be cut in order to suppress afferent responses.

The iterative stimulation (10-100 Hz, 0.1-1V, 200 μS) of the phrenic nerve centrally to the recording site i) evokes the direct response of the axon ii) evokes the appearence of sporadic recurrent responses of the same motoneuron. These responses are not randomly distributed, each stimulus is not efficient for eliciting the response (generally 4 to 6 responses for 100 stimuli), but each sporadic spike always follows the preceding stimulus with a very stable latency (fluctuations less than 200 μS). The recurrent responses are observed for 102 of the 158 tested motoneurons. The apparition of these responses is dependant on the spinal cord since

i) spinal section at C2 level does not abolish them ii) ventral root sections suppress them iii) the latency for the response includes some spinal delay (2 mS).

The recurrent responses depend on convergence phenomenon, they appear more frequently when stimulus intensity or frequency are increased. The excitability of phrenic motoneurons influences the occurrence of recurrent responses : they are more frequent 1) during inspiration than during expiration, 2) when the central respiratory drive is enhanced by hypercapnia or by lung collapse.

The recurrent responses are generally evoked by stimulation of the whole nerve (including the recorded axon), however responses can be evoked by stimulation of a nerve filament different from the recorded one, thus as previously reported for laryngeal motoneurons (1) the antidromic invasion of the soma is not necessary for the genesis of the recurrent response. The hypothesis involving a reexcitation of the initial segment by the somato-dendritic spike can be disregarded.

Systemic injections of antinicotinic drugs (DHBE, Mecamy-lamine) decrease or suppres the recurrent response ; thus the response probably depends on a mechanism involving a cholinergic synapse.

In order to obtain more accurate data on the mechanims of the recurrent response, an intracellular study of phrenic motoneurons was performed. By using stimulus parameters similar to those already described no somatic spikes coresponding to the recurrent responses recorded on the axons are seen.

Subthreshold stimulations for antidromic invasion of the recorded motoneuron sometimes evoke EPSPs on this motoneuron. These EPSPs depending on the activation of others motoneurons of the pool argue for the existence of excitatory relationships between phrenic motoneurons. The latencies of the EPSPs suggest that they coud be involved in the generation of the recurrent responses.

Similar results showing the existence of direct excita-tory relationships between motoneurons have also been obtained for intercostal motoneurons and suggest that the activity of a motoneuron could partly depend on the activity of the other members of the pool. These excitatory relationships would constitute a linkage mechanism synchronizing the motoneurons.

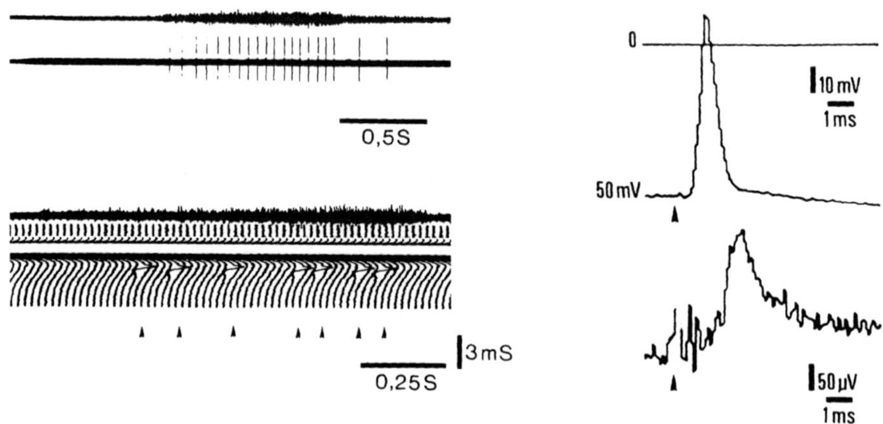

Fig 1 left : Top, spontaneous activities of phrenic nerve and motoneuron. Bottom, on vertical sweeps the arrows indicate the recurrent responses evoked by phrenic nerve stimulation.

rigth : intracellular recording of a phrenic motoneuron ; Top, antidromic invasion following phrenic nerve stimulation. Bottom, stimulation of the phrenic nerve below the invasion threshold for the recorded motoneuron, evokes an EPSP owing to the activation of the others members of the pool.

References

1 GAUTHIER P., BARILLOT J.C. and DUSSARDIER M. (1980). Mise en évidence d'interactions d'origine centrale entre motoneurones laryngés. J. Physiol. (Paris), 76, 647-661.

2 GOGAN P., GUERITAUD J.P., HORCHOLLE-BOSSAVIT G. and TYC-DUMONT S. (1977). Direct excitatory interactions between spinal motoneurones of the cat. J. Physiol., 272, 755-767.

3 MARTENSON A. (1967). Recurrent discharge studied in single units of laryngeal muscles. Acta Physiol. Scand., 70, 221-228.

4 NELSON P.G. (1966). Interaction between spinal motoneurones of the cat. J. Neurophysiol., 29, 275-287.

5 RENSHAW B. (1941). Central effects of centripetal impulses in axons of spinal roots. J. Neurophysiol., 4, 167-183.

34

Respiratory Motor Units in the Lateral Intercostal Muscles of the Cat

B. DURON and P. LE BARS

<u>INTRODUCTION</u>

The role played by the intercostal muscles in the ventilatory cycle remains questionable in spite of numerous electromyographic investigations (1) (7) (8). The lack of systematic studies performed on animals and the difficulty to accurately identify the electrical activities of each thoracic muscle on man are obviously the main cause of this uncertainty. Indeed, the external and internal intercostal muscles are two thin closely juxtaposed muscular layers entirely covered by more superficial muscles. Moreover, the thoracic muscular activities recorded in conscious man (7) (8), except during eupnea, may be the result of the voluntary motor control which can markedly differ from that generated by the respiratory centers. Thus, beyond the possible differences related to species, results obtained in conscious man or in animal are not unquestionably comparable. In the past, many divergent opinions have been expressed on the inspiratory or expiratory function of the intercostal muscles (2) and the greatest uncertainty concerns the lateral part of the intercostal muscles which participate not only in breathing but also in posture (3), (5), and in other non respiratory activities such as the thermal shivering (4). Are the external and internal

intercostal muscles respectively inspiratory or expiratory
in all the intercostal spaces ? We have aimed at answering
this question. The present study is a systematic analysis
of the electrical activities exhibited in the cat, during
eupnea or in the course of artificially induced dyspnea,
by both intercostal muscles and the intercostal nerve
filaments.

METHODS

In order to eliminate the influence of the
cerebral motor control, all the experiments were performed
on decerebrated preparations (twenty five adult cats
weighing between 2.2 and 3.8 kg). In such experimental
conditions the recorded "respiratory activities" are
assumed to be only due to the combined action of both the
central respiratory rhythm generators and the spinal
afferent segmental circuits.

Electromyography

Previously described (4) bipolar electrodes were
inserted under microscopic control into the external
intercostal muscles along the mid axillary line. The same
procedure was used for the internal intercostal muscles
after a small sector of the corresponding external
intercostal muscle was incised and reflected forward. In
some cases the electrodes inserted between the two
intercostal muscular layers could pick up the electrical
activity of both the internal and the external
intercostal muscles.

Nerve filaments activity

External and internal intercostal nerve filaments
of the same intercostal space were carefully dissected as
recommended by Sears (6) and placed on pairs of recording

electrodes (see schema) inserted in the corresponding
muscles to avoid nerve damage during the chest wall
displacements. Nerve and electrodes were then immersed
in a mineral oil bath.

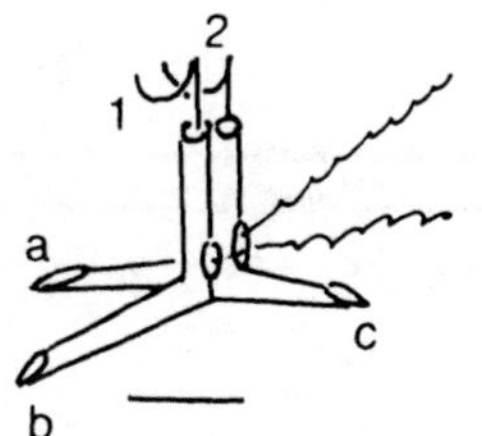

schema of the electrodes :

a. b and c are inserted in the

muscle.

1 and 2 receive the nerve filament

The inspiratory time was determined referring to
the electrical activity of the diaphragm. Recordings were
made both during quiet breathing and in the course of
artificially induced dyspnoeas. A bivalve tracheal canula
was used to selectively block either the inspiratory or
the expiratory flow and thus to reinforce the corresponding
respiratory central drive. Total tracheal occlusions have
also been performed to reinforce both the inspiratory
and expiratory activities. In some experiments the muscles
of both the paravertebral and parasternal regions have been
studies.

RESULTS

During quiet breathing

We observed a cephalo-caudal decrease of the
spontaneous electrical activity of the lateral external
intercostal muscles (fig. 1 A2 ; B2 ; C2). When observed,
this activity was most of the time devoid of respiratory
modulation. Nevertheless an inspiratory activity could
occasionaly be observed only on the first eight
intercostal spaces. The internal intercostal muscles were
generally silent except in the last four spaces where an
expiratory activity could be observed.

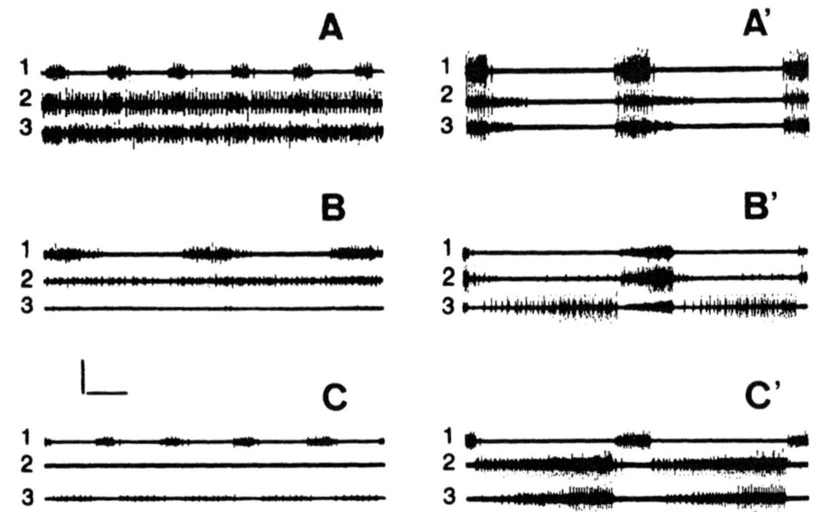

FIGURE 1 : Electrical activity of the lateral external and
           intercostal muscles in decerebrated cats during
           eupnea (A-B-C) and during the course of total
           tracheal occlusions (A'-B'-C'). On each sequence
           1 - diaphragm, 2 - lateral external intercostal,
           3 - lateral internal intercostal muscles of the
           third (A-A'), the sixth (B-B'), and the ninth
           (C-C') intercostal spaces. Note in A' the
           absence of any intercostal expiratory activity
           and in C' the absence of intercostal inspiratory
           activity. In the sixth intercostal space (B'),
           the external intercostal (2) exhibits an expira-
           tory activity. The small inspiratory activity
           recorded in B'3 is probably that of the corres-
           ponding external intercostal muscle.

During artificially induced dyspnoeas :

        In the cephalic part of the thorax (1 St - 5th
ribs) an inspiratory reinforcement of the electrical
activity of the lateral external or internal intercostal
muscles and nerve filaments occured in the course of the
tracheal occlusions interrupting the inspiratory flow
(fig. 1A') (fig. 2). At the same level, no expiratory
activity of any kind was observed during prolonged
interruptions of the expiratory flow (fig. 2). In the
caudal part of the thorax (9th - 13th ribs) whatever the
duration of the "inspiratory" tracheal occlusions was,
we never observed any inspiratory electrical activity
of the lateral intercostal muscles or nerves (external

or internal) (fig. 3). After interrupting the expiratory
flow, the same muscles and the corresponding nerve
filaments became active only during expiration (fig. 1 C')
(fig. 3). On the intermediate thoracic region (5th - 9th
ribs) a clear antagonism appeared between the lateral
external and internal intercostal muscles (fig. 1 B').

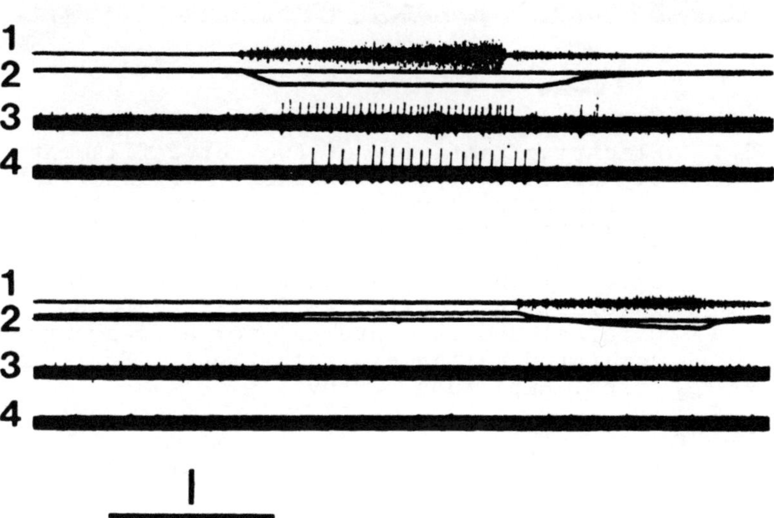

FIGURE 2 : Electrical activity of two nerve filaments of
the third intercostal space. A : interruption
of the inspiratory flow. B : interruption of
the expiratory flow. On each sequence : 1 -
diaphragm, 2 - oesophageal pressure, 3 - external intercostal nerve filament, 4 - internal
intercostal nerve filament. Both the external
and internal intercostal nerve filaments are
activated during the "inspiratory dyspnoea"
only. Scale time 1 sec. Amplitude 200 µV.

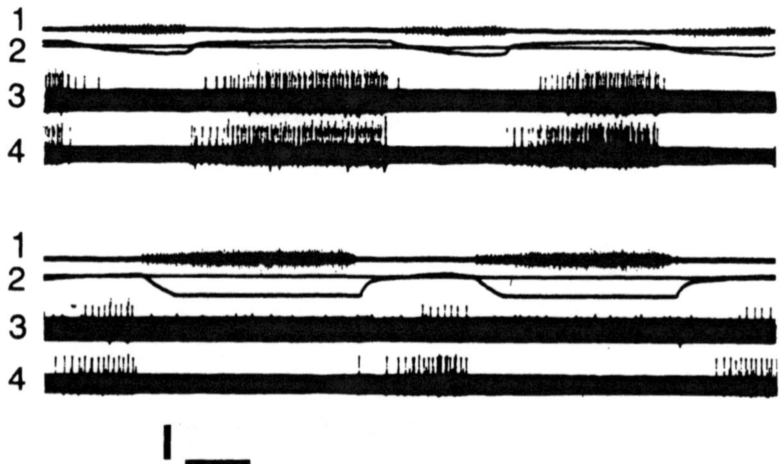

FIGURE 3 : Electrical activity of two nerve filaments of
          the ninth intercostal space. A : interruption
          of the inspiratory flow.B : interruption of the
          expiratory flow. On each sequence : 1 -
          diaphragm, 2 - oesophageal pressure, 3 - 9th
          external intercostal nerve filament, 3 - 9th
          internal intercostal nerve filament. Both the
          external and the internal nerve filaments are
          activated during expiratory only. Scale time
          1 sec. Amplitude 200 µV.

<u>CONCLUSION</u>

Our results obtained from two different technical
procedures demonstrate that, on the thorax, different
functionnal regions which can be characterized either by
the absence or the presence of an expiratory-inspiratory
antagonism. The activation of both the intercostal muscles
and the nerve filaments in our experimental conditions is
likely due to the combined action of both the central
respiratory rhythm generators and the spinal segmental
afferents. Under these influences, the lateral external
and internal intercostal muscles are synergist during either
inspiration (cephalic spaces) or expiration (caudal spaces).
The results we obtained in decerebrated cats during eupnea
are quite similar to those observed in conscious man during
quiet breathing by TAYLOR (7). Nevertheless, according to
this author, during the course of "a voluntary deeper

breathing" the external intercostal muscles are inspiratory and the internal ones expiratory at every level of the thorax. The discrepancy between these latters Taylor's data and the data we obtained in decerebrated cats during tracheal occlusions may be explained either by differences related to species or by the influence of the cerebral motor control.

REFERENCES

1. Campbell, E.J.M., (1958). The respiratory muscles and mechanics of breathing, Lloyd-Luke (Medical Books), London, 131 pp.

2. Duchenne de Boulogne, G.B., (1867). Physiologie des mouvements, Baillère, Paris, 872 pp.

3. Duron, B., (1981). Intercostal and diaphragmatic muscle endings and afferents. In T.F. Hornbein (Ed), Regulation of breathing, Volume 17, part. 1, Marcel Dekker, Inc., New-York, pp. 473-540.

4. Duron, B., and Caillol, M.C., (1971). Activité électrique des muscles intercostaux au cours du frisson thermique chez le chat anesthésié, J. Physiol. (Paris), 63, 523-537.

5. Massion, J., Meulders, M., and Colle, J., (1960). Fonction posturale des muscles respiratoires. Arch. Int. Physiol. Biochem., 68, pp. 314-326.

6. Sears, T.A., (1964). Efferent discharges in alpha and fusimotor fibres of intercostal nerves of the cat, J. Physiol., 174, pp. 295-315.

7. Taylor, A, (1960). The contribution of the intercostal muscles to the effort of respiration in man, J. Physiol. (London), 151, 390-402.

8. Tokizane, T., Kawàmata, K., and Tokizane, H., (1952). Electromyographic studies on the human respiratory muscles, Jpn. J. Physiol., 2, 232-247.

35

Cross-Correlation Analysis of the Connections between Bulbospinal Neurones and Respiratory Motoneurones

T. A. SEARS, P. A. KIRKWOOD and J. G. McF. DAVIES

The respiratory motoneurones innervating the rib cage and diaphragm are subject to a variety of influences in the awake state as metabolic, postural and other behavioural commands for respiratory movement are met through the integrative action of the respiratory motoneurones themselves. Under deep anaesthesia the commands are restricted to the metabolic system which exercises its control through respiratory bulbospinal neurones which couple the central pattern generator for respiratory rhythm to the spinal motoneurones and interneurones. There are two questions about the nature of this coupling, one concerning the extent to which the coupling to the phrenic and intercostal motoneurones is monosynaptic or polysynaptic, the other concerns the connectivity which determines the spatial distribution over the surface of the rib cage of the inspiratory and expiratory activities. Although these questions are inter-related only the first is considered here.

Despite published reports of both monosynaptic and polysynaptic connections from bulbospinal inspiratory neurones to spinal motoneurones [see 1,2,3], detailed descriptions of the strength and distribution of these connections are still lacking. We have used cross-correlation methods in anaesthetized, paralyzed cats to study the monosynaptic pathways to intercostal (T2-T9) and phrenic motoneurones. This method was chosen instead of spike-triggered averaging of PSPs because firstly it is specific to the motoneurones in receipt of the greatest drive (i.e., to those that are firing) and secondly, with suitable technology [4] a large sample of motoneurones can be studied. Analysis consisted of constructing cross-correlation histograms from the discharges of single antidromically identified inspiratory bulbospinal neurones (collision test used) and multi-unit motoneurone discharges recorded from either or both the contralateral phrenic nerve and filaments of the external intercostal nerves of several segments.

The ability to detect a monosynaptic connection with either spike-triggered averaging or cross-correlation methods depends on the availability of reliable criteria which can be used to separate the effects of other synaptic relationships from those which may be due to a monosynaptic one. This is especially the case if one wishes to make quantitative measurements, for instance of percentage connectivity. The two parameters which were used as criteria here are the duration and the latency of peaks in the cross-correlation histograms. A monosynaptic connection is expected to give a narrow peak, with a duration roughly corresponding to the rise-time of the underlying unitary EPSP in the motoneurone(s) [5,6] (typically less than 1ms) and a latency appropriate to the conduction velocity of the individual axon. A polysynaptic connection is expected to give mostly longer duration peaks and also longer latencies than the peaks from monosynaptic connections.

A third mechanism which would give a peak in the cross-correlation histograms is input synchronization, i.e., the firing of other neurones being synchronized to the firing

of the selected bulbospinal neurone and these other
neurones also exciting the motoneurones. Thus a cross-
correlation peak could exist between the discharges of the
bulbospinal neurone and the motoneurones even if they were
not connected. In a comparable situation where peripheral
afferent inputs were investigated and where the dynamic
sensitivity of muscle spindle primary endings readily
allowed afferent synchronization [7], this situation was
avoided by cutting most of the dorsal roots. With the
reference spikes being provided by a central neurone, as
here, this is not possible. Moreover presynaptic synchronization is more than just a theoretical possibility since
the inspiratory neurones of the medulla are known to be
capable of synchronization (see 8 for references). However,
there is one circumstance here in the experimenter's
favour, which is that there is a relatively long conduction
distance from the medulla to either the phrenic or the
intercostal motoneurones. Thus temporal dispersion among
the different descending axons, which are well known to
have a wide range of conduction velocities, would mean
that, even if the synchronization in the medulla was
constrained within a very short time span, the durations of
the resulting peaks in the cross-correlation peaks between
discharges of bulbospinal neurones and motoneurones would
be relatively long. A model set up to represent this
situation gave a minimum half-width of 1.1 ms for such
peaks. Latencies of these peaks depend on the timing of the
synchronization in the medulla and on the conduction
velocities of the different descending axons.

Experimentally we found two categories of peaks which
corresponded well with these expectations. Firstly there
was a group of narrow peaks (Fig. 1A,B) with half widths of
less than 1.1ms (mean 0.64 ms). These narrow peaks were at
the appropriate latency (0.8 - 2.2 ms after the arrival of
the bulbospinal impulse at the particular spinal cord
segment, as calculated from the collision test [cf. 10])

and their shapes corresponded to theoretical expectations
for a monosynaptic connection [5]. Secondly, there was a
group of peaks with broader and more variable half-widths
(Fig. 1C,D) which, in fact, were rather similar to others
previously described (e.g. ref. 9). The latencies of this
second group varied widely though, notably, some were much
too short for either a mono- or a polysynaptic connection
(e.g. Fig. 1C, conduction velocity of descending axon, 28
m/s; latency of arrival of axonal spike at C5, 2.0ms). For
these peaks with short latencies presynaptic synchronization was a necessary explanation. For the longer latency
peaks of this group presynaptic synchronization was a
sufficient explanation, although polysynaptic excitation
could also be involved. Only peaks belonging to the first
group (half-width less than 1.1ms and at appropriate
latencies) were taken as indicating a monosynaptic
connection.

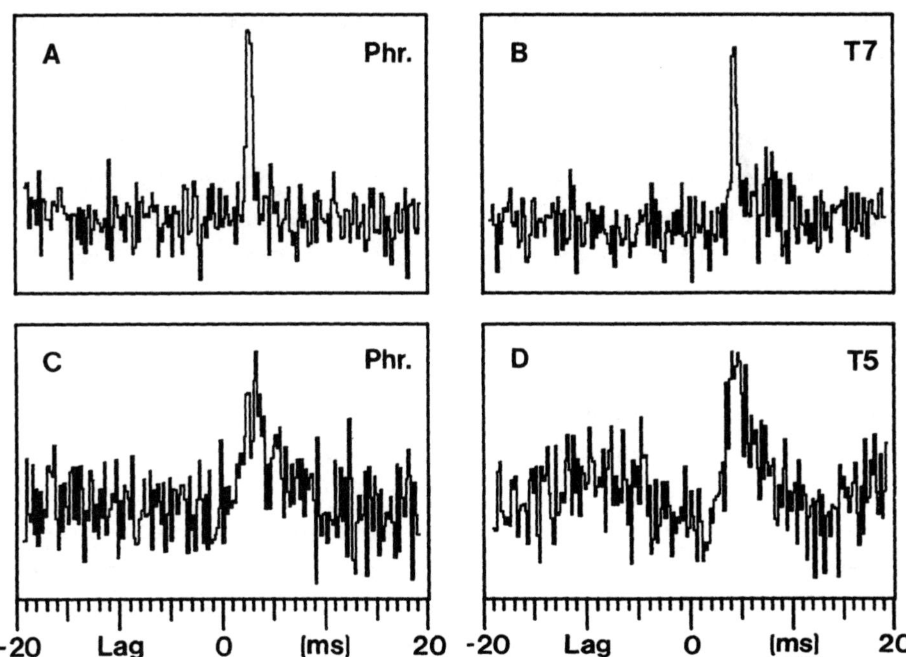

FIGURE 1. Cross-correlation histograms, reference spikes
from bulbospinal units. Ordinate: counts, full-scale
(A)5400-6400, (B)1400-1900, (C)2700-3150, (D)5800-6500.

26/59 bulbospinal neurones antidromically identified
from thoracic levels gave some monosynaptic connections to
intercostal motoneurones, whereas for the phrenic the
proportions were 8/20 neurones identified from T4-T5 and
5/7 identified from C3. For individual intercostal
segments, 50/295 unit/segment pairs showed a monosynaptic
connection. The connections were not randomly distributed;
a unit giving a connection in one segment was significantly
more likely to give one in another segment than the mean
connectivity would suggest. Part of this effect was related
to the location of the units in the medulla. The proportion
of connections from the dorsal respiratory group was 8/88
and from the ventral group caudal to obex 7/80. However in
the more rostral ventral group the proportion was 35/127.

The strength of the cross-correlation histograms was
measured by the ratio k [11] and a value of k = 1.0 was
assigned to histograms without a detectable peak. The mean
values of k were 1.019 for the intercostal connections and
1.055 for the phrenic. If k = 1.5 is taken as equivalent to
a 100uV EPSP, preliminary calculations [cf. 12] suggest
that if motoneurone depolarization is produced by these
monosynaptic connections alone, about 5200 bulbospinal
neurones would have to be active to excite the intercostal
motoneurones or about 1850 for the phrenic. These estimates
are of the maximum numbers of neurones, because the severe
criteria we have used for identifying monosynaptic
connections probably underestimate the connectivity by
rejecting some peaks (e.g. Fig. 1C) which could have
included monosynaptic effects concealed within those peaks
due to input synchronization. Nevertheless these required
numbers of bulbospinal neurones seem much larger than any
other estimates of actual numbers that may be made either
on the basis of general experience from recording in the
medulla or from what little anatomical evidence is
available. Thus we conclude, in contrast to our earlier

suggestions [13], that the majority of the excitation is
via interneurones [cf.1, 14]. The generalized description
derived from motoneurone synchronization measurements by
Kirkwood et al. [13] of a projection pattern of the
collaterals for any one presynaptic fibre to intercostal
motoneurones which is mostly restricted to 2 or 3 segments
may now be taken as a description of the projections of
interneurones which transmit the respiratory drive and
which themselves may be excited by collaterals of the axons
whose monosynaptic connections are here described.

References

1.   Fedorko, L., Merrill, E.G. and Lipski, J. (1983). Two
descending medullary inspiratory pathways to phrenic
motoneurones. Neurosci. Letters., 43, 285-291.
2.   Cohen, M.I. and Feldman, J.L. (1984). Discharge
properties of dorsal medullary inspiratory neurons:
relation to pulmonary afferent and phrenic efferent
discharge. J. Neurophysiol., 51, 753-776.
3.   Aminoff, M.J.and Sears, T.A. (1971). Spinal
integration of segmental, cortical and breathing inputs to
thoracic   respiratory motoneurones. J. Physiol., 215,
557-575.
4.   Davies, J.G., Forster I.C. and Sears, T.A. (1983). A
modular interface for the computer acquisition of
multichannel neural data. J.Physiol. 342, 7-8P.
5.   Kirkwood, P.A. (1979). On the use and interpretation
of cross-correlation measurements in the mammalian central
nervous system. J. Neurosci. Methods, 1, 107-132.
6.   Fetz, E.E. and Gustafsson, B. (1983). Relation between
shapes of post-synaptic potentials and changes in firing
probability of cat motoneurones. J. Physiol., 341, 387-410.
7.   Kirkwood, P.A. and Sears, T.A. (1982a). Excitatory
post-synaptic potentials from single muscle spindle
afferents in external intercostal motoneurones of the cat.
J. Physiol., 322, 287-314.
8.   Long, S.E. and Duffin, J. (1984). Cross-correlation of

ventrolateral inspiratory neurons in the cat. Exp. Neurol., 83, 233-253.

9.   Feldman, J.L. and Speck, D.F. (1983). Interactions among inspiratory neurons in dorsal and ventral respiratory groups in cat medulla. J. Neurophysiol., 49, 472-490.

10.   Kirkwood, P.A. and Sears, T.A. (1982b). The effects of single afferent impulses on the probability of firing of external intercostal motoneurones in the cat. J. Physiol., 322, 315-336.

11.   Sears, T.A. and Stagg, D. (1976). Short-term synchronization of intercostal motoneurone activity. J. Physiol., 263, 357-381.

12.   Sears, T.A. (1977). The respiratory motoneurone and apneusis. Fedn. Proc. 36, 2412-2420.

13.   Kirkwood, P.A., Sears, T.A., Stagg, D. and Westgaard, R.H. (1982). The spatial distribution of synchronization of intercostal motoneurones in the cat. J. Physiol., 327, 137-155.

14.   Lipski, J. and Merrill, E.G. (1983). Inputs to intercostal motoneurones from ventrolateral medullary respiratory neurones in Nembutal-anaesthetized cats. J. Physiol., 339, 25-26P.

36

Projection and Conduction Velocity of Single Inspiratory Bulbospinal Axons in Cervical Cord

T. E. DICK and A. J. BERGER

INTRODUCTION

The descending rhythmic drives directed to respiratory spinal motoneurons arise from many "premotor" or bulbospinal (BS) respiratory neurons of the dorsal and ventral respiratory groups. Mapping experiments [1] done by determining minimal stimulus threshold for antidromic activation (AA) have shown that BS axons lie in the lateral and ventral funiculi. Recent neuroanatomical experiments [2,3] support this concept of a wide distribution in the spinal cord.

Mean BS axonal conduction velocity (CV) has also been estimated using AA techniques [4]. However, mean CV estimates vary by more than a factor of two (26.2 to 59.6 m/s) [5]. Recently, spike-triggered averaging (STA) of extracellular field potentials has been used as an alternative method to AA for determining a neuron's projection [6] and its axon's CV [7,8]. In this study, data from each method were compared to determine: 1) the position of these BS axons in the cervical spinal cord, and 2) a precise estimate of axonal CV.

METHODS

In 19 anesthetized, paralyzed, and artificially ventilated cats, 26 inspiratory BS neurons (1 dorsal respiratory group and 25 ventral respiratory group cells) were recorded extracellularly with glass microelectrodes. A tungsten microelectrode (10-100 kΩ) was placed in the contralateral cervical spinal cord at C3 to antidromically activate these neurons. By

223

varying the position of the tungsten electrode, we mapped the stimulus thresholds of cathodal, constant-current pulses (200-μs duration at 10 Hz) that evoked BS neurons. Antidromic activation was confirmed by high-frequency following and collision tests.

At points where stimulus threshold was less than 30 μA, we recorded field potentials with the same tungsten electrode, averaging the field potentials for 5 ms following a trigger generated by the medullary spike (5-μs bin width, ≥ 2000 averages). Thus we located BS axons by determining the location of the smallest stimulus threshold for AA and the location of the largest recorded axonal potential by STA. Also we determined axonal CV from the latency of the antidromically evoked spike or from the averaged axonal potential latency. In many cases (10/26), this location procedure was done at both C3 and C4. Small lesions (100-μA DC, 10 s) were made to locate the position of the descending axon, using light-microscopic analysis of processed tissue.

RESULTS

Axonal Projection

The methods agreed in their determination of the projecting axon's location. Inspiratory BS axons were located in both the lateral and ventral funiculi of contralateral cervical spinal cord. However, the majority (81%, 21/26) projected in the ventral tracts. In ten cases where the descending axons were located at both C3 and C4, an axon's position remained approximately constant.

Five of six pairs of "neighboring" BS neurons — units recorded within 300 μm of one another — had axons that projected close, within 350 μm of each other in the cervical spinal cord. In the one exception where the axons from a pair were located 1700 μm apart, and the cells had very different firing patterns. One cell had a delayed onset (1.1 s after the onset of phrenic nerve activity), and its axon was located laterally. The other cell began to discharge simultaneously with the phrenic nerve and its axon was in the ventral tract.

Axonal Conduction Velocity

Estimates of CV were made with both methods. For AA, conduction time was the latency of the evoked spike's peak following stimulus onset. For STA, conduction time was the latency of the peak depolarization in the recorded axonal potential following the onset of the trigger spike. The estimate of the average axonal CV from AA at a single-point was significantly less (p<.001) than that determined by STA (Table 1). For example, mean CV, estimated at C3 with the AA method, was 42% less than the mean CV estimated at the same point by STA (Table 1). However, there was no significant difference (p>.2) in mean CV for two-point determinations. In such cases, estimates of axonal CV were calculated either from the difference in the latencies of the medullary spike evoked antidromically from two points in the spinal cord, or from the difference in the latencies of the averaged axonal potential recorded at two points in the spinal cord. The mean CV was 55.4 m/s estimated from AA data and 53.3 m/s estimated from STA data (Table 1).

TABLE 1   Mean ± standard deviation of conduction velocity for the 10 axons
that were located at both C3 and C4.

Methods	Single-point Determinations (m/s)		Two-point Determination (m/s)
	C3	C4	
Spike-triggered Averaging	44.0±8.2	46.9±8.9	53.3±13.1
Antidromic Activation	25.5±4.5	31.2±5.0	55.4±13.1
Paired t Test	p<.001	p<.001	N.S.

DISCUSSION

Previous investigations have shown similar axonal distributions of BS respiratory neurons in lateral and ventral funiculi of cervical spinal cord [1,2,3]. Particularly noteworthy in this study were: 1) ten axons located at both C3 and C4 maintained their positions in the descending tracts, and 2) axons arising from "neighboring" somata with similar firing patterns projected within 350 µm in the spinal cord. Even though the two methods (AA and STA) agreed in the location of an axon, each method has its advantages.

The AA method can determine whether or not an axon projected within a broad area, whereas the STA method can locate an axon more precisely because it involves maximizing an easily measured property of neurons.

Reported axonal CV estimates have varied [5]. The averages of two-point determinations of CV by both methods are at the high end of this range and were not significantly different. Single-point estimates using STA at C4 provided a closest estimate to the two-point determinations (Table 1). In determining CV, STA has a distinct advantages over AA. With STA, utilization time is not a component of conduction time and many problems associated with electrical stimulation are avoided [8].

ACKNOWLEDGMENTS

This work was supported by USPHS grant NS 14857.

REFERENCES

1.   Merrill, E.G. (1974). Finding a respiratory function for the medullary respiratory neurons. In: Bellairs, R. and Gray, E. (eds). Essays on the Nervous System. pp. 451-86. (Oxford: Claredon Press)
2.   Feldman, J.L., Loewy, A.D., and Speck, D.F. Projections from the ventral respiratory group to phrenic and intercostal cat: An autoradiographic study. [submitted]
3.   Holstege, G. and Kuypers, H.G.J.M. (1982). The anatomy of brain stem pathways to the spinal cord in cat. A labeled amino acid tracing study. In: Kuypers, H.G.J.M. and Martin, G.F. (eds). Descending Pathways to the Spinal Cord, Progress in Brain Research, 57. pp. 145-75. (New York: Elsevier Biomedical Press)
4.   Lipski, J. (1981). Antidromic activation of neurones as an analytic tool in the study of the central nervous system. J. Neurosci. Meth., 4, 1-32
5.   Feldman, J.L. Neurophysiology of breathing in mammals. In: Bloom, F.E. (ed). Handbook of Physiology, The Nervous System, Intrinsic Regulatory Systems of the Brain. (Bethesda: American Physiological Society) [submitted]
6.   Berger, A.J. and Averill, D.B. (1983). Projection of single pulmonary stretch receptors to solitary tract region. J. Neurophysiol., 49, 819-30
7.   Kirkwood, P.A. and Sears, T.A. (1975). Spike triggered averaging for the measurement of single unit conduction velocities. J. Physiol. (London) 245, 58P-59P
8.   Lemon, R. (1984). Methods for Neuronal Recording in Conscious Animals. (New York: Wiley & Sons)

37

Physiological Properties of Motor Units in the Diaphragm

G. C. SIECK, M. FOURNIER and M. J. BELMAN

In hind limb muscles 4 types of motor units are distinguished based on contractile and fatigue properties : 1) Fast-twitch, fatigable (FF) ; 2) Fast-twitch, intermediate fatigability (FI) ; 3) Fast-twitch, fatigue resistant (FR) and 4) Slow-twitch, non-fatigable (S) (Burke et al, 1973). Generally, relatively few FI motor units are present (i.e., 5 % or less). Thus, the overall population of motor units usually form 2 major groups based on fatigability, i.e., fatigable (FF) and fatigue resistant (FR and S). Two recent studies (Nemeth et al., 1981 and Kugelberg and Lindegren, 1979) have reported a direct correlation between motor unit fatigue resistance and the oxidative activity of muscle fibers belonging to the motor unit. With the segregation of motor units into fatigable and fatigue resistant groups, the oxidative activities of muscle fibers should also separate into 2 populations. In a recent study (Sieck et al., 1983), we found that the oxidative activities of diaphragmatic muscle fibers were unimodally distributed. This suggested that either the correlation between fatigue resistance and oxidative activity was absent in the diaphragm, or that motor units did not show the same separation of fatigue properties. The purpose of this study was to characterize the contractile and fatigue-related properties of motor units in the cat diaphragm.

Adult cats (2.5 to 3.5 kg) were anesthetized with sodium pentobarbital and blood pressure, heart rate, end-tidal CO_2 and rectal temperature monitored. Following laporatomy, EMG electrodes were inserted into the sternal and costal regions of the diaphragm. The costal margin of the rib cage was freed and attached via a non-compliant string to a force transducer. In preliminary studies, we determined that the C5 ventral root innervated a restricted area of the sternal and ventral costal regions of the diaphragm. Thus, we optimized the orientation of the force transducer in the line of pull of these fibers. The central tendon of the diaphragm was fixed, so that force measurements were made under isometric conditions. Muscle fiber length was adjusted to achieve optimal force/length relationships during the experiment. After cervical laminectomy, the spinal cord was transected at C3 and the animal was mechanically ventilated. The C5 ventral root was isolated and single filaments teased and placed on bipolar stimulating electrodes. Filaments were sampled systematically in a rostral to caudal direction. Verification of the isolation of a single motor unit was based on varying the stimulus intensity above threshold and noting : 1) a constant latency and waveform of the EMG evoked response ; and 2) a constant twitch force. Isolated motor units were usually restricted to either the sternal or ventral costal region of the diaphragm based on the evoked EMG response. There appeared to be a rostral to caudal distribution of motor units within the C5 ventral root with rostral axons innervating the sternal region and more caudal axons innervating the costal region. Time to peak tension (contraction time, CT) and twitch force (Ptw) were measured for each unit. Subsequently, force/frequency relationships were assessed at 1, 5, 10, 15, 20, 30, 40, 50, 75 and 100 Hz stimulation. The fatigue-related properties of each motor unit was assessed using a fatigue test (Burke et al., 1973) which consisted of trains of 13 pulses presented at a rate of 40 Hz, with a train duration of 330 msec and 1 train per sec. The force produced at 2 min was divided by the initial force to derive a fatigue index. In hind limb muscles, fatigue indices above 0.75 are considered fatigue-resistant units, while indices below 0.25 are considered fatigue units (Burke et al., 1973). Contractile and fatigue-related properties of 70 diaphragmatic motor units have been assessed. Contraction times ranged between 28 and 115 msec, with the predominance of faster contraction times (66 % with CT's of 45 msec or less). Twitch tensions (Ptw) ranged from 0.3 to 8.5 g and tetanic tensions (Po) ranged from O.9 to 35 g. Twitch/tenanic ratios (Ptw/pO) ranged from O.8 to O.62. Generally, no consistent correlation was present betweem CT ant Ptw, Po or Ptw/Po. However, diaphragmatic motor units with slower CT's did show fusion at lower frequencies. Thus, there were differences between fast and slow motor units in the shape of the force/frequency curves. In contrast to hind limb muscles, we found that the majority of diaphragmatic motor units showed intermediate fatigue indices (53 % of diaphragmatic motor units with fatigue indices between O.25 and O.75). Approximately 28 % of motor units were fatigable and 19 % fatigue resistant. These findings indicate that diaphragmatic motor units have contractile properties which are similar to those found in other skeletal muscles. However, the high proportion of motor units with intermediate fatigability suggests that the diaphragm differs significantly in its resistance to fatigue. This might reflect the influence of the continuous rhythmic pattern of activation of the diaphragm which contrasts with that of other skeletal muscles.

Research supported by NIH Grant HL29999-O2

References

1. BURKE, R.E., LEVINE, D.N., TSAIRIS, P. and ZAJAC, F.E. (1973). Physiological type and histochemical profiles in motor units of the cat gastrocnemius. J. Physiol., 234, 723-748.

2. KUGELBERG, E. and LINDEGREN, B. (1979). Transmission and contraction fatigue of rat motor units in relation to succinate dehydrogenase activity of motor unit fibers. J. Physiol., 288, 285-300.

3. NEMETH, P.M., PETTE, D. and VRBOVA, G. (1981). Comparison of enzyme activities among single muscle fibers within defined motor units. J. Physiol., 311, 489-495.

4. SIECK, G.C., SACKS, R.D., BLANCO, C.E. and EDGERTON, V.R. (1983). Quantification of succinate dehydrogenase activity of diaphragmatic muscle fibers. Soc. for Neuroscience, 9, 1164.

38

Evidence for Differential Inputs to Phrenic Motoneurons based on Dendritic Morphology

W. E. CAMERON, D. B. AVERILL and A. J. BERGER

INTRODUCTION

Two mechanisms have been proposed to explain the orderly activation of phrenic motoneurons (PMs) during a normal inspiration. The first hypothesis, the size principle (1,2), predicts that excitability is inversely related to cell size. The second hypothesis proposes that early and late recruited PMs receive different inputs from various subpopulations of respiratory neurons located in the brain stem (3-5). The purpose of the present study was twofold: i) to provide a direct test of the size principle in the phrenic nucleus and ii) to determine whether early and late recruited PMs are morphologically distinct as a result of the receipt of different afferent inputs.

METHODS

Phrenic motoneurons of the C5-C6 spinal cord were studied in the chloralose-urethan-anesthetized, paralyzed and artificially ventilated cat. Antidromically identified PMs were impaled with beveled glass microelectrodes and injected by iontophoresis with horseradish peroxidase (HRP). The cervical cord was fixed, frozen, sectioned and reacted with diaminobenzidine (DAB), and neuronal reconstructions made from serial sections. A PM was classified based upon its extra- and intracellularly recorded, spontaneous discharge as well as the amplitude and trajectory of its central respiratory drive potential (6). Early recruited units initiated discharge within the first quarter of the inspiratory cycle, while late recruited units fired during the second half of the cycle. A third group of cells had no spontaneous discharge; these were classified as high threshold units.

RESULTS

Figure 1 shows three PMs reconstructed from sagittal sections, arranged according to order of recruitment, top to bottom: early, late and high threshold cells as indicated by their discharge (inset). As predicted by the size principle, axonal conduction velocity (CV), axon diameter and cell body size increase as one moves down the recruitment sequence. Although each cell shown is from a different experiment, the central drive was assumed to be roughly equivalent by maintaining a 5% end-expired CO_2. In two experiments, pairs of cells (late:late and early:early) were labeled within the same experiment. Comparison of these pairs confirmed similarity between PMs with the same discharge pattern. Similar trends were found when the total sample of 29 PMs was analyzed (7 early, 12 late, and 10 high threshold).

In addition to the difference in sizes, the shapes of cell bodies and the distribution of dendrites also were found to differ between early and late recruited units. The cell bodies of late and high threshold PMs were more ellipsoid than early recruited cells, with the major axes of their somas oriented parallel to the longitudinal axis of the phrenic nucleus and spinal cord. From the poles of these elongate cells, stem dendrites with largest diameters emerged to generate a large, rostrocaudally extending, dendritic surface area within the confines of the motor column. The early PMs, on the other hand, generated more laterally directed dendrites in the vicinity of the cell body. Once again, the analysis of the early and late pairs from the same experiments supported these observations.

DISCUSSION

The theory of neurobiotaxis (7) states that "the growth of the chief dendrite, and eventually the displacement of the cell body itself, takes place in the direction whence the majority of stimuli proceed to the cell." Therefore, we suggest that dendrites of late and high threshold cells are arranged so as to optimize receipt of input from those medullary respiratory neurons whose axons terminate predominantly within the confines of the phrenic nucleus (8). From these data we conclude that early and late

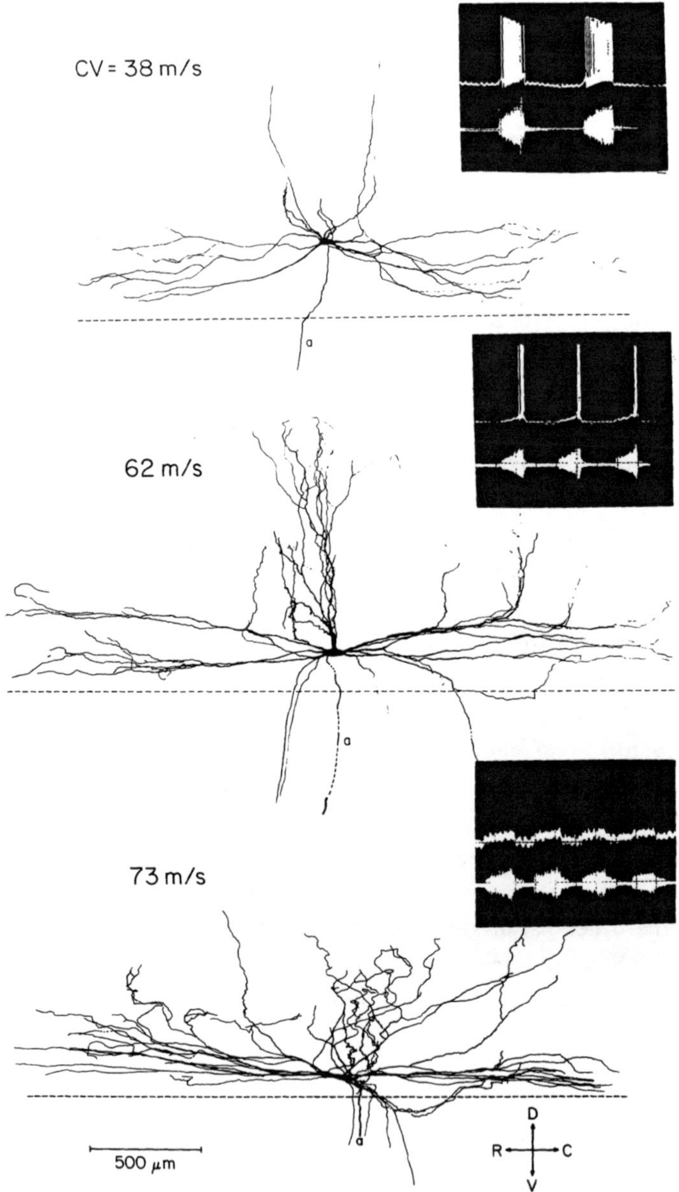

FIGURE 1   Sagittal reconstructions of three PMs.   The axonal velocity (CV) is given on the left of each reconstruction.   The IC record (top) and phrenic neurogram (bottom trace) are presented in the inset at the upper right.   The broken line indicates the ventral margin of the gray matter and a indicates the axon.   D - dorsal; V - ventral; R - rostral; C - caudal.

recruited PMs are morphologically distinct, based not only on their size, but also on the distribution of their membrane surface area. This latter distinction may reflect the mechanism by which these cells receive varying proportions of synaptic contacts from the different subpopulations of respiratory bulbospinal neurons.

ACKNOWLEDGMENTS

Dr. Cameron's present address is: Departments of Anatomy and Pediatrics, University of Pittsburgh School of Medicine, 3550 Terrace Street, Pittsburgh, Pennsylvania 15261, U.S.A.

This work was supported by USPHS grant NS 14857, NIH postdoctoral fellowships to W.E.C. and D.B.A. and a Muscular Dystrophy Association fellowship to W.E.C.

REFERENCES

1. Henneman, E. (1957). Relation between size of neurons and their susceptibility to discharge. Science, 126, 1345-46
2. Henneman, E., Somjen, G. and Carpenter, D.O. (1965). Functional significance of cell size in spinal motoneurons. J. Neurophysiol., 28, 560-80
3. Hilaire, G., Monteau, R. and Dussardier, M. (1972). Modalités du recrutement des motoneurones phréniques. J. Physiol. Paris, 64, 457-78
4. Hilaire, G. and Monteau, R. (1979). Facteurs déterminant l'ordre de recrutement des motoneurones phréniques. J. Physiol. Paris, 75, 765-81
5. Hilaire, G., Gauthier, P. and Monteau, R. (1983). Central respiratory drive and recruitment order of phrenic and inspiratory laryngeal motoneurones. Resp. Physiol., 51, 341-59
6. Berger, A.J. (1979). Phrenic motoneurons in the cat: Subpopulations and nature of respiratory drive potentials. J. Neurophysiol., 42, 76-90
7. Kappers, C.U.A. (1917). Further contributions on neurobiotaxis. J. Comp. Neurol., 27, 261-98
8. Feldman, J.L., Loewy, A.D. and Speck, D.F. Projections from the ventral respiratory group to phrenic and intercostal motoneurons in the cat: An autoradiographic study. J. Neurosci. (submitted)

Section 4
Afferent Systems and their Terminations at the Brain Stem Level

39

Central Projections of Phrenic Afferent Fibers

D. MARLOT, J.-M. MACRON and B. DURON

INTRODUCTION

An histological study of the phrenic diaphragm complex in the cat (1) has shown than 25 % of the myelinated fibers of the phrenic nerve (PN) are afferents. Moreover, only 2 % (about 15 nervous fibers) could have a spindle origin. During electrophysiological recordings, most of the PN afferents are active during expiration or inspiration and are certainly related to either muscle spindles or Golgi tendon organ (2, 3). According to previous data (4, 5), the present report deals with projections of phrenic afferents to the cerebellar cortex (CCx) and to the external cuneate nucleus (ECN) and the lateral reticular nucleus (LRN) which are two of the main relay nuclei of the spinocerebellar tracts.

MATERIAL AND METHODS

Experiments were performed on spontaneously breathing adult cats anesthetized with intraperitoneal injection of sodium pentobarbital (35 mg/kg). The trachea and the femoral vein were cannulated. A bipolar electrode (copper wires) was inserted through an abdominal approach to the left lateral part of the dome of the diaphragm. The right PN (C5 and C6 branches) was isolated, desheathed and cut. The skin was then sutured and the animal placed in a prone position in a stereotaxic apparatus. An occipital craniotomy and a cervical laminectomy extending from C1 to C7 segments were performed. The dura was cut and removed and the nervous tissue was covered with warm saline. The central end of the right PN was then exposed via a dorsal approach and covered with warm paraffin oil. The right deep radial nerve was also isolated, cut and its central end was immersed in a paraffin oil pool. The rectal temperature was maintened between 36 and 38°C with a heating pad.

The central end of the right PN was placed on a bipolar silver wire electrode either for stimulating (0.1 to 5V, 20 to 50 µs, 3 Hz) or for recording (bandpass 30 Hz to 10 kHz). One or two tungsten microelectrodes (resistance = 3-5 megohms, 1 kHz) were introduced at different levels of the brain stem by a micromanipulator for recording (bandpass 30 Hz to 3 kHz), stimulating (1 to 8V, 20 to 50 µsec, 3 Hz) or microlesioning (DC current : 30 µ A, 30 sec). The accessible surface of the anterior lobe of the CCx (lobules IV and V) was explored with a single 100 µm diameter insulated steel wire electrode. For records obtained in the ECN, LRN or CCx, the reference electrode was placed on temporal muscles or bone. Evoked responses recorded in the ECN or on the CCx were amplified, averaged (32 to 512 sweeps) and directly photographied from an oscilloscope. Responses in the LNR were only amplified and stored on tape for further analyzis. Sites with PN responses in the brain stem were lesioned for subsequent histological verifications and the extent of the different spinal cord lesions performed in some animals was also confirmed histologically.

<u>RESULTS</u>

1 - <u>Phrenic potentials on the cerebellar cortex</u>

Stimulation of the central end of the right PH evoked a potential on the CCx (figure 1 A). The initial component of this response was a positive wave of 13.4 µv in average. This positive deflexion had a mean latency of 9.5 m sec and a mean duration of 19.5 m sec. The differentiation between

mossy and climbing fibers input is essentially related to
3 criteria (depth profile and latency of the response,
repetitive stimulation). With our recording technic, the
depth profile criterion has not been tested. The latency
of the evoked phrenic response may correspond to activation of either indirect mossy fiber pathway (spinoreticu-
lar pathway) or climbing fiber pathway. Nevertheless, the
phrenic response followed repetitive stimulation above
20/sec which was in agreement with activation of mossy
fibers. In these conditions, the positive phase could be
attributed to the synaptic activation of granule and Golgi
cells by mossy fibers inputs. A negative deflexion
followed the initial positive wave and was related to the
post-synaptic activations of Purkinje cells dendrites by
parallel fibers.

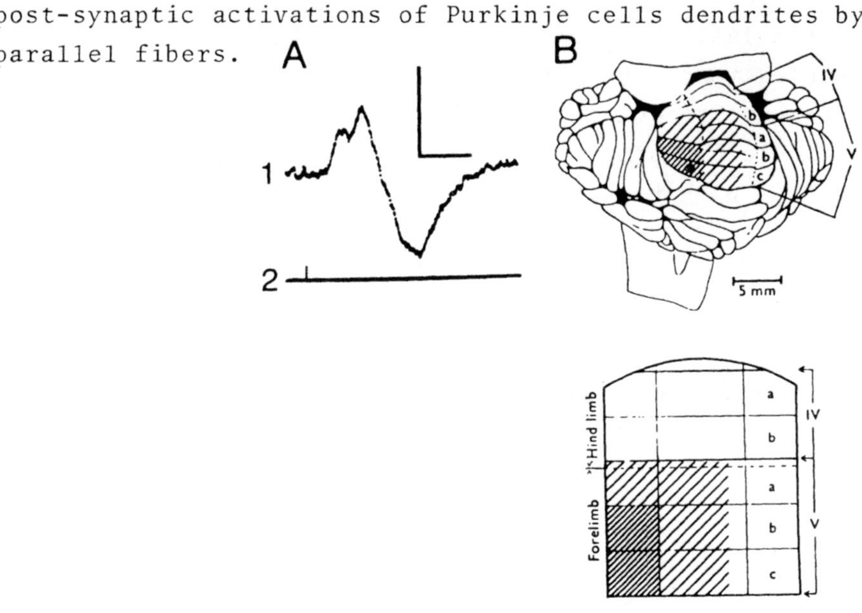

Figure 1 : A : cerebellar evoked phrenic response,
recorded in the intermediate cortex (black point of the
cerebellum schema, lobule V zone c) ; 2, stimulator
output. Scale = 20 ms and 20 µV ; average of 512 sweeps.
B : hatched areas indicate the distribution of the phrenic
responses in the anterior lobe using schemas from Coffey
et al. (5). Sparse hatching indicates areas with small
responses.
Reproduced from Marlot et al. (6).

The distribution of these evoked responses on the CCx is

shown on figure 1B. During experiments, we investigated
all the accessible surface of the anterior lobe of the CCx
(lobules IV and V according to Larsell, 7). Evoked phrenic
responses were recorded in the lobule V including the
ipsilateral intermediate cortex and a large portion of
the vermis.
Responses of highest amplitude were obtained in zones b
and c of the intermediate cortex. Different spinal cord
sections have shown that the main part of the phrenic
afferents to the CCx were conduced in the contralateral
ventro-lateral sector of the spinal cord.

2 - Evoked_phrenic_responses_in_the_external_cuneate
    nucleus

Stimulation of the central end of the right PN evoked
a complex response in the ipsilateral ECN (Figure 2 A).
Conversely, stimulation of the ECN only evoked a single
antidromic potential in the ipsilateral PN (Figure 2 B).
Responses with highest peack amplitude were preferentially

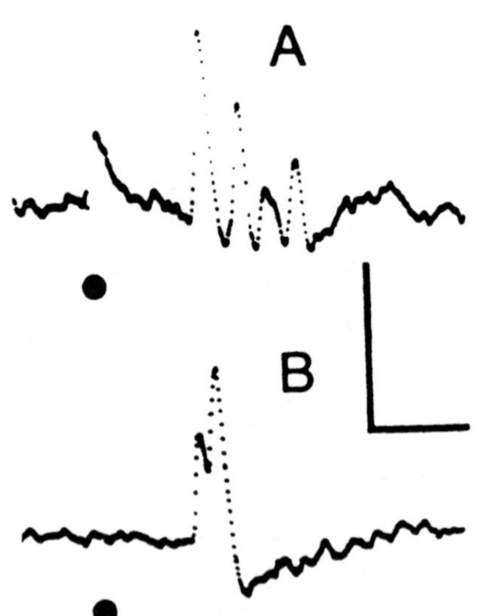

Figure 2 : In A, stimulation
of the PN evoked an orthodromic response in the ipsilateral ECN with 4 positive waves
(average of 512 sweeps).
In B, stimulation of the ECN
evoked an antidromic single
potential in the PN (average
of 128 sweeps) corresponding
to the early waves of the
record A.
Scale : 2 msec, 25 µV (A) and
50 µV (B). Black dots indicate time of stimulation.

found for the following stereotaxic coordinates in the
ECN : P = 12 to 13, L = 4 to 5 and H = -4.2 to -6.3. The
latency of the cuneate response ranged from 1.40 to 2.50
msec with a mean value (+/- SD) of 2.01 +/- 0.14 msec.

The latency from the site of stimulation in the phrenic
nerve to the spinal dorsal roots C5-C6 ranged from 0.3
to 0.4 msec. The latency from C5-C6 level to the external
cuneate nucleus ranged from 1.7 to 1.6 msec, corresponding
to the activation of fibers having conduction velocities
equal to 38-41 m/sec. It was possible to separate 2
different groups of waves in the global response recorded
in the ECN. The absolute refractory period (ARP), tested
by stimulating the PN with paired supraliminar (2 times the
threshold) shocks, was different according to which group
was examined. The ARP of the early waves ranged from 0.8
to 1.5 msec similar to that of nervous fibers (8) whereas
that of the later waves ranged from 3 to 4 msec similar to
that of cells (9, 10). The early group had always a peak
amplitude higher than that of the later group ranging
from 10 to 50 µv. The threshold stimulus in the PN for
an evoked response in the ECN was similar for the 2
components and ranged from 0.2 to 0.3 V (duration = 20 µ
sec).

The 3 criteria (ARP, threshold, and antidromic stimulation
of the ECN) indicate that the early component may be
related to terminal potentials of direct PN afferents and
the late group to postsynaptic activities. Sections of the
ipsilateral dorsal columns or of the C5 and C6 spinal
dorsal roots abolished the response recorded either in the
ECN (PN stimulation) or in the PN (ECN stimulation). The
stimulation of the central end of the right PN evoked
also an orthodromic response in the ipsilateral dorsal
column (figure 3). The threshold stimulus of this response
was similar to that previously found for the response in
the ECN.

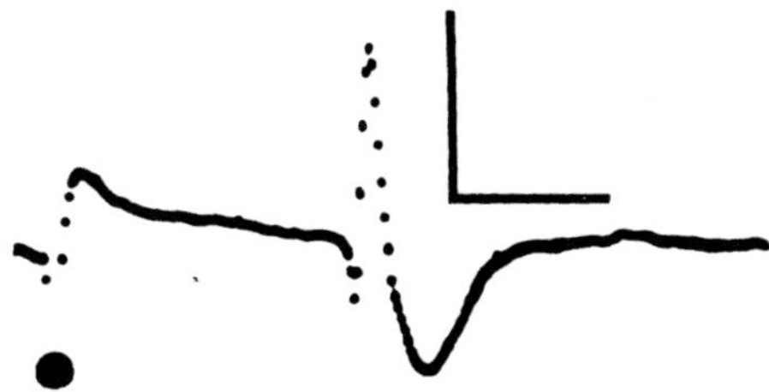

Figure 3 : Response
obtained in the ipsilate-
ral dorsal column (avera-
ge of 256 sweeps) after
stimulation of the right
PN. Scale = 1 msec, 20µV.
Black dot indicates time
of stimulation.

3 - <u>Evoked phrenic responses in the lateral reticular nucleus</u>

Excitatory and inhibitory responses to stimulation of the right PN have been recorded in the ipsi- and controlateral LRN. The mean threshold stimulus for these responses was 0.9 +/- 0.6 V ranging from 0.3 to 3.0 V for square wave pulses with a duration of 50 μsec.

Excitatory responses : Single shock stimulation of the PN evoked a unit activity consisting of a single spike or a burst of 2-17 spikes (figure 4). Increasing the intensity of the stimulus elicited an increase in the number of spikes. In few cases, repetitive stimulation of the phrenic afferents provoked a tonic discharge that stopped with a variable delay after cessation of the stimulation. The mean latency of the responses measured for a stimulus intensity equal to 2 times the threshold was 17.7 +/ - 6.5 msec ranging from to 40 msec.

Inhibitory responses : Spontaneous tonic reticular activities were totally or partially inhibited by stimulation of phrenic afferent fibers. Tonic reticular activities were also transitory inhibited (Figure 4) by each single shock of the repetitive stimulation of the PN. In these cases, the duration of the inhibition ranged from 80 to 120 msec.

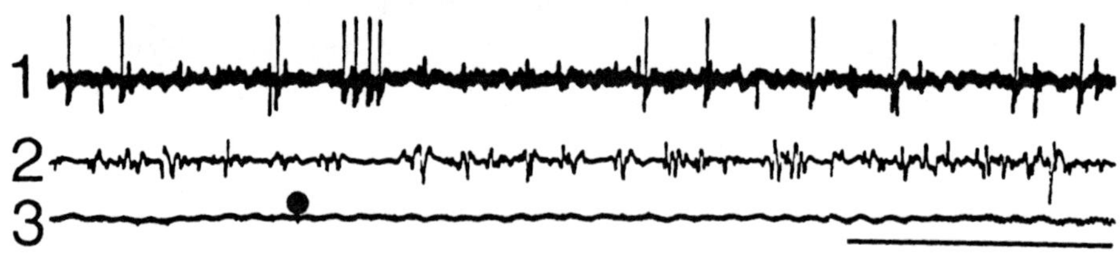

Figure 4 : 1 : unit activity in the LRN ; 2 : EMG of the diaphragm ; 3 : stimulator output (black dot indicates time of stimulation) stimulation of the PN firstly excitaded and secondly inhibited the spontaneous activity of the LRN unit. Scale = 100 msec.

During these experiments, we also recorded phrenic responses in other regions of the lateral bulbar reticular formation. Indeed, excitatory responses (burst of spikes)

were found in the nucleus ambiguus and in a zone dorsal to
this nucleus. Activities of the respiratory related units
recorded in the nucleus ambiguus were not modified by PN
stimulations except in one case where an expiratory related
unit has been inhibited during stimulation of the phrenic
afferent fibers.

CONCLUSION

The present results show that some phrenic afferents
project to the CCx in regions where projections of afferents from forelimbs and upper intercostal muscles have
been previously demonstrated (4, 5). The latency of the
cerebellar potentials does not correspond to activation
of direct tracts. Indirect spinocerebellar paths (spinore-
ticulo- and spino-olivocerebellar tracts) may be involved
in the conduction of the phrenic afferents projecting to
the CCx. In these conditions, it is surprising to find
responses to PN stimulation in the ECN. These potentials
correspond to activation of large afferent fibers,
certainly group I and II muscle afferents of the diaphragm.
It has been impossible to demonstrate a projection of this
type of afferents to the CCx. This negative result is
perhaps due to the small number of these afferents in the
PN which may be unable to evoke a surface potential on the
CCx but could only modify the activity of few Purkinje
cells. Moreover, these afferents could also project to the
thalamus as has been shown for some neurones of the ECN in
the primate (11). The phrenic projections to the LRN
correspond to activation of high threshold muscle afferents.
According to the latency of the responses, these phrenic
afferents could project to the CCx. The reticular neurones
are influenced by numerous and different systems especially by cardiorespiratory and somatosensory inputs (12).
Then, the phrenic afferent fibers projecting to the brain
stem may interact with other peripheral inputs, and the
overlapping between respiratory and non-respiratory informations could be important for the adaptation of respiratory and somatomotor functions during different behaviors

such as exercise.

<u>REFERENCES</u>

1. Duron, B., Jung-Caillol, M.C. and Marlot, D. (1978). Myelinated fiber supply and muscle spindles in the respiratory muscles : quantitative study. <u>Anat. Embryol</u>., <u>152</u>, 171-192.

2. Duron, B. and Caillol, M.C. (1973). Investigation of afferent activity in the intact phrenic nerve with bipolar electrodes. <u>Acta Neurobiol. Exp</u>., <u>33</u>, 427-432.

3. Corda, M., von Euler, C. and Lennerstrand, G. (1965). Proprioceptive innervation of the diaphragm. <u>J. Physiol</u>. (London), <u>178</u>, 161-177.

4. Oscarsson, O. (1973). Functional organization of spino-cerebellar paths. In : Iggo, A. (ed.). <u>Handbook of sensory physiology</u>. pp. 339-380 (Springer Verlag).

5. Coffey, G.L., Godwin-Austen, R.B., Mac gillivary, B.B. and Sears,T.A. (1971). The form and distribution of the surface evoked responses in cerebellar cortex from intercostal nerves in the cat. <u>J. Physiol</u>. (London), <u>212</u>, 129-145.

6. Marlot, D., Macron, J.M. and Duron, B. (1984). Projections of phrenic afferences to the cat cerebellar cortex. <u>Neurosci. Lett</u>., <u>44</u>, 95-98.

7. Larsell, O. (1953). The cerebellum of the cat and the monkey. <u>J. Comp. Neurol</u>., <u>99</u>, 135-200.

8. Paintal, A.S. (1966). The influence of diameter of medullated nerve fibers of cats on the rising and falling phases of the spike and its recovery. <u>J. Physiol</u>. (London) <u>184</u>, 791-811.

9. Gill, P.K. and Kuno, M. (1963). Properties of phrenic motoneurones. <u>J. Physiol</u>. (London), <u>168</u>, 258-273.

10. Brock, L.G., Coombs, J.S. and Eccles, J.C. (1953). Intracellular recording from antidromically activated neurones. <u>J. Physiol</u>. (London), <u>122</u>, 429-461.

11. Boivie, J. and Bomas, K. (1981). Termination of a separate (proprioceptive ?) cuneothalamic tract from the external cuneate nucleus in monkey. <u>Brain Res</u>., <u>224</u>, 235-246.

12. Langhorst, P., Schulz, B., Schulz, G. and Lambert, M. (1983). Reticular formation of the lower brainstem. A common system for cardiorespiratory and somatomotor functions : discharge patterns of neighboring neurons influenced by cardiovascular and respiratory afferents. <u>J. Autom. Nerv. Syst</u>., <u>9</u>, 411-432.

40

A Role for Phrenic Afferents?

D. T. FRAZIER, W. R. REVELETTE, D. FRYMAN and L. JEWELL

In 1962 Sant'Ambrogio, Wilson and Frazier (1) reported an absence of proprioceptive drive in the reflex regulation of diaphragmatic function. Three years later Corda et al. (2) reported that there were very few muscle spindles and tendon organs within the diaphragm. In addition, the reflex augmentation of diaphragm activity induced by tracheal occlusion was abolished following bilateral vagotomy (3). They suggested that the absence of the load-compensation reflex was related to the scarcity of proprioceptors in the diaphragm. A recent study by Duron et al. (4) has confirmed that the diaphragm is supplied by very few muscle spindles and tendon organs. Also, Breslav (5) and Arita and Bishop (6) have similarly reported that vagal receptors appear to solely mediate the diaphragmatic load compensation reflex.

The hypothesis that the diaphragm has a very weak or absent auto-genic load compensation reflex mechanism remained unchallenged until recent reports by Banzett et al. (7) and Fryman and Frazier (8). Banzett et al. (7) studied the effects of negative lower torso pressure on the electrical activity of the diaphragm in quadraplegic men. They found that application of a negative pressure within a cuirass fitted around the lower torso of their patients produced an increase in the moving time-averaged diaphragm EMG. Using spinalized-vagtomized cats, Fryman and Frazier (8) reported that lower body negative pressure induced augmentation in diaphragm electrical activity. This reponse was elminated by C_3 - C_5 bilateral dorsal rhizotomy suggesting that phrenic afferents were involved. These findings, plus the mounting evidence (9, 10) that diaphragm afferents may be involved in detection of added mechanical loads to breathing, has prompted us to reconsider the role of phrenic afferents in the control of respiration.

Three separate experiments demonstrating the reflex potential of diaphragm afferents have been carried out.

I. Diaphragmatic Stretch Reflex

Anesthetized cats were bilaterally vagotomized and spinalized at C-7 and placed in the supine position. A pair of hook electrodes were placed in the left crus through an abdominal incision. Stretch of the diaphragm was accomplished by positioning a latex balloon between the liver and diaphragm. The device was held in place by tethers sutured to the abdominal wall. Inflation was made through a catheter connected to a syringe. Balloon pressure was monitored from a branch of the catheter. The abdominal incision was closed and the animal's breathing was allowed to stabilize with the balloon deflated. Rapid inflations had the most profound effect during expiration when crural activity is normally silent. Diaphragmatic stretch was associated with a triphasic EMG response which began 60 msec, and persisted for approximately 100 msec, following stimulation. These data indicate that the diaphragm, like other skeletal muscles, has a stretch reflex apparently mediated by muscle spindle fibers. However, the latency of the response to stretch suggests that supraspinal connections are required.

II. Abdominal Expansion

In these experiments the animals were spinalized and bilaterally vagotomized, mounted in a spinal suspension frame with their abdomen supported. Release of the abdomen (unloading) was performed during expiration and the electronically-averaged costal and crural emg's generated during the loaded and first unloaded inspirations were compared. A sharp rise in the activity of both regions was noted beginning at approximately 1.5 seconds into the first unloaded inspiration. The procedure in this study involved a rapid shortening of the diaphragm. With the operating length decreased, the mechanical efficiency of the diaphragm is suddenly compromised. Afferents capable of responding to diaphragmatic unloading apparently alter phrenic nerve motor outflow by a mechanism involving either disinhibition and/or facilitation.

III. Effect of Diaphragmatic C-Fiber Stimulation on Ventilation in the Dog

This study tested the possible influence of diaphragmatic C-fibers on the control of ventilation. Dogs were anesthetized and intubated with a tracheal cannula. The left femoral artery and vein were also cannulated. The circulation to the left crus, the left phrenico-abdominal artery and vein, was exposed and cannulated. Crural perfusion was maintained by coupling the left femoral arterial and phrenico-abdominal catheters. Myoelectric activity of the left crus and left posterior cricoartytenoid (PCA) was recorded. C-fiber stimulation was accomplished by injection of capsaicin into the phrenic arterial catheter (0.125, 0.25 and 1.0 mg) while recording tracheal airflow, % CO_2 and both raw and averaged crural and PCA electromyograms. Capsaicin infusions of .25 and 1.0 mg were associated with immediate apneusis lasting approximately five seconds consisting of a tonic increase in crural and PCA emgs. Following apneusis, inspiratory airflows and volumes returned to control levels with low level tonic crural and PCA activity still present. Bilateral vagotomy did not eliminate apneusis. Systemic capsaicin injection (1 mg) was associated with a short period of apnea followed by hyperpnea, with no increase in tonic crural or PCA activity. These data suggest that crural C-fiber excitation facilities medullary inspiratory neuronal activity.

Overall Conclusions

1. Diaphragm afferents are capable of altering motor drive to the diaphragm and other respiratory muscles.

2. The diaphragm, like other skeletal muscles, has a stretch reflex.

REFERENCES

1. Sant'Ambrogio, G., Wilson, M. F. and Frazier, D. T. (1962). Somatic afferent activity in reflex regulation of diaphragmatic function in the cat. J. Appl. Physiol. 17: 829-832

2. Corda, M., Von Euler, C. and Lennerstrand, G. J. (1965). Proprioceptive innervation of the diaphragm. J. Physiol. 178: 161-177

3. Corda, M., Eklund, G. and Von Euler, C. (1965). External intercostal and phrenic alpha motor responses to changes in respiratory load. Acta Physiol. Scand. 63: 391-400

4. Duron, B., Jung-Caillol, M. C. and Marlot, D. (1978). Myelinated nerve fiber supply and muscle spindles in the respiratory muscles of cat: quantitative study. Anat. Embryol. 152: 171-192

5. Breslav, I. A., Klyvera, N. Z. and Konza, E. A. (1980). Mechanisms of regulation of respiration under a resistive load. Bull. Exp. Biol. Med. 89: 408-411

6. Arita, H. and Bishop, B. (1983). Firing profile of diaphragm single motor unit during hypercapnia and airway occlusion. J. Appl. Physiol. 55: 1203-1210

7. Banzett, R. B., Inbar, G. F., Brown, R., Goldman, M., Rossier, H. and Mead, J. (1981). Diaphragm electrical activity during negative lower torso pressure in quadriplegic men. J. Appl. Physiol. 51: 654-659

8. Fryman, D. and Frazier, D. T. (1983). Modulation of diaphragm activity by phrenic afferents. Fed. Proc. 42: 1014

9. Guz, A., Nolde, M. I. M., Widdicombe, J. G., Trenchard, D., Mushin, W. W. and Makey, A. R. (1966). The role of vagal and glossopharyngeal afferent nerves in respiratory sensation, control of breathing, and arterial pressure regulation. Cli. Sci. 30: 161-170

10. Eisele, J., Trenchard, D., Burki, N. and Guz, A. (1968). The effect of chest wall block on respiratory sensations and control in man. Cli. Sci. 35: 23-33

41

Voltage-Dependent Currents in Rat Nodose and Petrose Neurons Maintained in Culture

J.-L. BOSSU, A. FELTZ and J.-M. THOMANN

The understanding of the ionic basis of action potentials is essential in an analysis of how chemosensory information is carried via the primary sensory neurones; the signal can be modulated at the periphery (mainly by dopamine which is generally assumed to be released by glomic cells) and at the central level (GABA, 5HT presynaptic actions...). In a population of rat nodose and petrose neurones maintained in culture (1), using "patch-clamp" technique in whole cell configuration (2), we recorded separately the various cationic currents normally involved in spike generation.

Fig. 1 typifies the three inward currents which in a single cell are generally evoked in presence of tetrodotoxine (TTX) 3-15 µM after cae-sium loading and alternatively superfusing with 5 mM Ca solution (0 Na, 10 mM Mg) and then with a 50 mM Na solution (0 Ca, 10 mM Mg). The current elicited in Na free medium just above resting potential was a transient wave which successively activated and inactivated in a volt-age dependent manner and was over in a few tens of milliseconds. Step commands above -10 mV induced a large sustained current. These two currents are distinct since certain conditions will block one or the other. They were both only altered when varying Ca concentration and

behave as typical "Ca" currents. Both currents appear upon on I-V curve

as more or less distinct peaks.

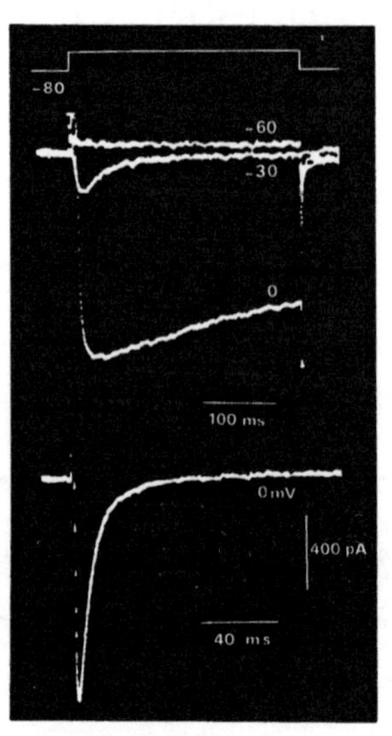
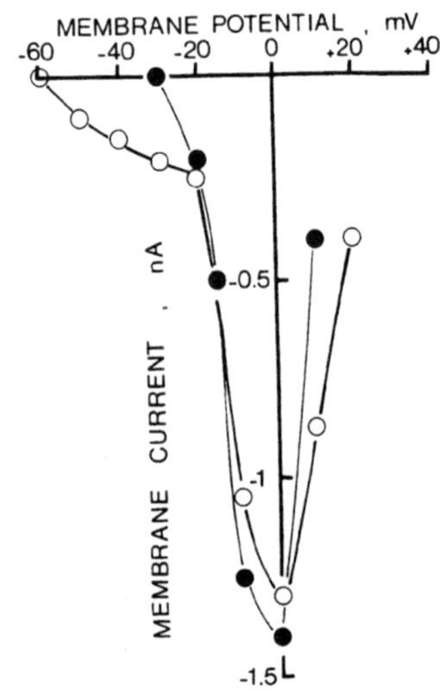

FIGURE 1   *Upper left:* Samples inward Ca currents evoked by 300 ms depolarising commands from holding -80 mV increasing to 0 mV.
The external solution contained (in mM): choline chloride
130, $CaCl_2$ 5,TEA 7.5, $MgCl_2$ 2, Hepes/Tris 5 pH 7.4 and the
pipette (or internal) solution: CsCl 110, TEA 20, $MgCl_2$ 2,
pCa 8, Hepes/NaOH 10 pH 7.2. *Bottom left:* TTX-resistant Na
current recorded in the same cell, and evoked by 300 ms
depolarising jump to 0 mV from -80 mV. The external solution
contained in mM: NaCl 50, choline chloride 80, $MgCl_2$ 10,
TEA 7.5, Hepes/NaOH 5 pH 7.4, TTX 10 µM. *Right:* Current-
voltage relationships for inward calcium current, o and inward slow Na current, ● recorded in the same cell. Current
values at peak.

In "sodium" medium, a TTX sensitive component and a TTX resistant

component (TTX up to 3-15 µM) were observed (Fig. 1, bottom trace). The

latter depending on command voltage peaks in 3-6 ms and inactivates in

20-50 ms; we therefore referred to it as the slow Na current. Inter-

estingly the slow Na current appears in the same voltage span as the

sustained Ca current from -25 mV upwards (but see 3) , and in physiological conditions is 5-10 larger in amplitude.

The outward currents when carried by K as sole cation are certainly

of multiple origin. A typical current-voltage relationship (as measured

at the end of a 150 ms plateau current) is illustrated in Fig. 2 (see

traces of records on upper left). In our culture conditions the large

conductance channels as recorded in cell attached conditions can only

be elicited above +100 mV (Fig. 2, bottom trace),   therefore a minor

contribution to the outward K current is certainly linked to a previous

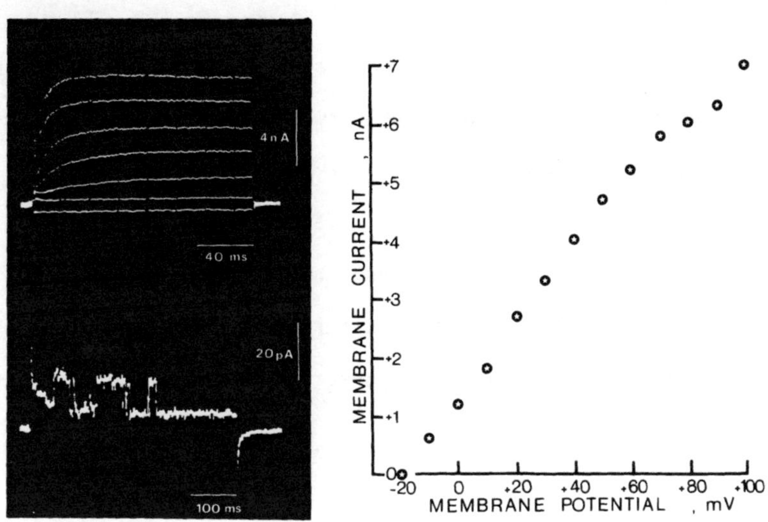

FIGURE 2   *Upper left:* Samples outward K currents evoked by 150 ms de-
         polarizing commands increasing from the holding potential -
         60 mV to -20, 0, +20, +40, +60, +100 mV, plus leak current by
         a step of -40 mV from holding potential. The external sol-
         ution contained (in mM):   choline chloride 130,   KCl 5,
         $CaCl_2$ 2, $MgCl_2$ 10, Hepes/Tris 5 pH 7.4 and the pipette sol-
         ution: KCl 130, $MgCl_2$ 2, pCa 7, Hepes/Tris 10 pH 7.4. *Bottom
         left:* Large unitary K current in cell attached recording
         configuration: response to +100 mV jumps from -20 mV. The
         pipette   solution   contained   KCl 130 mM   which   imposes
         E_K = 0 mV. *Right:* Same cell as in upper left. Current-voltage
         relationship for outward K currents. Current values at pla-
         teau.

Ca entry, and the current is mainly carried through the delayed K rectifier.

Our results reveal a population of Ca channels which close slowly as indicated by noise analysis: since they are elicited just above resting potential, *i.e.* "threshold channels" (4), they are certainly of functional importance in triggering the action potential. They could control a repetitive firing activity. Once elicited, the action potential results from the successive activation of two distinct Na currents. The TTX-sensitive Na current from -40 mV upwards supplies a large current which accounts for most of the initial part of the spike. The TTX-resistant current has slower kinetics and higher threshold. For neurones maintained in culture, this current carries most of the charge transfer occurring during the hump of the spike, along with a small Ca component which is only revealed in TEA prolonged spike. It appears to be also a characteristic of these cranial sensory neurones in adult animals (Gallego, 1983). This Na current is reminiscent of the sodium response to serotonin which is observed in these sensory neurones of the C type (6,7).

1. Baccaglini, P.I. and Cooper, E. (1982). *J. Physiol.* *324*, 429-439

2. Hamill, O.P., Marty, A., Neher, E., Sakmann, B. and Sigworth, F.J. (1981). *Pflügers Arch.* *391*, 85-100

3. Kostyuk, P.G., Veselovsky, N.S. and Tsyndrenko, A.Y. (1981). *Neuroscience 6*, 2423-2430

4. Gilly, W.F. and Armstrong, C.M. (1984). *Nature 309*, 448-450

5. Gallego, R. 1983. *J. Physiol.* *342*, 591-602

6. Higashi, H. and Nishi, S. (1982). *J. Physiol.* *323*, 543-567

7. Stansfeld, C.E. and Wallis, D.T. (1984). *J. Physiol.* *352*, 49-72

42

The Chemoreceptor and Lung Stretch Input to the Brainstem

K. M. SPYER

Considerable interest has been focussed in recent years on the organisation of the reflex control of respiration. In this respect neuroanatomical studies have indicated that the major central projections of the afferents of the IXth and Xth cranial nerves, and their various branches, are to the nucleus of the tractus solitarius (NTS) (Kalia and Mesulam, 1980a,b ; Ciriello, Hrycyshyn and Calaresu, 1981). The precise regions of the NTS innervated by the arterial chemoreceptors and slowly adapting lung stretch afferents have been unresolved until recent studies although various speculations had derived from the neuroanatomical investigations. We have used an electrophysiological approach, the antidromic mapping technique (Lipski, 1981), to determine the central projections of individual slowly adapting lung stretch afferents in cat and rabbit (Donoghue, Garcia, Jordan and Spyer, 1982) and carotid body chemoreceptors in the cat (Donoghue, Felder, Jordan and Spyer, 1984).

Slowly adapting lung stretch afferents

The activity of 18 lung stretch afferents has been recorded from nodose ganglia in cat and rabbit (Donoghue et al., 1982) and their antidromic responses on microstimulation within the dorsomedial medulla observed. In the cat these neurones had axonal conduction velocities in the range 16-33 m/s (7 afferents) and in the rabbit in the range 17.5-39 m/s (11 afferents). They were routinely activated by stimulation within the ipsilateral NTS. Most afferents were antidromically excited on stimulation within the dorsomedial NTS at levels rostral to the obex as well as the medial and ventromedial areas. Conversely, stimulating within the lateral and ventrolateral divisions of the NTS was effective in only 25 % of afferents. These observations are compatible with those of another neurophysiological study (Berger and Averill, 1983).

Carotid body chemoreceptors

The activity of 13 arterial chemoreceptors was recorded within the petrosal ganglion (axonal conduction velocity 0.6-3.5 m/s). All were activated antidromically on stimulating within the dorsomedial and medial subnuclei of the ipsilateral NTS and in the majority from the commissural subnucleus (Donoghue et al., 1984). The innervation of the commissural nucleus extended to the contralateral NTS. In 4 cases stimulating within the lateral subnucleus, and particularly its dorsolateral aspect, was also effective.

These studies suggest that these groups of afferents relay primarily within the medial regions of the NTS rather than in the immediate vicinity of the respiratory neurones of the ventro-lateral NTS (Merrill, 1974). These observations will be discussed in relation to the morphology of these respiratory neurones. Further, the results of recent neurophysiological studies investigating the control of afferent transmission through the NTS will be reviewed.

This investigation has been supported by grants from the Medical Research Council.

REFERENCES

BERGER, A.J. and AVERILL, D.B. (1983). Projection of single pulmonary stretch receptors to solitary tract region. Journal of Neurophysiology, 49 (3), 819-830.

CIRIELLO, J., HRYCYCHYN, A.W. and CALARESU, F.R. (1981). Glosso-pharyngeal and vagal afferent projections to the brain stem of the cat : a horseradish peroxidase study. Journal of Autonomic Nervous System, 4, 63-79.

DONOGHUE, S., FELDER, R.B., JORDAN, D. and SPYER, K.M. (1982). The central projections of carotid baroreceptors and chemo-receptors in the cat : a neurophysiological study. Journal of Physiology, 347, 397-410.

DONOGHUE, S., GARCIA, M., JORDAN, D. and SPYER, K.M. (1982). The brain-stem projections of pulmonary stretch afferent neurones in cats and rabbits. Journal of Physiology, 322, 353-363.

KALIA, M. and MESULAM, M.M. (1980a). Brain stem projections of sensory and motor components of the vagus complex in the cat. I. The cervical vagus and nodose ganglion. Journal of Comparative Neurology, 193, 435-465.

KALIA, M. and MESULAM, M.M. (1980b). Brain stem projections of sensory and motor components of the vagus complex in the cat. II. Laryngeal, tracheobronchial, pulmonary, cardiac and gastrointestinal branches. Journal of Comparative Neurology,, 193, 467-508.

LIPSKI, J. (1981). Antidromic activation of neurones as an analytic tool in the study of the central nervous system. Journal of Neuroscience Methods, 4, 1-32.

MERRILL, E.G. (1974). Finding a respiratory function for the medullary respiratory neurons. In Essays on the Nervous System (Editors : R. BELLAIRS and E.G. GRAY) pp 451-486. Oxford, Clarendon Press.

43

Role of the Dorsal Respiratory Group (DRG): Processing of Vagal and Superior Laryngeal Nerve Afferent Input

D. R. McCRIMMON, D. F. SPECK and J. L. FELDMAN

The DRG, centered about the ventrolateral subnucleus of the tractus solitarius, is one of two medullary regions where respiratory neurons are concentrated. Respiratory related afferents, arriving via the IXth and Xth cranial nerves, terminate in the DRG as well as the medial and dorsolateral subnuclei of the tractus solitarius (1,2,3). Electrical stimulation of the superior laryngeal nerve elicits a short latency excitation of the contralateral phrenic nerve (4,5,6,7). Berger has proposed that the phrenic nerve excitation involves a disynaptic pathway and that the second order neurons are bulbospinal inspiratory neurons which have cell bodies located in the DRG. Activation of pulmonary stretch receptors (PSRs) is known to influence ventilatory control in several ways including: i) inspiratory shortening, ii) expiratory prolongation, and iii) augmentation of integrated phrenic neural discharge (see reviews 8,9). A subset of DRG inspiratory neurons receive monosynaptic excitation from PSR afferents (1,10,11) and many investigators have hypothesized that these neurons may be of crucial importance in the inspiratory-shortening reflex. Extensive electrolytic (12,13,14) or aspiration lesions of the NTS have been found to attenuate the inspiratory-shortening reflex. These studies implicate the NTS as a necessary component in the inspiratory-inhibitory reflex; the lesions, however, were quite extensive and could have produced their effects by damaging the afferent fibers intracranially or disrupting terminal fields in NTS regions other than the DRG.

Despite the anatomical and electrophysiological demonstration of projections from these receptors to the nucleus tractus solitarius (NTS), very little is known about the neuronal circuitry underlying specific respiratory reflexes. This study was designed to test whether DRG neurons play an obligatory role in respiratory reflexes mediated by cranial nerve afferents. To address this question we made focal uni- or bi-lateral lesions of the DRG and determined the effect on the phrenic nerve response to activation of vagal or superior laryngeal nerve afferents. Experiments were conducted on anesthetized, vagotomized, paralyzed and artificially ventilated cats. We verified that 1) single electrical pulses (5 to 30 uA, 0.1 ms duration) administered to the superior laryngeal nerve during inspiration elicit a short onset latency (4 to 6 ms) excitation of the contralateral phrenic nerve followed by a bilateral inhibition, and 2) stimulus trains delivered to the vagus nerve (20 to 60 uA, 100 Hz, 0.1 ms pulse duration; intensities chosen to maximize activation of pulmonary stretch receptors) throughout inspiration produce a current dependent shortening of inspiratory duration (T_I). We recorded and lesioned in the DRG with a linear array of 2 to 4 tungsten microelectrodes (tip separation of 0.7 to 1.0 mm). When DRG respiratory neurons were recorded on an electrode it was moved to the approximate center of this activity and a lesion was made by passing cathodal current (20 uA for 60 sec). If respiratory neurons were recorded on an adjacent electrode, then lesions were made between these electrodes (20 uA for 60 seconds each polarity). The array was moved in 500 um steps (in the horizontal plane) and we iterated the above procedure until respiratory neurons (in the DRG) were no longer detected. Lesions were located ventral and lateral to the tractus solitarius and extended from about 1 mm caudal to 2 mm rostral to the obex. Uni- or bi-lateral DRG destruction (verified to be at least 80% complete by cell counts) severely attenuated or abolished the short latency phrenic nerve excitation to superior laryngeal nerve stimulation; this finding is consistent with the interpretation that the second order neurons mediating this response are bulbospinal DRG neurons. The longer latency bilateral phrenic nerve inhibition was unaffected. In no case did the lesions disrupt respiratory rhythm (15). In addition, the relative shortening of T in response to vagus nerve stimulation of a given intensity was the same before and after lesioning. We conclude: 1) the DRG is an obligatory component in the short latency phrenic nerve excitation elicited by superior laryngeal nerve stimulation but is not required for the longer latency bilateral inhibition; 2) neuronal pathways exclusive of the DRG are sufficient to produce inspiratory termination in response to activation of pulmonary stretch receptor afferents; and, 3) the monosynaptic excitatory input from PSRs to DRG I(+) neurons (16) many of which project to phrenic motoneurons, underlies another aspect(s) of the PSR reflex, perhaps the inspiratory-augmenting reflex.

This work was supported by NIH grant HL-23820, D.R.M. is a Parker B. Francis Fellow of the Puritan-Bennett Foundation. The current address for D.F.S. is: Department of Physiology and Biophysics, University of Kentucky, Lexington, KY.

REFERENCES

1). AVERILL, D.B., CAMERON, W.E. and BERGER, A.J. Monosynaptic excitation of dorsal medullary respiratory neurons by slowly adapting pulmonary stretch receptors. J. Neurophysiol. In Press.

2). DONOGHUE, S., GARCIA, M., JORDAN, D. and SPYER, K.M. (1982). The brain-stem projections of pulmonary stretch afferent neurons in cats and rabbits. J. Physiol. (Lond.) 322: 352-364.

3). KALIA, M. and MESULAM, M.M. (1980). Brain stem projections of sensory and motor components of the vagus complex in the cat: II. Laryngeal, tracheobronchial, pulmonary, cardiac, and gastrointestinal branches. J. Comp. Neurol. 193: 467-508.

4). BERGER, A.J. (1977). Dorsal respiratory group neurons in the medulla of cat: spinal projections, responses to lung inflation and superior laryngeal nerve stimulation. Brain Res. 135: 231-254.

5). BERGER, A.J. (1978). Respiratory gating of phrenic motoneuron responses to superior laryngeal nerve stimulation. Brain Res. 157: 381-384.

6). BERGER, A.J. and MITCHELL, R.A. (1976). Lateralized phrenic nerve responses to stimulating respiratory afferents in the cat. Am. J. Physiol. 230: 1314-1320.

7). ISCOE, S., FELDMAN, J.L. and COHEN, M.I. (1979). Properties of inspiratory termination by superior laryngeal and vagal stimulation. Respir. Physiol. 36: 353-366.

8). COHEN, M.I. (1979). Neurogenesis of respiratory rhythm in the mammal. Physiol. Rev. 59: 1105-1173.

9). PACK, A.I. (1981). Sensory inputs to the medulla. Ann. Rev. Physiol. 43: 73-90.

10). BACKMAN, S.B., BALLANTYNE, D., MIFFLIN, S.W., ANDERS, C., JORDAN, D.,, SPYER, M. and RICHTER, D.W. Responses of inspiratory Beta-neurones to activation of lung stretch receptors investigated by the spike triggered averaging technique. Pflugers Arch. In Press.

11). RICHTER, D.W., CAMERER, H. and ROHRIG, N. (1979). Monosynaptic transmission from lung stretch receptor afferents to R[beta]-neurones. In: Central Nervous Control Mechanisms in Breathing, Physiological and clinical aspects of regular periodic and irregular breathing in adults and in the perinatal period. Proceedings of the International Symposium held at the Wenner-Gren Center, Stockolm, ed. C. Von Euler and H. Lagercrantz, Oxford: Pergamon Press, p. 267-271.

12). ANDEREGGEN, P., OBERHOLZER, R.J.H. and WYSS, O.A.M. (1946). Le mechanisme central des reflexes respiratories d'origine vagale. II. La localisation du centre expirateur. Helv. Physiol. Acta 4: 213-232.

13). OBERHOLZER, R.J.H., ANDEREGGEN, P. and WYSS, O.A.M. (1946). Le mechanisme central des reflexes respiratoires d'origine vagale. IV. Localisation du centre reflexe inspiratoire. Helv. Physiol. Acta 4: 495-512.

14). RICHARDSON, C.A. and MITCHELL, R.A. (1982). Power spectral analysis of inspiratory nerve activity in the decerebrae cat. Brain Res. 233: 317-336.

15). SPECK, D.F. and FELDMAN, J.L. (1982). The effects of microstimulation and microlesions in the ventral and dorsal respiratory groups in cat. J. Neurosci. 2: 744-757.

16) BACKMAN, S.B., ANDERS, C., BALLANTYNE, D., ROHRIG, N., CAMERER, H., MIFFLIN, S. JORDAN, D., DICKHAUF, H., SPYER, K.M. and RICHTER, D.W. (1984). Evidence for a monosynaptic connection between slowly adapting pulmonary stretch receptor afferents and inspiratory β neurons. Pflügers Arch. 402: 129-136.

44

Analysis of the Connection between Slowly Adapting Pulmonary Stretch Receptor Afferents and Inspiratory Neurones within the Tractus Solitarius Region in Cats

S. B. BACKMAN, D. BALLANTYNE, S. MIFFLIN, K. ANDERS,
D. JORDAN K. M. SPYER and D. W. RICHTER

Inspiratory neurones in a region ventral and ventrolateral to the tractus solitarius have been differentiated into two types on the basis of their response to activation of slowly adapting pulmonary stretch receptors during lung inflation. One type, inspiratory beta neurones, discharge action potentials in phase with phrenic nerve activity and are also excited by lung inflation while the other type, inspiratory alpha neurones, discharge with phrenic nerve activity but are not excited by lung inflation (1). While indirect evidence suggests that the synaptic connectivity mediating the excitation of inspiratory beta neurones by slowly adapting pulmonary stretch receptor afferents is monosynaptic (3), this has not yet been firmly established.

The purpose of the present study was to determine the nature of this connection by averaging the synaptic noise recorded from inspiratory beta neurones using the spike of a single slowly adapting pulmonary stretch receptor afferent as a trigger (2).

Experiments were done on pentobarbital-anaesthetized, paralyzed, thoracotomized and artificially ventilated cats. Extracellular single unit spikes of slowly adapting pulmonary stretch receptors (SARs) were recorded in the nodose ganglion and identified by their response (10-100 spikes/sec) to moderate lung inflation (2-6 mmHg tracheal pressure). Intracellular recordings were obtained from inspiratory neurones in the ventral and ventrolateral region of the tractus solitarius using glass micropipettes filled with 2 M potassium citrate. Inspiratory neurones were classified as beta if they received a wave of e.p.s.p.s in response to moderate lung inflation (Fig. 1A), and if a well-defined, constant latency e.p.s.p. was evoked by electrical stimulation of the vagus nerve at an intensity (0.05 ms, 0.2-0.7 V) which elicited a Hering-Breuer reflex (Fig. 1B). Inspiratory neurones which did not fulfill these criteria were classified as alpha.

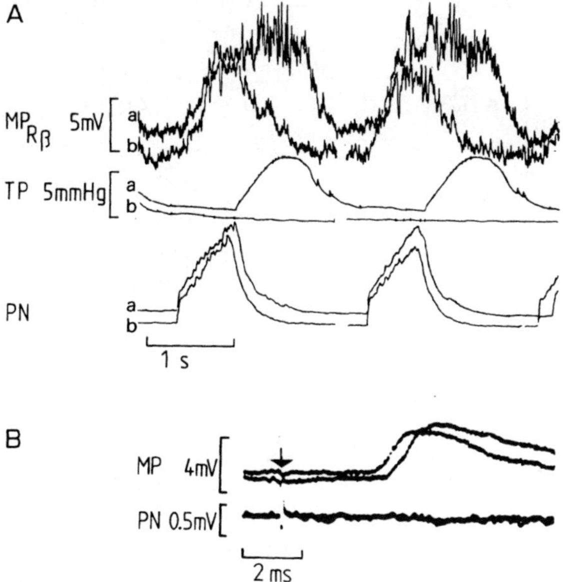

FIG. 1 Identification of an inspiratory beta neurone. Inspiratory beta neurones were identified by: (A) Their inspiratory associated depolarization of membrane potential (MP) and by their depolarization associated with lung inflation. In condition "a" changes in MP are shown when the lungs were inflated (as seen by the rise in tracheal pressure, TP) shortly after peak phrenic nerve activity (PN). In "b" changes in MP in the absence of lung inflation are shown. (B) Intracellular records of MP showing short latency e.p.s.p.s evoked by vagus nerve stimulation (arrows, 0.7V, 0.05 msec pulses). When the vagus nerve was stimulated at two different positions seperated by 16 mm, the difference in latencies of the e.p.s.p.s was 0.5 msec corresponding to an afferent vagal fibre conduction velocity of 32 m/s.

By back-averaging the mass-activity recorded from the vagus nerve at two recording sites separated by a fixed distance using the spike

discharge of a single SAR afferent as a trigger, the absence of
synchronous activation of afferents was verified and the extra-
cephalic conduction velocity was calculated to range from 24-62 m/s
(Fig. 2A). SAR spike triggered averages of synaptic noise (amplified x
1,000; filtered: 1 Hz-10 kHz) were performed on inspiratory neurones
when spike discharge had ceased (membrane potential -30 to -45 mV). An
averaged response was considered genuine if it appeared at the same
latency using consecutive series of spike triggers and if a negative
response was obtained with an independent trigger source (Fig. 2B).

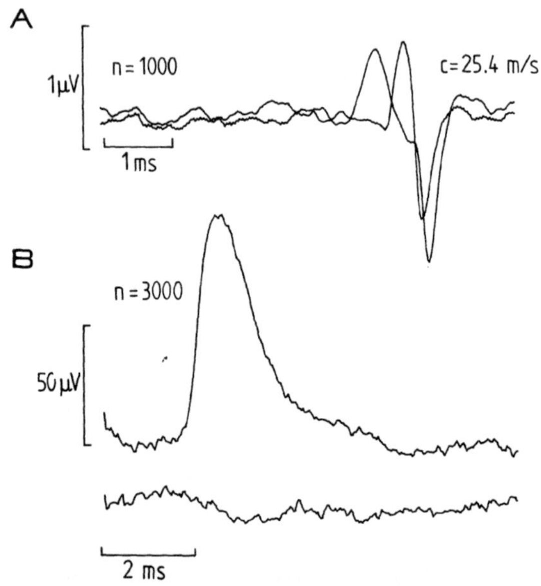

FIG. 2: Spike triggered average of synaptic noise
recorded from an inspiratory beta neurone. (A) Back-average
recorded from two sites on the vagus nerve (seperated by
10 mm) using the discharge of a single SAR unit as a
trigger. The average of 1000 sweeps at each recording site
on the vagus nerve reveals an unitary spike not
superimposed on any other waveform, verifying the absence
of synchronization of afferent SAR activity. The difference
in latencies of the unitary spikes suggests an orthodromic
conduction velocity (c) of the SAR fibre of 25.4 m/s. (B)
Synaptic noise recorded from an inspiratory beta neurone
was averaged over 3000 sweeps using the spike discharge of
the SAR unit illustrated in A as a trigger indicating a
prominent synaptic depolarization. The lower trace is the
average of 3000 sweeps using an independent trigger and
reveals a flat response.

Averages of 100-10000 sweeps of 10 ms duration revealed a clear wave of depolarization in 11 inspiratory beta neurones tested (Fig. 2 B). Moreover, in 3 cases, there was a positive correlation between the spike discharge recorded from one SAR afferent and the synaptic noise recorded from two inspiratory beta neurones. Averaged responses had an amplitude from 67 μV to 265 μV and a latency of 1.6 ms to 3.0 ms. In every instance, this depolarization was clear after averaging less than 100 sweeps. With 10 beta neurones, averaging over a larger number of sweeps revealed two clear components to this depolarization. The initial component was rapidly rising, with rise times from 0.27 ms to 1.1 ms, followed by a second more slowly rising component. In the remaining neurone, the rising phase showed no obvious division into fast and slow components (rise time: 0.62 ms). In each neurone, the repolarizing phase exhibited a number of small inflections and in 10 cases was followed by a series of small waves of depolarization. The distribution of interspike intervals of the SAR activity showed no peaks corresponding to these waves and the first clear wave of depolarization did not recur at intervals less than the modal interspike interval of the SAR discharge.

Four inspiratory alpha neurones were tested for connectivity with SAR afferents. These tests were made only under conditions where the Hering-Breuer reflex was intact. Averages of 1000-3000 sweeps of 10 ms duration were essentially flat.

The findings provide strong evidence for a direct and powerful connection between SAR afferents and inspiratory beta neurones. They also suggest lack of a direct connection between SAR afferents and inspiratory alpha neurones.

(Supported by the DFG and the Canadian Heart Foundation)

References:

1. Baumgarten von, R., and E. Kanzow:
            Arch. ital. Biol. 96, 361-373, 1958
2. Backman, S. B., C. Anders, D. Ballantyne, N. Röhrig,
    H. Camerer, S. W. Mifflin, D. Jordan, H. Dieckhaus,
    K. M. Spyer, and D. W. Richter: Pflügers Arch., in press
3. Richter, D. W., H. Camerer, and N. Röhrig: in "Central
            nervous control mechanisms in breathing",
            C. von Euler, and H. Lagercrantz (Eds.),
            Pergamon Press, Oxford, pp.267-270, 1979

45

Antidromic Mapping of the Brainstem Projection of Pulmonary Rapidly Adapting Receptor Neurons in the Cat

L. KUBIN and R. O. DAVIES

The intrapulmonary rapidly adapting receptors (RAR) are airway mechanoreceptors that respond to rapid inflations or deflations of the lung with an irregular discharge that adapts rapidly when the volume stimulus is maintained (1,2). They are also stimulated during hyperpnea, pneumothorax and anaphylaxis, and by a variety of noxious agents. However, their central connections, reflex pathways and role in the control of ventilation in normal and pathological states are not well understood (1).

Previous neuroanatomical investigations (3,4) have shown that the vagal visceral afferent fibers terminate principally in the nucleus of the tractus solitarius (NTS); however, the results of such studies do not allow one to distinguish the projection patterns to the various subnuclei of fibers of different sensory modalities. Therefore, we have used the antidromic mapping technique (5) to determine the central pathway and presumed terminal sites within the NTS of single, functionally identified RAR neurons recorded in the nodose ganglion. This same method proved successful for the mapping of the brainstem projections of pulmonary stretch receptor (PSR) neurons (6).

The present experiments were done on decrebrate, paralyzed cats, ventilated with a cycle-triggered pump. The first step in our protocol was to locate inspiratory neurons of the ventrolateral NTS (vlNTS) by extracellular recording of strong, multi-unit activity using a stainless steel microelectrode (later used for stimulation).

Then the activity of single units were recorded extracellularly in the nodose ganglion and RAR were identified by their rapid adaptation (greater than 70% within 2 s) (7) to a fast-rising, then maintained, hyperinflation of the lung. The medulla was systematically explored with a monopolar stimulating electrode (using 100 µs constant current pulses not exceeding 150 µA) to activate the RAR antidromically. The first stimulations were made in the vlNTS. Because in our first experiments we found a contralateral projection, in subsequent experiments, after the stimulations in the vlNTS, penetrations were made on the contralateral side. There, we proceeded (usually in 0.2 mm steps) from rostral to caudal levels, and then in the opposite direction on the ipsilateral side. This minimized possible damage of some axonal branches by the multiple, closely spaced penetrations (33-68 in a single mapping session). In each penetration, the electrode was lowered in 50-100 µm steps and the threshold current and antidromic latency were recorded at each point; depth-threshold curves were then determined for each penetration. At each site, changes in the latency of the response were also assessed as the stimulus intensity was increased above threshold. Since, in certain penetrations, antidromic responses having up to nine distinct latencies could be recorded, the collision test was used to confirm that each response latency resulted from stimulation of a branch of the RAR under study. At the end of the mapping session, the chest was opened and the pulmonary location of the receptive field of the RAR verified by local, mechanical probing of the surface of the lung. A single, low threshold, medullary site was marked by depositing ferric ions from the electrode.

The location of each medullary penetration was determined on the basis of its stereotaxic coordinates with respect to the obex, the location of the prussian blue mark and inspection of the transverse serial sections of the medulla. The position of each penetration was then superimposed on a corresponding camera lucida drawing of a transverse section. The location of each track was also plotted on a map showing a dorsal view of the medial and lateral borders of the NTS (drawn from measurements taken from the serial sections). The medullary course of each RAR axon and its main collaterals within the NTS was deduced from: (1) the latency of the antidromic response at each low threshold point; (2) an estimation of the conduction velocity along a supposed course of a given branch; and (3) the experimental depth-threshold curves for each penetration and for each antidromic response characterized by a distinct latency.

For ten of the eleven cells studied, the reconstructions showed that the densest branching was located on the ipsilateral side in the ventral portion of the medial part of the caudal NTS. In these ten cells, a continuation of at least one major branch was traced to the contralateral side, to the medial part of the caudal NTS. There were also a few branches stimulated with low intensity current in the lateral NTS; in only two cells these branches reached the vlNTS where we were able to record inspiratory neuronal activity. One of the reconstructions showing the course of the main axon and its major branches and, schematically, the relative density of the presumed terminal branches of a single RAR is shown in Figure 1.

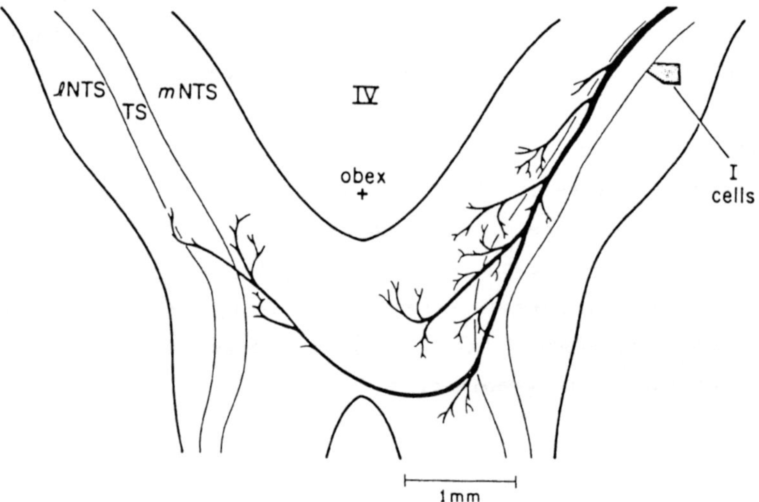

Figure 1: Horizontal view of the medulla showing the medial and lateral borders of the NTS and the tract itself. A reconstruction of the course of the axon within the tract and its main collaterals is shown. The relative density of presumed terminal branches is characterized schematically by the number of fine branches attached to the main collaterals. The hatched area shows the location where inspiratory neuronal activity was recorded.

One RAR neuron, in contrast to the others, had a branching pattern limited to the ipsilateral medial, lateral and vlNTS, with the latter

projection being most pronounced. This cell, although classified as
an RAR on the basis of its adaptation rate, was possibly an extreme
example of a high threshold PSR; this was suggested by the unchar-
acteristic, pronounced repeatability of its responses to lung
inflations. An additional two cells, both unequivocally classified as
PSR, were tested for contralateral projections; no evidence for this
was found. One of these PSR cells was further studied in sufficient
detail to determine that its densest projection was localized in the
vlNTS.

This study has delineated the region that presumably contains the
second-order neurons of the RAR pathway. Subsequent characterization
of the relay neurons and the determination of their connections should
facilitate our understanding of the role of RAR in ventilatory
control. (Supported by USPHS Grants HL-08805 and BRSG 5-S07-RR-5465.)

REFERENCES

1. Pack, A.I. (1981). Sensory inputs to the medulla. Ann. Rev.
    Physiol., 43, 73-90.

2. Sant'Ambrogio, G. (1982). Information arising from the tracheo-
    bronchial tree of mammals. Physiol. Rev., 62, 531-569.

3. Cottle, M.K. (1964). Degeneration studies of primary afferents of
    IXth and Xth cranial nerves in the cat. J. Comp. Neurol., 122,
    329-345.

4. Kalia, M. and Mesulam, M.-M. (1980). Brain stem projections of
    sensory and motor components of the vagus complex in the cat: I.
    The cervical vagus and nodose ganglion. J. Comp. Neurol., 193,
    435-465.

5. Lipski, J. (1981). Antidromic activation of neurones as an
    analytic tool in the study of the central nervous system. J.
    Neurosci. Meth., 4, 1-32.

6. Donoghue, S., Garcia, M., Jordan, D. and Spyer, K.M. (1982). The
    brain-stem projections of pulmonary stretch afferent neurones in
    cats and rabbits. J. Physiol. (London), 322, 353-363.

7. Widdicombe, J.G. (1954). Receptors in the trachea and bronchi of
    the cat. J. Physiol. (London), 123, 71-104.

46

Responses of Inspiratory Neurones of the DRG to Maintained Lung Inflation

S. ISCOE and S. LONG

INTRODUCTION

Changes in lung volume are common, occurring most often during alterations of posture. Since the brainstem respiratory centre(s) receive input from slowly adapting pulmonary stretch receptors (PSR), changes in lung volume would be expected to affect breathing pattern. This is indeed the case (1,2,3,4). The most recent studies (3,4) indicate a prolongation of expiratory duration (Te), an increase in tidal volume (V_T) but little or no change in inspiratory duration (Ti) when breathing occurs at an elevated functional residual capacity (FRC). This absence of change in Ti is surprising since the resulting increases in both PSR activity and chemical drive would be expected to shorten Ti. Complete PSR adaptation, however, requires at least 30 min (5); only partial adaptation is apparent over shorter durations of breathing at increased FRC (4,6,7). Increases in arterial PCO_2 during breathing at elevated FRC appear to be ineffective in reducing Ti (4).

The object of this study was to examine the effects of breathing at increased lung volumes on the discharge patterns of inspiratory neurones of the dorsal respiratory group (DRG). In particular, the responses of I β neurones (those receiving input from PSR) to increases in FRC were studied to determine if habituation of these second order afferents occurs when ventilation occurs at an elevated FRC. Such habituation might explain the lack of change of Ti observed in previous studies.

METHODS

Experiments were conducted on cats, anaesthetized with pentobarbital
sodium (35 mg/kg i.v., supplemented as required). Surgical procedures
included tracheostomy and cannulation of both femoral arteries and a
vein. Arterial pressure was continuously monitored (Statham P23Db) as
was rectal temperature; body temperature was maintained at 37°C with a
heating pad.

The cats were placed in a stereotaxic frame, the head ventroflexed
45°. Both C5 branches of the phrenic nerves were isolated, cut
distally and desheathed. An occipital craniotomy was performed; when
the obex was not visible, it was exposed by moving the cerebellum
rostrally. The phrenic nerves were then placed on silver hook
electrodes and electrical activity amplified. This activity was
processed (8) and used to drive a cycle triggered pump (CTP, 9). Once
the ventilatory pattern on the CTP had stabilized, the cat was
paralyzed and bilaterally thoracotomized. An expiratory threshold
load (ETL) of 1-2 cm H_2O was used to prevent collapse of the lungs.

Recordings of brainstem inspiratory neurones were made with
tungsten microelectrodes placed in the DRG, the ventolateral nucleus
of the solitary tract. All units from which recordings were made were
found between 0.5 and 2.2 mm lateral and 0.5 and 2.5 mm rostral to the
obex at depths of between 1 and 2 mm. No histological verification of
recording sites was made. After successful isolation of an
inspiratory neurone, its activity was gated and standard pulses passed
to an instantaneous frequency meter (10). The cell was identified as α
(no PSR input) or β (PSR input) on the basis of changes in discharge
patterns between respiratory cycles with and without lung inflation.
FRC was then increased by applying a larger expiratory threshold load
(+5 cm H_2O) for 5 min. The unit's discharge was then examined again
with and without lung inflation.

RESULTS

The discharge patterns of both $I\alpha$ and $I\beta$ neurones were remarkably
unaffected by increases in lung volume sustained for 5-10 min.
Examples of the discharge patterns of an $I\alpha$ and $I\beta$ neurone during
inflations at control FRC and at the elevated FRC due to application
of ETL are shown in Figs. 1 and 2, respectively. The $I\alpha$ neurone fired

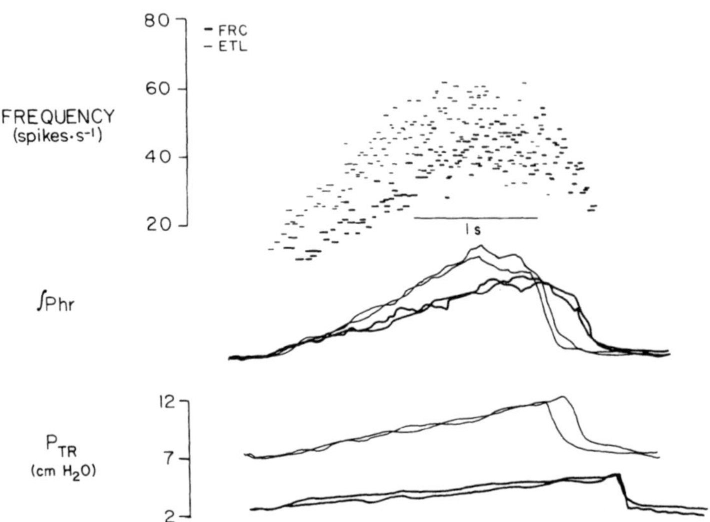

FIGURE 1 Response of an Iα neurone to 4 inflations, 2 at control FRC and 2 during ETL. Traces from top: instantaneous discharge frequency, integrated phrenic activity, and tracheal pressure. All traces synchronized to start of phrenic activity. PaCO$_2$ at control FRC, 48.7 torr; ETL, 54.3 torr.

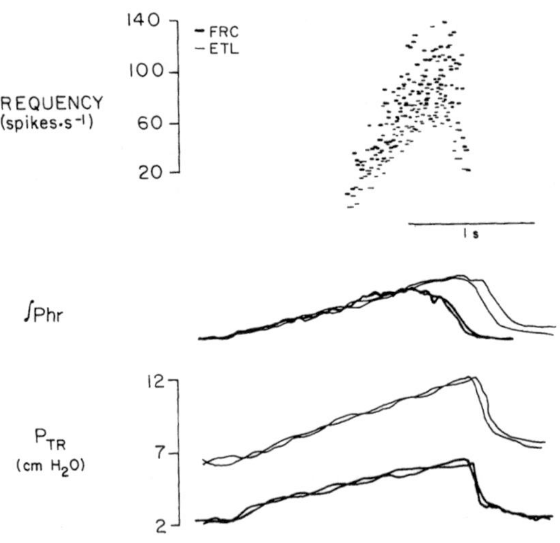

FIGURE 2 Response of an Iβ neurone to 4 inflations, 2 at control FRC, 2 during ETL. Traces as in Fig. 1. PaCO$_2$ at control FRC, 32.2 torr; ETL, 31.4 torr.

earlier and at an increased frequency, reaching a slightly higher peak frequency during inflations at the elevated lung volume. This may reflect, however, the increased rate of rise and peak amplitude of phrenic activity associated with a 5.6 torr increase in $PaCO_2$.

The Iβ neurone's discharge was unaffected by the change in FRC (Fig. 2). The onset of discharge did not change, nor the rate at which the unit increased its frequency, nor the peak frequency of discharge. Unlike the phrenic activity depicted in Fig. 1, in this case there was no difference in the rate of rise of integrated phrenic activity at the two lung volumes, although there was a slight ($\sim$350 ms) prolongation of Ti. In this case, $PaCO_2$ decreased by 0.8 torr.

The responses of all individual Iα and Iβ neurones at the two lung volumes are shown in Fig. 3. Overall, there was no significant change

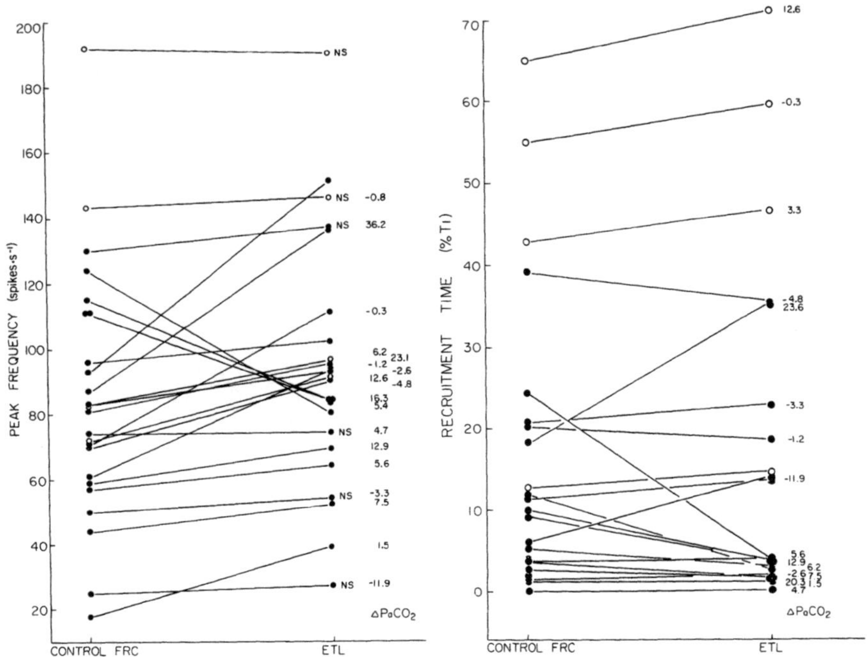

FIGURE 3 Discharge patterns of Iα ($\bullet$) and Iβ (o) neurones during ventilation at control FRC and on ETL. Left: peak discharge frequency. Each symbol average of 6–15 observations. Symbols marked NS represent neurones in which no significant change occurred. Change in arterial PCO_2 ($\Delta PaCO_2$) provided beside symbols for which blood gas data are available. Right: time of onset of discharge (recruitment time as % of Ti) for each neurone at control FRC and during ETL. Arterial PCO_2 indicated.

in either peak frequency (left) or the time of onset of discharge (right) although individual neurones did show significant changes. Such changes were, however, unassociated with changes in arterial CO_2 which, on average, increased significantly (6.5 torr, P < .05).

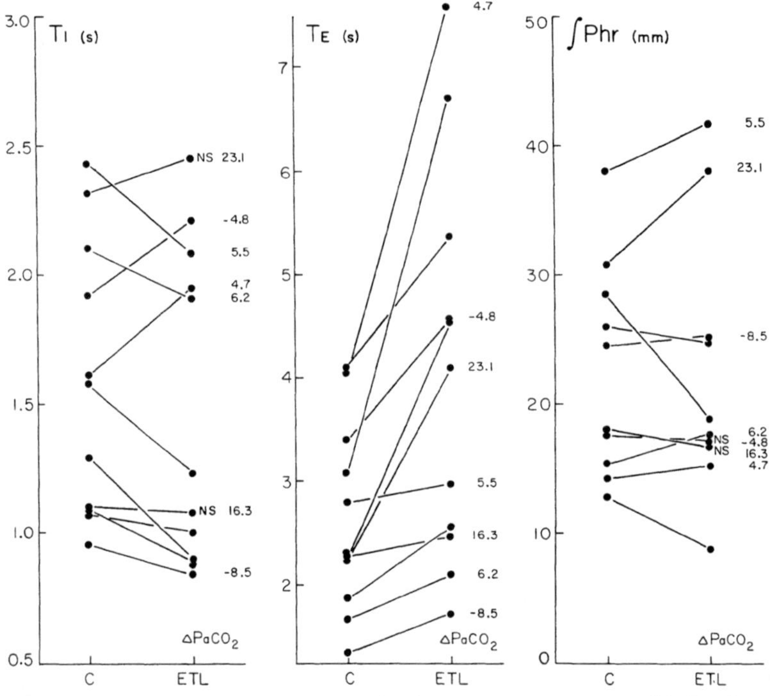

FIGURE 4   Changes in inspiratory (Ti) and expiratory (Te) duration, and peak amplitude of integrated phrenic activity ($\int$Phr) during ventilation at control FRC (C) and on ETL. Each symbol from a different cat. Only non-significant changes (NS) indicated. Changes in arterial PCO_2 as indicated.

The changes in Ti, Te, and peak amplitude of integrated phrenic activity ($\int$ Phr) for individual cats at the two lung volumes are shown in Fig. 4. There was no significant change in either Ti or $\int$Phr but there was a significant increase in Te (P<.01).

DISCUSSION

Application of ETL of 5 cm H_2O in these paralyzed, thoractomized cats presumably increases FRC by a volume equal to that produced by an ETL of 10 cm H_2O in a spontaneously breathing cat with intact chest wall. In the latter, such changes were associated with a doubling of peak amplitude of integrated phrenic activity and a 27% increase in peak discharge frequency of PSR (4).   We assume that similar increases in PSR frequency occurred in the cats of this study.

Despite this presumed increase in PSR activity, inspiratory neurones, both $I\alpha$ and $I\beta$, showed no overall change in either peak frequency of discharge or time of onset of discharge during inflations at elevated FRC (Fig. 3).   This lack of change is consistent with the hypothesis that habituation of central inspiratory neuronal activity occurs during breathing at elevated lung volumes. These results support those of Stanley et al., (6) who observed a recovery of phrenic activity in dogs ventilated at elevated lung volumes with constant blood gases, when PSR adaptation was prevented, and of Jammes et al., (11) who observed a gradual disappearance of the apnoeic response to vagal (PSR) stimulation in dogs breathing on an ETL.

Unlike spontaneously breathing cats (4) and dogs (3), our paralyzed ventilated cats and the paralyzed ventilated dogs of another study (12) did not increase peak amplitude of integrated phrenic activity when breathing occurred at an elevated FRC, even in the presence of occasional large increases in $PaCO_2$ (Fig. 4). Although Jammes et al., (3) observed that the increase in V_T during loaded breathing was abolished by vagal blockade, such increases are still present in spontaneously breathing vagotomized cats (13, 14) and absent in paralyzed, ventilated, non-vagotomized dogs (12). These results strongly suggest that activation of non-pulmonary (and non-vagal) afferents is responsible for the increase in phrenic activity when breathing occurs at an elevated lung volume.

Inspiratory (both $I\alpha$ and $I\beta$) and phrenic activities are inhibited by activation of lower intercostal and lumbar nerve afferents (15). Moreover, lesions of the ventrolateral cervical spinal cord reduce V_T but have no effect on respiratory frequency of decerebrate, spontaneously breathing cats (16). Contraction of accessory respiratory muscles appears to be necessary to elicit increased phrenic activity at elevated FRC since paralysis alone, without thoracotomy, abolishes the increase in phrenic activity (Iscoe, unpublished observations).

In conclusion, our data show that increases of FRC in paralyzed ventilated cats do not result in any net change in $I\alpha$ and $I\beta$ activity. Nor is there any change in Ti although Te is prolonged.   Unlike spontaneously breathing cats, there is no increase in phrenic activity when end-expiratory lung volume is increased by ETL in paralyzed, ventilated cats. It is unclear what receptors are responsible for this increase in phrenic activity. Most intercostal and abdominal afferents when stimulated produce inhibition of phrenic activity (15) but phrenic afferents may have a prominent role in the response (14). The results strongly suggest that non-pulmonary receptors have a prominent role in determining ventilatory pattern at elevated lung volumes.

REFERENCES

1. Bishop, B. (1977). Vagal control of diaphragm timing in cat while breathing at elevated lung volumes. Respir. Physiol. 30, 169-184.

2. D'Angelo, E. and Agostoni, E. (1975). Tonic vagal influences on inspiratory duration. Respir. Physiol. 24, 287-302.

3. Jammes, Y., Bye, P.T.P., Pardy, R.L., Katsardis, C., Esau, S. and Roussos, C. (1983). Expiratory threshold load under extracorporeal circulation: effects of vagal afferents. J. Appl. Physiol.: Respirat. Environ. Exercise Physiol. 55, 307-315.

4. Finkler, J., and Iscoe, S. (1984). Control of breathing at elevated lung volumes in anesthetized cats. J. Appl. Physiol.: Respirat. Environ. Exercise Physiol. 56, 839-844.

5. Muza, S.R. and Frazier, D.T. (1983). Response of pulmonary stretch receptors to shifts of functional residual capacity. Respir. Physiol. 52, 371-386.

6. Stanley, H.M., Altose, M.D., Cherniack, M.S., and Fishman, A.P. (1975). Changes in strength of lung inflation reflex during prolonged inflation. J. Appl. Physiol. 38, 474-480.

7. Farber, J. (1982). Pulmonary receptor discharge and expiratory muscle activity. Respir. Physiol. 47, 219-225.

8. Cohen, M.I. (1968). Discharge patterns of brain-stem respiratory neurons in relation to carbon dioxide tension. J. Neurophysiol. 31, 142-165.

9. Feldman, J.L. and Gautier, H. (1976). Interaction of pulmonary afferents and pneumotaxic center in control of respiratory pattern in cats. J. Neurophysiol. 39, 31-44.

10.   Iscoe, S. and Young, R.A. (1981). Microprocessor-based instantaneous frequency meter. <u>Electroencephalogr.</u> <u>clin.</u> <u>Neurophysiol</u>. <u>51</u>, 443-445.

11.   Jammes, Y., Bye, P.T.P., Pardy, R.L., and Roussos, C. (1983). Vagal feedback with expiratory threshold load under extracorporeal circulation. <u>J</u>. <u>Appl</u>. <u>Physiol</u>.: <u>Respirat</u>. <u>Environ</u>. <u>Exercise</u> <u>Physiol</u>. <u>55</u>, 316-322.

12.   Bartoli, A., Bystrzycka, E., Guz, A., Jain, S.K., Noble, M.I.M., and Trenchard, D. (1973). Studies of the pulmonary vagal control of central respiratory rhythm in the absence of breathing movements. <u>J</u>. <u>Physiol</u>. (<u>London</u>). <u>230</u>, 449-465.

13.   Bishop, B. (1969). Diaphragm and abdominal muscle responses to elevated airway pressures in the cat. <u>J</u>. <u>Appl</u>. <u>Physiol</u>. <u>22</u>, 959-965.

14.   Fryman, D. and Frazier, D.T. (1983). Modulation of diaphragm activity by phrenic afferents. (Abstract). <u>Federation</u> <u>Proc</u>. <u>42</u>: 1014.

15.   Shannon, R. (1980). Intercostal and abdominal muscle afferent influence on medullary dorsal respiratory group neurons. <u>Respir</u>. <u>Physiol</u>. <u>39</u>, 73-94.

16.   Krieger, A.J., Christensen, H.D., Sapru, H.N., and Wang, S.C. (1972). Changes in ventilatory patterns after ablation of various respiratory feedback mechanisms <u>J</u>. <u>Appl</u>. <u>Physiol</u>. <u>33</u>, 431-435.

Supported by the Medical Research Council of Canada and the Ontario Thoracic Society.
We thank S. Gordon, L. Royko, and M. Rupert for their technical assistance.

47

Nonmyelinated Vagal Afferents from the Lungs: Studies on the Tonic Vagal Sensory Influences

Y. JAMMES, S. DELPIERRE, N. MEI and Ch. GRIMAUD

<u>INTRODUCTION</u>

Since the principle work by Hering and Breuer (8), numerous studies have been devoted to the functional role played by the pulmonary vagal receptors phasically activated during tidal breaths. In particular, Clark and Von Euler(1) have clearly demonstrated that this phasic vagal information is responsible for the inspiratory "off-switch" mechanism. However,we must wait for the work by Hammouda and Wilson(7) which first presented evidence that small vagal fibres supplied a tonic input to the respiratory centres. Neurophysiological studies have shown that small myelinated fibres connected to rapidly adapting receptors and nonmyelinated vagal endings (12-15) supply sensory information not related to tidal lung volume changes. An extensive review on nonmyelinated bronchopulmonary vagal afferents (called also C fibres) has been recently put forth by Coleridge and Coleridge (2). Most studies have been devoted to the ventilatory and bronchomotor effects of the stimulation of lung C fibres, but little information exists concerning the possible roles of their resting discharge. In the present study we present a short overview of published results obtained by our group on the spontaneous activity of lung C fibres in the cat and on their influences on the bronchoconstrictor vagal tone. New data on the role played by these afferents on the spontaneous breathing pattern are added.

1 - <u>COMPOSITION OF AFFERENT BRONCHIAL VAGAL BRANCHES</u>

In cats, motor vagal fibres constitute distinct superficial bundles at
the ventral side of the nodose ganglion. Therefore, at this level it is
possible to suppress selectively either the motor fibres(motor vagotomy) or the sensory neurones(sensory vagotomy)(9).In order to determine
the number and the diameter of fibres in the afferent vagal component,
a unilateral motor vagotomy was performed (10). Pulmonary vagal branches
were studied using electron microscopy. This work reveals that nonmyelinated fibres constitute more than 90 % of the total population of the
afferent vagal component as well as of the entire nerve.We also observed a preferential compartmentalization by Schwann cell envelopment of
fibres having comparable diameters; it was called a "packing effect".
Therefore,nonmyelinated bronchopulmonary fibres constitute the most
important quantitative component of the vagal sensory pathway.The other
fibres are connected to lung stretch receptors and irritant receptors.

2 - <u>LOCALIZATION OF BRONCHOPULMONARY C FIBRES</u>

This was assessed either by pulling the lung parenchyma or the main
bronchi in the open chest preparation or by injecting drugs as phenyldiguanide or capsaicine into the pulmonary or the bronchial circulation (2-5). Lung C fibres are associated with endings located in the
larynx and the trachea (5), in lower bronchi (2-5) and in the alveoli
(2-5-12). The latter endings were called J receptors by Paintal (12).

3 - <u>RESTING DISCHARGE OF BRONCHOPULMONARY C FIBRES</u>

Under the particular conditions of neurophysiological studies on lung
receptors,i.e. during artificial ventilation and tracheotomy,the spontaneous firing rate of bronchopulmonary C fibres is low(less than one
impulse per second).However,when physiological conditions are restored
(i.e. normal alveolar CO_2 content, $F_{ET}CO_2$, tidal volume V_T, and airway
temperature) the resting firing rate of lung C fibres doubled, thus
being far from negligible (5) (Table 1). This spontaneous activity of
lung receptors constitutes a tonic neurophysiological background without
any respiratory rhythmicity.

These receptors seem to detect preferentially changes in inspired CO_2
content and they adapt when a high CO_2 level is maintained (5).There-
fore,peak spike frequency can reach 8 to 10 impulses.s^{-1} during step
changes in $F_{ET}CO_2$.Eighty per cent of these afferent fibres increase

their discharge frequency when $F_{ET}CO_2$ reaches 0.04, i.e. its physiological value.

Table 1 : Resting discharge frequency of bronchopulmonary C fibres (from data reported in J. Physiol.(London), 1981, <u>316</u>, 61-74).

	Extraphysiological conditions		Physiological conditions	Resting discharge (impulses.s^{-1})
$F_{ET}CO_2$	0.02	$\longrightarrow$	0.05	$1.1\pm0.2 \rightarrow 2.4\pm0.5$
V_T	100 ml	$\longrightarrow$	40 ml	$0.8\pm0.3 \rightarrow 1.6\pm0.4$
Tracheal temperature	28°C	$\longrightarrow$	35°C	$0.7\pm0.2 \rightarrow 1.5\pm0.3$

4 - BRONCHOMOTOR AND VENTILATORY RESPONSES TO INCREASE IN SPONTANEOUS DISCHARGE OF LUNG C FIBRES

Activation of C fibres by hypercapnia is responsible for the early increase in lung resistance measured in cats during inhalation of CO_2 enriched gas (4) and the reflex increase in respiratory frequency in response to hypercapnia (13).These fibres also detect the changes in lung volume beyond the normal range of tidal volume, since they are stimulated by hyperdeflation of the lung and strongly inhibited by overinflation(5).These findings agree with the definition of "deflation" receptors proposed by Paintal (12) for type J receptors. The respiratory response to mechanical stimulation of C fibres appears to be mostly characterized by changes in ventilatory timing.In fact, enhancement of the resting discharge of bronchopulmonary C fibres seems to always result in an increase in respiratory frequency (13-15) and in lung resistance (2). In addition, nonmyelinated vagal afferents from the trachea and main bronchi act also as sensors of thermal changes in the airway lumen (11).Thus, the bronchomotor response to cold air breathing in cats is a reflex mostly mediated by small afferent fibres in the vagus and superior laryngeal nerves. This response is associated with a significant decrease in amplitude of the integrated phrenic discharge and with modifications in the recruitment of phrenic motoneurones. Surprisingly, the ventilatory timing does not vary.

5 - FUNCTIONAL ROLES OF THE SPONTANEOUS DISCHARGE OF LUNG C FIBRES

a) <u>Origin of the bronchoconstrictor vagal tone</u>:In cats the selective

blockade of conduction in C fibres(and also in small myelinated fibres)
by local application of procaine solution on both vagal trunks induced
a decrease in pulmonary resistance similar to that produced by the selective sensory vagotomy(9)(Table 2).Further total bivagotomy did not
change lung resistance.

Table 2. Changes in lung resistance (R_L) after sensory vagotomy or procaine block of conduction in small vagal fibres.All changes are significant for $2p < 0.001$.From data published in J.Physiol.(London), 1979,
291, 305-316.

	Control	Sensory Vagotomy	Control	Procaine Block
R_L,cmH$_2$O.L^{-1}.s	16.5 $\pm$ 0.6	11.0 $\pm$ 0.6	17.0 $\pm$ 1.3	10.3 $\pm$ 1.5
ΔR_L, %		- 33		- 39

From these results, it appears that bronchoconstrictor vagal tone had
an exclusive peripheral origin and that pulmonary endings, in particular those connected with nonmyelinated vagal fibres, are involved in
this mechanism. Any change in alveolar CO_2 content, in thermal environment of lung receptors and perhaps in lung volumes may modulate the
resting tonic vagal sensory pathway resulting in reflex bronchomotor
responses.

(b) <u>Breathing pattern</u>. The influence of the resting discharge of bronchopulmonary C fibres on breathing pattern is controversial (2). Some
authors attribute the control of the ventilatory timing during eupnea
to both phasic (pulmonary stretch receptors) and tonic vagal sensory
pathways (3),but others attribute the adjustment of the spontaneous
respiratory frequency only to volume related vagal informations(1-6).

In an attempt to elucidate the possible role of the tonic vagal
sensory component in the control of tidal breaths,we performed further
experiments in seven adult cats. They were anesthetized with sodium
pentobarbital, tracheotomized and allowed to breath spontaneously. End
tidal CO_2 and O_2 concentrations were measured with rapid analyzers and
maintained in their physiological range. Tidal volume (V_T) was measured
with a Fleisch pneumotachygraph (n° 00). A phrenic root was dissected
free in the neck and electrical activity recorded with platinum elec-

trodes. Raw and integrated phrenic output (called Ephr) were displayed with tidal volume on a multichannel ink recorder (Siemens, Mingograph). Ephr was measured at 1.0 s (Ephr 1.0) and also every 0.1 s for each breath. This permitted us to obtain the Ephr vs. time profile and thus to calculate the b coefficient of the relationship Ephr= a.Time^b. This moving time average approaches the firing rate and recruitment pattern of phrenic motoneurones by inspiratory cells(14). The postinspiratory phrenic activity (PIA) and its ratio to the expiratory period (PIA/T_E) were also considered. In addition, both vagus nerves were dissected in the neck. One nerve was placed on two pairs of electrodes: the distal electrodes were used for peripheral stimulation while the proximal electrodes for recording of vagal action potentials. Teflon sheets containing procaine solution (2°/oo) were placed around both vagus nerves. Thus, the compound C wave of evoked potentials can be selectively suppressed as indicated during recording of nerve potentials. Breathing pattern and phrenic discharge were analyzed for ten consecutive breaths under control conditions (vagus nerves unblocked), after suppression of conduction in small vagal fibres and finally following section of the vagus nerves.

Procaine block significantly increased V_T (42 % of its maximal increase measured after bivagotomy) but did not change the inspiratory period (T_I) nor the total breath duration (Ttot) (Figure 1). Ephr 1.0 markedly increased after procaine block and was similar to the value measured after bivagotomy (Figure 2). The b coefficient of Ephr vs. time relationship was also significantly modified by the procaine block, which also induced a lengthening of PIA with increase in PIA/T_E ratio (PIA/T_E was respectively 6.7 $\pm$ 0.7 % in intact animals, and 10.9 $\pm$ 0.4% after procaine block).

Further bivagotomy increased V_T but not Ephr 1.0 (Figures 1 and 2). PIA,T_I and Ttot were prolonged and PIA/T_E ratio also significantly increased (15.7 $\pm$ 0.6 %). Moreover, total vagotomy always restored the control value of the b coefficient (Figure 2, legend).

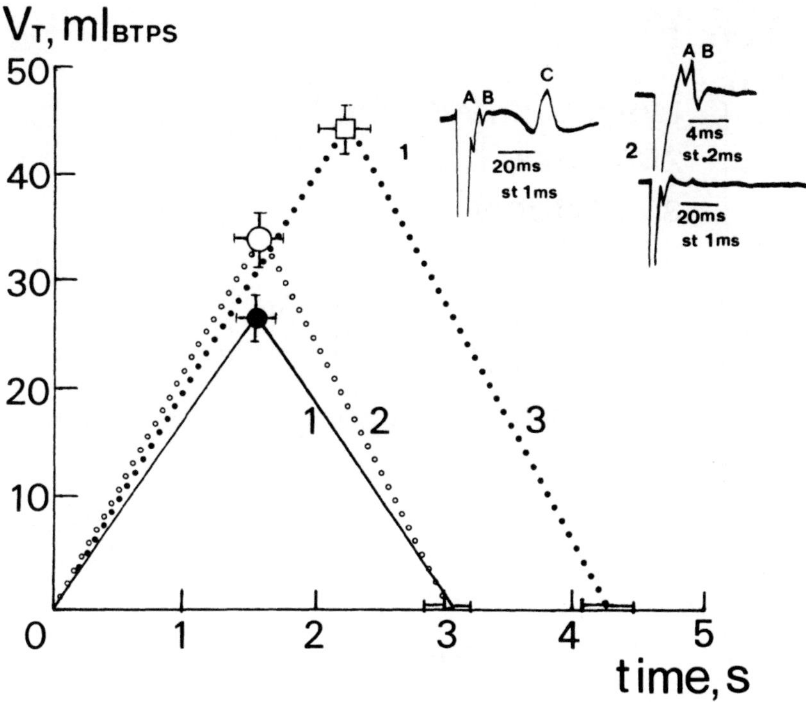

Figure 1. Mean spirograms (+ one standard error) measured during eupnea in anesthetized cats whose vagus nerves were intact (1), blocked by procaine (2), then severed (3). Recording of evoked vagal potentials proofs the integrity of A, B and C waves (1) and the selective suppression of C wave during procaine block (2). Two different single pulse durations were used for stimulation in order to identify A (and B) or C waves (respectively : 0.2 and 1.0 ms).

Actually, the tonic sensory pathway supplied by small pulmonary vagal fibres appears to mainly influence the onset of phrenic discharge (marked changes in Ephr 1.0 and the b coefficient) but does not control the ventilatory timing. Changes in breathing pattern and phrenic discharge observed after total bivagotomy cannot be interpreted only in terms of suppression of afferents from pulmonary stretch receptors, which control the inspiratory duration, but also as a consequence of the removal of tonic vagal sensory inputs, which mainly act on the early recruitment of phrenic motoneurones.

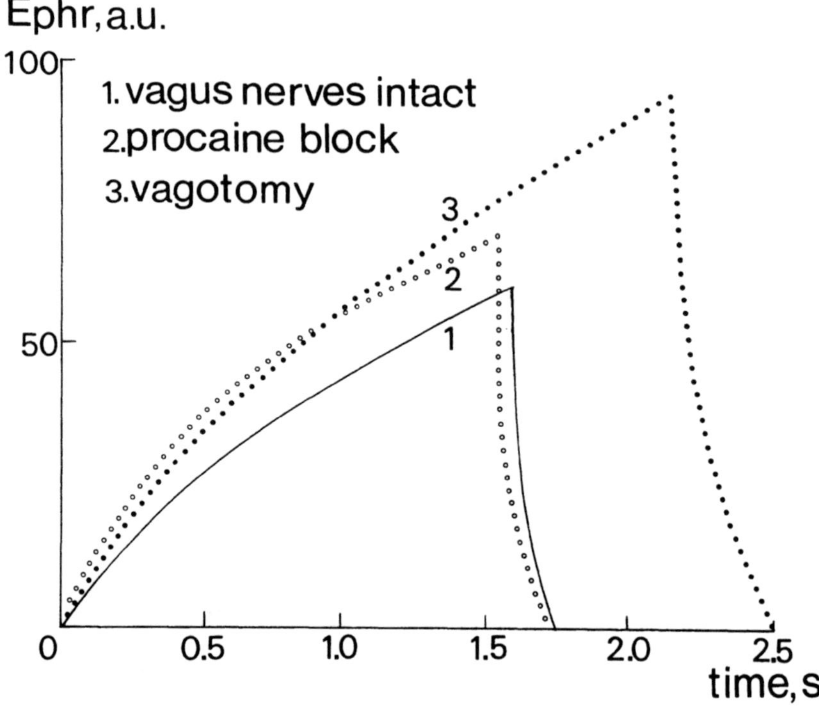

Figure 2. Mean shape of time average spontaneous phrenic activity recorded under three experimental conditions: (1) Ephr = 41.9 . Time 0.75;
(2) Ephr = 52.6 . Time 0.63 ; (3) Ephr = 52.9 . Time 0.75.

<u>CONCLUSIONS</u>

Present data clearly show that tonic sensory influences from the
lungs, carried by quantitatively the most important vagal pathway, control both the activity of vagal and phrenic motoneurones. Therefore,
pulmonary information not related to the amplitude of tidal breaths
seem to play a major role in respiratory control.

References :

1. Clark,F.J. and Von Euler,C. (1972). On the regulation of depth and
   rate of breathing. <u>J. Physiol. (London)</u>, <u>222</u>, 267-295.

2. Coleridge,J.C.G. and Coleridge,H.M. (1984). Afferent vagal C fibre
   innervation of the lungs and airways and its functional significance.
   In: Adrian,R.H.(ed.), <u>Physiology, Biochemistry and Pharmacology,</u>
   pp. 1-110 (Springer Verlag).

3. D'Angelo,E. and Agostoni,E. (1975).Tonic vagal influences on inspiratory duration. Respir. Physiol.,24, 287-302.

4. Delpierre,S.,Jammes,Y.,Mei,N.,Mathiot,M.J. and Grimaud,Ch. (1980). Mise en évidence de l'origine vagale réflexe des effets bronchoconstricteurs du CO_2 chez le Chat. J. de Physiol. (Paris), 76, 889-891.

5. Delpierre,S.,Grimaud,Ch.,Jammes,Y. and Mei,N.(1981). Changes in activity of vagal bronchopulmonary C fibres by chemical and physical stimuli in the cat. J. Physiol. (London), 316, 61-74.

6. Grunstein,M.M.,Younes,M. and Milic-Emili,J.(1973).Control of tidal volume and respiratory frequency in anesthetized cats. J. of Appl. Physiol., 35, 463-476.

7. Hammouda,M. and Wilson,W.H. (1939). Reflex acceleration of respiration arising from excitation of the vagus or its terminations in the lungs. J. Physiol. (London), 94, 497-524.

8. Hering,E. and Breuer,J.(1868).Die Selbststeuerung der Athmung durch den Nervus vagus. Sitzber Akad.Wiss., Wien, 57, 672-677.

9. Jammes,Y., and Mei,N.(1979).Assessment of the pulmonary origin of bronchoconstrictor vagal tone. J. Physiol. (London), 291, 305-316.

10. Jammes,Y.,Fornaris,E.,Mei,N. and Barrat E. (1982).Afferent and efferent components of the bronchial vagal branches in cats. J. Auton. Nerv. Sys., 5, 165-176.

11. Jammes,Y.,Barthelemy,P. and Delpierre,S.(1983).Respiratory effects of cold air breathing in anesthetized cats. Respir. Physiol., 54, 41-54.

12. Paintal,A.S. (1955).Impulses in vagal afferent fibres from specific pulmonary deflation receptors.The response of the receptors to phenyldiguanide,potato-starch,5-hydroxytryptamine and nicotine and their role in respiratory and cardiovascular reflexes. Q.J. exp. Physiol., 40, 89-111.

13. Russel,N.J.W.,Raybould,H.E. and Trenchard,D.(1984). Role of vagal C-fiber afferents in respiratory response to hypercapnia. J. Appl. Physiol. R.E.E.P., 56, 1550-1558.

14. Siafakas,N.M.,Chang,H.K.,Bonora,M.,Gauthier,H.,Milic-Emili,J. and Duron,B.(1981). Time course of phrenic activity and expiratory pressures during airway occlusion in cats.J.Appl.Physiol.,R.E.E.P.,51,99-108.

15. Winning,A.J. and Widdicombe,J.G.(1976).The effects of lung reflexes on the pattern of breathing in cats.Respir.Physiol., 27, 153-266.

48

The Effects of Pulmonary Denervation on Respiratory Timing in the Awake Rat

R. L. MARTIN-BODY and J. D. SINCLAIR

INTRODUCTION

The respiratory role of impulses carried in the vagus has been infrequently studied in awake animals. Of particular importance is the dependence on vagal integrity of the increase in respiratory frequency in response to inhalation of CO_2. There is no doubt that after functional vagotomy in awake animals and man substantial increases occur in VT and VE in response to CO_2. However, respiratory frequency and phase duration alterations are equivocal. For example, in the rabbit, respiratory frequency may either increase or decrease (1, 2); variable changes in TI and TE are responsible. In studies on 8 cats TI and TE decreased only marginally less than before vagotomy in 4 animals, did not change in 2 animals, and increased in the remaining animals (3). Unfortunately these results are difficult to interpret because some cats were tracheostomised whereas others had partial resection of their vocal cords to prevent laryngeal obstruction consequent to the vagotomy. In 3 of 4 dogs studied f still increased after vagal block (4) whereas in another study on one dog f decreased (5). In man, vagal block by local anaesthetic severely reduced but did not abolish the f response to CO_2 (6).

We have reinvestigated the variability of the effects of vagotomy on respiratory patterning using two chronic preparations. In both preparations unilateral innervation of the vocal cords remained.

METHODS

Chronic pulmonary denervation was achieved in rats either by left pneumonectomy in combination with right cervical vagotomy (PNVG), or by left intrathoracic vagotomy below the origin of the recurrent laryngeal nerve in combination with right cervical vagotomy (IVCV). Rapid development of pulmonary oedema in awake rats after bilateral cervical vagotomy prevents the use of a more direct technique. The efficacy of the denervation procedures was tested by measurement of the duration of apnoea in response to maintained lung inflation in the anaesthetised state (chloral hydrate, 34 mg kg-1).

Ventilation was measured in 12 rats three weeks after left pneumonectomy and 1 week after right cervical vagotomy. Their responses were compared with those of 12 age-matched rats which had undergone left pneumonectomy alone (PNAM). In the second series of experiments the respiratory patterns of 8 rats were measured 28-40 days after left intrathoracic vagotomy and on the day of right cervical vagotomy at least 1 hour after discontinuation of halothane anaesthesia. These were compared with 8 age-matched intact rats (IVAM). Respiratory measurements were made in the awake state using a barometric plethysmograph while the rats breathed air and after 10 minute exposures to CO_2 concentrations between 0.5 and 10%.

RESULTS

PNVG produced a significant reduction in the apnoea of lung inflation. IVCV abolished this response.

During air breathing PNVG resulted in a significant rise in VT and increase in TI so that the mean inspiratory flow rate was unaltered. TE failed to change. This pattern of response was also observed after IVCV but with larger changes.

When breathing hypercapnic gas mixtures PNVG rats had an enhanced VT response to CO_2 but a substantial reduction in the f response. This was due to attenuation of the responses of both TI and TE to CO_2. As a result the relationship between VT and TI was steeper and shifted to the right compared with PNAM rats.

IVCV rats only tolerated exposure to mild and moderate

concentrations of CO2.   Tests at 3.6% and 5.7% CO2 revealed an enhancement of the VT response to CO2 which was greater than that noted in PNVG rats.   TI was observed to decrease with increasing concentrations of CO2 whereas TE lengthened.   This unexpected observation occurred in all 8 rats studied.   The lengthening of TE offset the increase in VT and decrease in TI so that mean expiratory flow rate (VT/TE) and f were unchanged during hypercapnia.

DISCUSSION

Chronic pulmonary denervation by either PNVG or IVCV produced the classically described increases in VT and TI during air breathing. Unexpectedly, TE failed to increase. Possible explanations include a lack of vagal control of TE; a significant influence of the remaining laryngeal function; or a balance between the opposing actions of pulmonary stretch receptors and rapidly adapting receptors (7).

Measurement of respiratory frequency alone would have suggested that awake completely denervated animals respond to CO2 in a manner similar to anaesthetised animals since in the latter state f varies little after bilateral cervical vagotomy.   However, examination of the respiratory phase durations showed that TI decreased at least as much as in intact control rats and that TE increased with higher levels of CO2. The changes in TE must be explained by the mechanisms which are known to influence expiration.   In the intact state there are a number of mechanisms which assist expiration. These include a reduction in airways resistance and post-inspiratory contraction of the diaphragm, and an increase in expiratory muscle activity. However, such facilitatory mechanisms may be offset to some extent by inhibition of the activity of the main glottal abductor, the posterior cricoarytenoid muscle, by hypercapnia (8).   It is hypothesised that when IVCV rats were breathing 3.6% CO2 some or all of the facilitatory mechanisms were operative and their effects successfully outweighed the increase in TE which would result from an increased VT.   The lengthening of TE with 5.7% CO2 may reflect a reduction in the effectiveness of these mechanisms after vagotomy; increased time for lung emptying would then result from the still further increased VT and any chemically mediated expiratory inhibition.

REFERENCES

1. Richardson, P. S. and Widdicombe, J. G. (1969). The role of the vagus nerves in the ventilatory responses to hypercapnia and hypoxia in anaesthetized and unanaesthetized rabbits. Resp. Physiol., 7, 122–35

2. Glogowska, M. (1975). The frequency response to CO_2 in chronically vagotomised conscious rabbits. Bull. Pathophysiol. Respir. (Nancy), 11, 91–2P

3. Gautier, H. (1975). Effects of hypoxia or hypercapnia on ventilatory pattern of chronic cats before and after vagotomy. Bull. Pathophysiol. Respir. (Nancy), 11, 89–90P

4. Phillipson, E. A. (1974). Vagal control of breathing pattern independent of lung inflation in conscious dogs. J. Appl. Physiol., 37, 183–89

5. Guz, A., Noble, M. I. M., Eisele, J. H. and Trenchard, D. (1970). The role of vagal inflation reflexes in man and other animals. In: Porter, R. (ed). Breathing: Hering–Breuer Centenary Symposium., pp. 17–40. (London: Churchill)

6. Guz, A., Noble, M. I. M., Widdicombe, J. G., Trenchard, D. and Mushin, W. W. (1966). The effect of bilateral block of vagus and glossopharyngeal nerves on the ventilatory response to CO_2 of conscious man. Resp. Physiol., 1, 206–10

7. Davies, A., Dixon, M., Callanan, D., Huszczuk, A., Widdicombe, J. G. and Wise, J. C. M. (1978). Lung reflexes in rabbits during pulmonary stretch receptor block by sulphur dioxide. Resp. Physiol., 34, 83–101

8. Bartlett, D. (Jr), Knuth, S. L. and Knuth, K. V. (1981). Effects of pulmonary stretch receptor blockade on laryngeal responses to hypercapnia and hypoxia. Resp. Physiol., 45, 67–77

49

Efferent Modulation of Pulmonary Stretch Receptor Sensitivity

C. A. RICHARDSON, D. A. HERBERT, R. A. MITCHELL

In 1966, Widdicombe (1) reported that nerve fibers in the pulmonary branches of the vagus nerve and in small strands of the recurrent laryngeal nerve fired with respiratory rhythm and suggested that these nerve fibers innervated airway smooth muscle. Subsequently, McAllen and Spyer (2) identified neurons in the nucleus ambiguus with inspiratory rhythm that projected to the airways. More recently Mitchell et al. (3) demonstrated that tracheal smooth muscle contracted with inspiration and relaxed with expiration. Moreover they demonstrated that transpulmonary pressure, which they suggested was an index of airway resistance, increased during inspiration. These observations were consistent with those reported by Nadel and Widdicombe (4,5) who demonstrated that tracheal volume and lower-airway resistance reflexes paralleled the level of breathing. In other words, reflexes that caused an increase in breathing caused contraction of airway smooth muscle, diminishing tracheal volume and increasing lower-airway resistance; whereas reflexes that inhibited ventilation caused relaxation of airway smooth muscle, increasing tracheal volume and diminishing lower-airway resistance.

The effect of the natural contraction of smooth muscle upon pulmonary receptors has not been reported. Many investigators have studied afferent fibers from the receptors after cutting the vagus nerve thus denervating the lung, or have recorded from a small strand of an otherwise intact vagus during spontaneous breathing of the animal. In both cases the effect of the smooth muscle contraction upon the threshold and firing of pulmonary receptors would not be apparent. We determined the effect of inspiratory related contractions of airway smooth muscle upon pulmonary receptors by recording afferent activity with microelectrodes in the intact vagus nerve in anesthetized, paralyzed cats (6).

Firing rates of pulmonary stretch receptors were modulated in step with the inflation-deflation cycle of the mechanical respirator, as expected. Firing rates of most slowly adapting receptors (SARs), but not rapidly adapting receptors (RARs), were also strongly modulated in step with the phrenic nerve activity even when the respirator was turned off and the cat motionless (Figure). The modulation of some receptors' firing rates by the inspiratory motor output was as great as the change in firing rate in response to a lung inflation of 18 ml of air (one tidal volume). Atropine blocked the inspiratory-related modulation of SAR firing rates; it did not block the inflation-related modulation.

We conclude that, through the mechanical link with airway smooth muscle, stretch receptor sensitivity depends on inspiratory output and thus, with efferent modulation of a sensory input, forms a closed loop.

References:
1. Widdicombe J.G. (1966), Action potentials in parasympathetic and sympathetic efferent fibres to the trachea and lungs of dogs and cats, J. Physiol. (Lond) 186:56-88.
2. McAllen R.M., K.M. Spyer (1978), Two types of vagal preganglionic motoneurones projecting to the heart and lungs, J. Physiol. (Lond) 282:353-364.
3. Mitchell R.A., D.A. Herbert, D.J. Baker, Parasympathetic bronchoconstrictor fibers fire with inspiratory rhythm to evoke rhythmic fluctuations in airway smooth muscle tone, (Submitted to J. Appl. Physiol. August 1984).
4. Nadel J.A., J.G. Widdicombe (1962), Effect of changes in blood gas tensions and carotid sinus pressure on tracheal volume and total lung resistance to airflow, J. Physiol. (Lond) 163:13-33.
5. Widdicombe J.G., J.A. Nadel (1963), Reflex effects of lung inflation on tracheal volume, J. Appl. Physiol. 18:681-686.
6. Richardson C.A., D.A. Herbert, R.A. Mitchell, Modulation of pulmonary stretch receptor activity and pulmonary airway resistance by parasympathetic efferents in the lung, (Submitted to J. Appl. Physiol. June 1984).

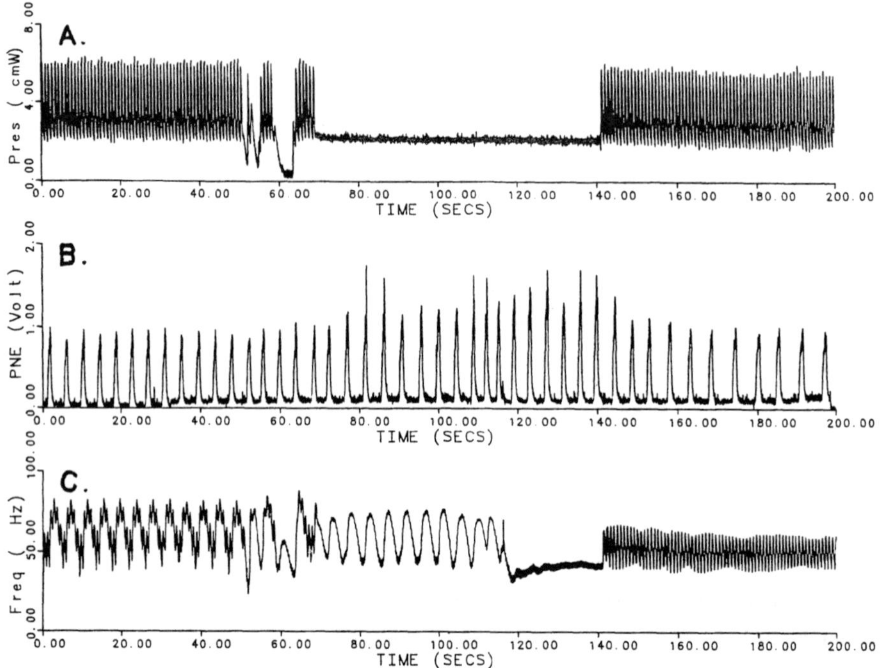

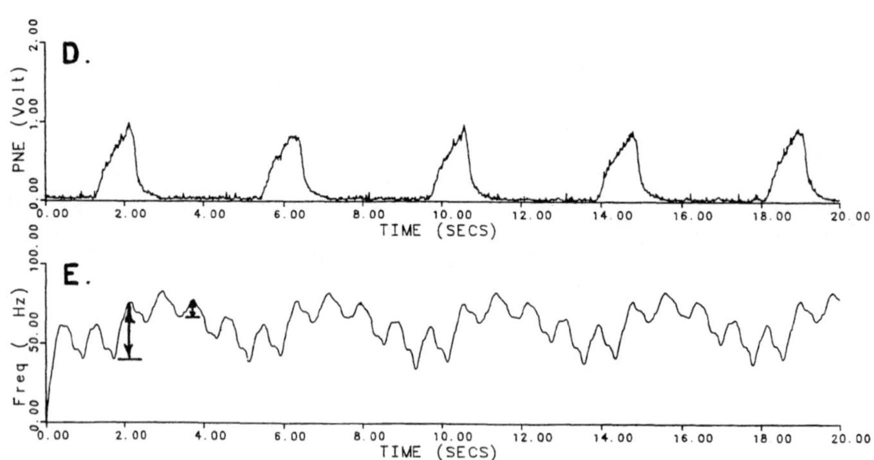

<u>Figure</u>. Modulation of the firing rate of a pulmonary stretch receptor
by lung inflation and inspiratory motor output.
The tracheal pressure (A) and phrenic nerve activity (B,D) were
recorded in an anesthetized, paralyzed cat, ventilated with a fixed
tidal volume of 18ml. Simultaneously, a single fiber was recorded in
the intact cervical vagus. The instantaneous firing rate of the single fiber was computed (C,E). In the interval from 0 to 50 s (A,B,C),
the pulmonary stretch receptor's firing rate was modulated by the
inflation/deflation cycle of the respirator with a change of about
50 Hz for each tidal volume. The lowest rate after each deflation (a
valley) was itself modulated in step with the phrenic nerve activity;
the range of this modulation was on the order of 50-60 Hz. The
highest rate after each inflation (a peak) was also modulated in step
with the phrenic nerve activity; the range of this modulation was on
the order of 20-30 Hz. From 70 to 110 s, the respirator was turned
off and the cat was motionless. The firing rate of the pulmonary
stretch receptor varied from 35 to 80 Hz with each burst of inspiratory activity on the phrenic nerve. Atropine was slowly injected
(0.5 mg/kg i.v.) in the 20 s interval from 100 to 120 s and the
inspiratory related modulation was eliminated (only the higher frequency cardiac related modulation remained). When the respirator was
turned back on at 140 s, the stretch receptor firing rate had only
volume modulation and no inspiratory modulation of its peaks and val-
leys compared with the interval from 0 to 40 s.
Variation of the sensitivity of this stretch receptor to the 18ml
inflation (D,E) ranged from 40Hz/tidal volume (t=2s, neural inspiration) to 10Hz/tidal volume (t=4s, neural expiration).

50

Activation of Bulbar Post-Inspiratory Neurons by Upper Airway Stimulation

J. E. REMMERS, D. W. RICHTER, D. BALLANTYNE, C. R. BAINTON and J. P. KLEIN

Changes in the duration of mechanical expiration constitute the major means by which the respiratory frequency is altered in response to chemical stimulation and end-expiratory volume is controlled. Accordingly, factors regulating the time that elapses between bursts of inspiratory motor output play a pivitol role in setting the respiratory rhythm and lung volumes. Analysis of excitatory and inhibitory post-synaptic potentials (EPSPs and IPSPs) in bulbar respiratory neurons has revealed that the neural respiratory cycle consists of inspiratory (1), post-inspiratory (PI), and neural expiratory (nE) intervals (1). Each interval is associated with depolarization of at least one type of respiratory neuron, namely: inspiratory, post-inspiratory or expiratory neurons. This means that the expiratory phase of the mechanical breathing cycle encompasses two neural intervals, PI and nE. The former may correspond to the first portion of mechanical expiration when expiratory "braking" retards airflow, and the latter occurs later in expiration when contraction of expiratory muscles may facilitate airflow.

Whether or not these two neural intervals occurring during mechanical expiration constitute separate neural "phases" is unknown at present. Here, a phase of the neural respiratory cycle is taken to signify an interval associated with characteristic activities of various respiratory neurons whose duration is variable and controlled independently from other phases. Slowly adapting pulmonary stretch receptors, via the Hering Breuer reflex, prolong the nE period. We sought to identify inputs that lengthen the PI period. For this purpose we used electrical stimulation of the superior laryngeal nerve (SLN), the carotid sinus nerve (CSN), and areas in the rostral pons (RP) that terminate inspiration and prolong expiration. Stimulation of laryngeal receptors evokes powerful airway protective reflexes that prolong the interval between phrenic bursts (2), and stimulation of carotid chemoreceptors early in mechanical expiration delays the onset of inspiration (3).

Experiments were carried out on chloralose anesthetized, paralyzed,

vagotomized cats. Phrenic efferent activity was recorded and the upper airway was cannulated from below through a tracheostomy. PI
neurons of the ventral respiratory group at the level of the obex
were impaled using 5-10 megohm glass micropipettes filled with 3
M KCI. None of these neurons could be antidromically activated by
stimulating the spinal cord or the vagus nerves. For the purposes of
this study, we define the PI period as the interval between the
abrupt depolarization of PI neurons, usually coincident with the
rapid decline in the integrated phrenic neurogram from its peak
value, and the onset of IPSP induced hyperpolarization indicative
of the nE phase.

Single shock (0.5-1.5 V, 0.05 ms) stimulation of the ipsilateral
SLN, CSN, or RP evoked short latency (3-4 msec) compound EPSPs. For
each stimulus, brief (0.1-0.2 s) tetanic stimulation (100 Hz) at the
beginning of the post-inspiratory (PI) period increased depolarization
of PI neurons throughout the PI period and lengthened this period
without significantly changing the durations of inspiratory and
neural expiratory periods. Changes in subglottic pressure,
insufflation of smoke into the upper airway or application of
water to the larynx produced the same effects. Sustained tetanic
stimulation of both SLNs usually arrested the respiratory rhythm
for a considerable time (20-30 s); sometimes a respiratory rhythm
reappeared consisting of alternating I and PI periods but with no
apparent nE period. Augumenting IPSPs indicative of the nE period did
not appear before stimulation ended but always preceded the next I
period. The duration of the nE phase was not greatly altered.

Our results reveal a dissociation of the durations of the PI and
nE periods suggesting that the interval between inspiratory bursts
encompass two distinct neural phases. The first phase is prolonged
by afferent inputs from the larynx, and by stimulating the CSN and
the RP. The "apnea" caused by SLN afferent activity can be
characterized as a post-inspiratory apneusis.

REFERENCES:

1.) D.W. Richter, J. Exp. Biol., 100: 93, 1982.
2.) E.E. Lawson, J. Appl. Physiol., 50: 874, 1981.
3.) F.L. Eldridge, J. Physiol., 222: 297, 1972.

51

Localization of Synaptic Inputs of Central and Peripheral Origin on Laryngeal Respiratory Motoneurones

J. C. BARILLOT, A. L. BIANCHI and P. GOGAN

Intracellular recordings from either somata, proximal axons or distal axons of laryngeal motoneurones, have been obtained in decerebrated cats. These neurones were identified by antidromic stimulation of the Recurrent Laryngeal Nerve (RLN), and were synaptically driven by stimulation of the ipsilateral Superior Laryngeal Nerve (SLN). Some of these motoneurones were injected intracellularly with horseradish peroxidase (HRP) for morphological reconstruction. To differentiate between the recordings obtained from somata or axons, we used the following criteria : (I) refractory period, tested by twin shocks applied to the Recurrent Nerve at 5 x threshold, (II) duration, shape and rise time of action potentials, and (III) localization of the cell compartment of recording, either after HRP labelling (Fig. 1B, C) or with the aid of stereotaxic coordinates (in cases of impalment distant from the Nucleus Ambiguus : Fig. 1A).

Respiratory variations of membrane potential could be recorded in all somata, and along axons for at least 2 mm. The laster signals were attenuated.

Effects of supra-liminal stimulation of SLN were different according to the functional nature of the recorded laryngeal, inspiratory (ILM) or expiratory (ELM) motoneurone.

In ELMs, stimulation of the SLN was followed by an excitatory response (min. latency 4.2 ms +- 0.2 S.D.) in both somata and

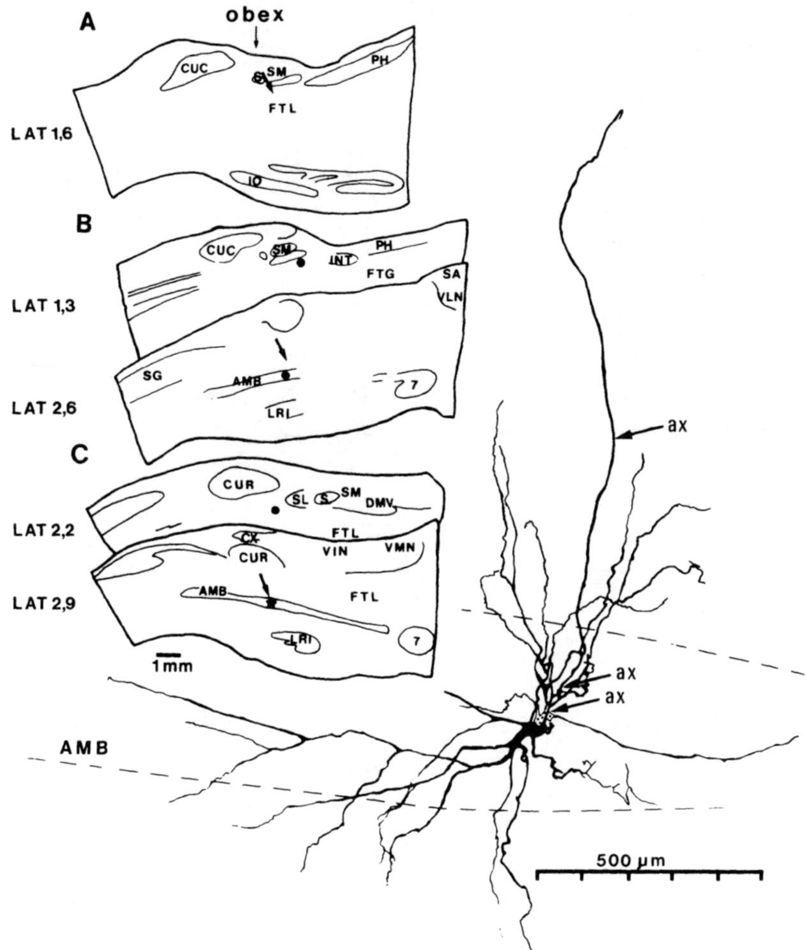

FIGURE 1 Localization, in parasagittal sections of the medulla, of three sites of recording (arrows) : distal axon (A), proximal axon (B), soma (C) in three different ELMs. In B and C, HRP injection allowed the reconstruction of the recorded region : axon in B (dots show limits of labelling, situated in two different planes), soma in C (star for the soma itself, dot for limit of axonal labelling). This last ELM is detailed on the right (for place convenience, a dendritic expansion in caudal direction has been excluded).
AMB : nucleus ambiguus ; CUC, CUR : cuneate nucleus, caudal and rostral division ; DMV : dorsal motor nucleus of the vagus ; FTL : lateral tegmental field ; FTG : giganto cellular tegmental field ; INT : nucleus intercalatus ; IO : inferior olive ; LRI : lateral reticular nucleus, internal division ; PH : nucleus praepositus hypoglossi ; S : solitary tract ; SM, SL : medial and lateral nuclei of the solitary tract ; SA : stria acustica ; VIN, VLN, VMN : inferior, lateral and medial vestibular nuclei ; 7 : facial nucleus.

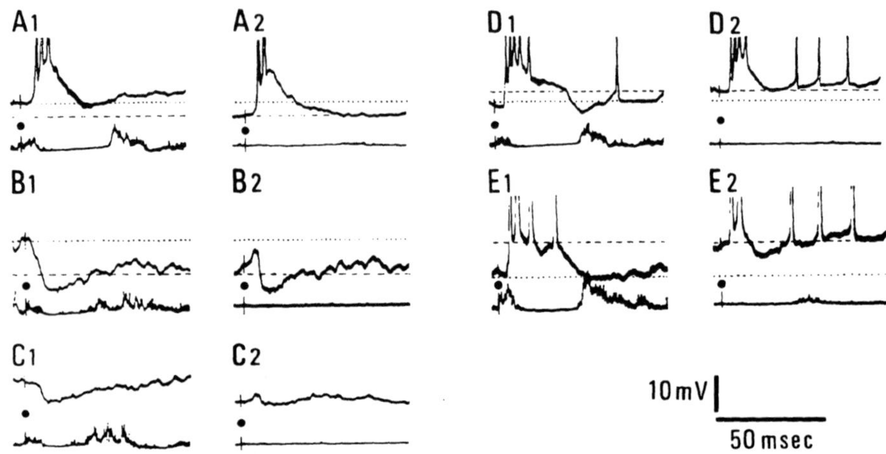

FIGURE 2 Effects of SLN stimulation on two inspiratory laryngeal motoneurones (A1-C2) and two expiratory ones (D1-E2). Upper traces : intracellular recordings ; lower traces : intergrated phrenic nerve activity. Dotted line and dashed line : mean level of membrane potential during inspiration and expiration, respectively. A1-A2 : axonal recordings from one ILM during inspiration (A1) and expiration (A2). Action potentials (clipped on the figure) are superimposed on large EPSPs (12 mV). The amplitude of these EPSPs is not affected by the changes of membrane potential related to the respiratory cycle. B1-B2 : somatic recordings from another ILM. A hyperpolarizing response is observed in both inspiration (B1) and expiration (B2). An hyperpolarizing current of 5 nA reduced this response during inspiration (C1) and abolished it during expiration (C2). D1-D2 : axonal recordings of one ELM during inspiration (D1) and expiration (D2). Action potentials (clipped on the figure) are superimposed on large depolarizing potentials. The amplitude of these depolarizing responses varies with the respiratory change of membrane potential. E1-E2 : somatic recordings from another ELM during inspiration (E1) and expiration (E2). Events are similar to those observed in D1-D2. Resting membrane potentials : - 84 mV (A1, A2), - 44 mV (B1 C2), - 46 mV (D1, D2), - 58 mV (E1, E2). In all cases the extracellular field potentials on SLN and RLN stimulation had an amplitude not greater than 0.5 mV. Reproduced from BARILLOT, J.C., BIANCHI, A.L. and GOGAN, P., Neurosci. Lett., 47 (1984), 107-112, with permission.

axons, the amplitude of which depended on the level of the membrane potential. In axons, this response was recorded at least two millimeters from somata with an attenuation of 1.4 to 2 times (Fig. 2 D1, D2, E1, E2).

In ILMs, when recordings were obtained from somata, stimulation of the SLN evoked an inhibitory response (min. latency 4.3 ms +- 0.3 S.D.) while in recordings from axons the same stimulation evoked an excitatory response (min. latency 4.2 ms +- 0.2 S.D.) (Fig. 2 A1, A2, B1, B2). The amplitude of the inhibitory response in somata depended on the membrane potential, i.e. was larger during inspiration than during expiration. Injection of hyperpolarizing currents revealed that two post-synaptic events were exerted on the somato-dentritic membrane : direct inhibitory post-synaptic potentials (IPSP) during expiration, and IPSPs combined with a decrease of excitatory input during inspiration (Fig. 2 C1, C2). On the contrary, the amplitude of the depolarizing response in axons did not depend on respiratory variations of membrane potential, suggesting that this phenomenon was not of somato-dentritic origin.

From our results, it appears that for both ELMs and ILMs, the rhythmical variations of membrane potential induced in somato-dentritic region by the central respiratory activity spread electrotonically in the axons.Such a close coupling between somata and axons has already been described for spinal motoneurones (1) and respiratory bulbospinal neurones (2).

However, the most striking finding of our study is the difference between the ELMs and the ILMs in response to SLN stimulation. For ELMs, a coherent excitatory effect was produced in both somata and axons. For ILMs, inhibitory responses developed in somata, while axons propagated excitatory responses. Analysis of these responses suggests that excitatory synaptic effects of SLN afferents take place mainly on, or close to, the initial segment of the ILMs, while inhibitory effects reach the somato-dentritic region, together with the excitatory influences of central respiratory drive (3).

This work was supported by CNRS (LA 205) and INSERM (82 6002) grants. We are grateful to Mrs J. ROMAN and A. ROUVIERE for technical assistance.

REFERENCES

1 RICHTER, D.W, SCHLUE, W.R., MAURITZ, K.H., NACIMIENTO, A.C. (1974). Comparison of membrane properties of the cell body and the initial part of the axon of the phasic motoneurones in the spinal cord of the cat. Exp. Brain Res., 20, 193-206.

2 MERRILL, E.G. (1974). Finding a respiratory function of the medullary neurones. In : R. Bellairs and E.G. Grays (Eds), Essays on the Nervous System, Clarendon Press, Oxford, 451-486.

3 BARILLOT, J.C., BIANCHI, A.L. and GOGAN, P. (1984). Laryngeal respiratory motoneurones : morphology and electrophysiological evidence of separate sites for excitatory and inhibitory synaptic inputs. Neurosci. Lett., 47, 107-112.

52

Intercostal Muscle Mechanoreceptor Influence on Medullary Respiratory Activity

R. SHANNON, D. C. BOLSER and B. G. LINDSEY

Intercostal muscle spindle endings and tendon organs are known to reflexly regulate intercostal muscle activity through spinal segmental pathways. There are numerous reports in the literature suggesting that these mechanoreceptors can also influence respiratory motor activity through their effects on brainstem respiratory neuron activity. For example, increasing the length and/or tension of the intercostal muscles or electrical stimulation of mechanoreceptor afferent fibers (Group I) can alter the breathing pattern. These studies did not resolve the specific actions of the muscle spindle endings (MSE) and tendon organs (TO) on brainstem respiratory activity.

The objectives of the experiments reported here were a) to determine if MSE and TO in external and internal intercostal muscles could be selectively stimulated, and if so, b) test their effects on medullary inspiratory drive and respiratory timing. The experiments were performed on anesthetized (Dial) and mid-collicular decerebrate cats that were thoracotomized, vagotomized and artifically ventilated.

It was possible to increase the activity of intercostal muscle TO without increasing the activity of MSE [1]. The technique is based on the well known response of MSE and TO during active muscle shortening, elicited by electrical stimulation of the muscle motor nerves. During this type of muscle contraction there is an increase in TO firing rate and a decrease in MSE firing rate. Muscle contraction in a single intercostal (T6) space was elicited by electrical stimulation of the peripheral cut end of the spinal ventral root innervating that space.

This space was surgically separated from the muscles of the adjoining spaces. In order to study MSE and TO in either external or internal intercostal muscles, the muscle not being studied was denervated. All other muscles innervated by the main intercostal nerve of this space was also denervated. In addition, two intercostal spaces rostral and caudal to this space were denervated bilaterally. The activity in single dorsal root filaments was monitored during active muscle shortening in order to identify fibers from MSE and from putative TO. The response of these identified fibers was then monitored during a pseudo isometric contraction (ribs were clamped). All external (154 in 29 cats) and internal (68 in 14 cats) intercostal MSE decreased their firing rate or ceased firing during the isometric contraction. Other dorsal root fibers (85 from external and 47 from internal intercostal muscles) were recruited or increased their firing rate during isometric contraction and arise from tension receptors, some of which are TO. Isometric twitches generate more muscle force than isotonic twitches, which results in more TO activity for studying their reflex effects on respiratory activity.

Additional studies [1] showed that medullary dorsal respiratory group I-neuron and phrenic motor activity could be transiently reduced or prematurely terminated (fig. 1) during isometric twitches of external (5 cats) or internal (5 cats) intercostal muscles. We know from our previous electrical stimulation studies [2,3] that medullary inspiratory drive can be reduced or terminated by an increase in external or internal intercostal Group I afferent fiber activity. Since MSE activity (Ia fibers) decreases and TO activity (Ib fibers) increases during the ventral root elicited intercostal muscle contraction, it is concluded that an increase in TO activity is responsible for the reduced medullary I-neuron and phrenic activity.

It was also possible to stimulate intercostal MSE without stimulating TO [4]. The muscles in a single intercostal space (T6) were vibrated with a vibrator attached to the caudal rib of the space; the rostral rib was secured by clamps. This space was surgically separated from the muscles of adjoining spaces, and denervated as described for the TO experiments. Dorsal root fibers from MSE and tension receptors (putative TO) were identified by their response during ventral root elicited (electrical stimulation) muscle shortening. No

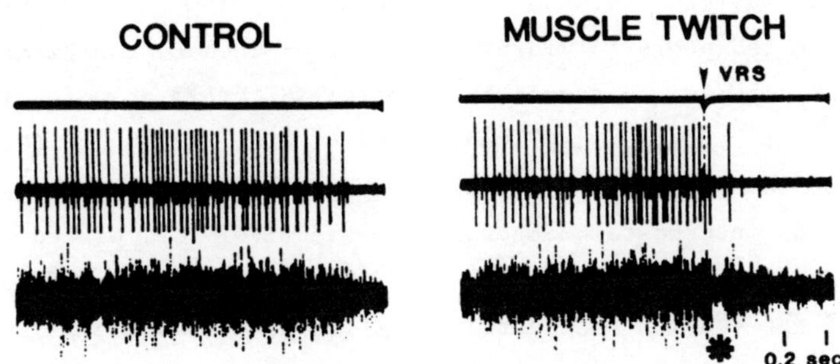

FIGURE 1   Response of dorsal respiratory group I-neuron (middle
trace) and phrenic motoneurons (bottom trace) to intercostal muscle contraction elicited by ventral root stimulation (VRS). I-neuron activity is prematurely terminated and phrenic activity is transiently reduced (*).

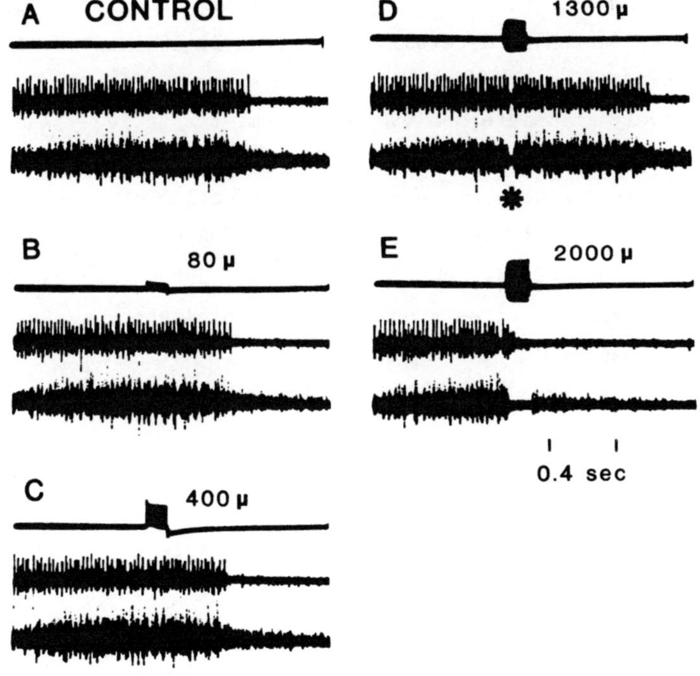

FIGURE 2   Response of caudal ventral respiratory group I-neurons
(middle trace) and phrenic motoneurons (bottom trace) to
intercostal muscle vibration (200 Hz). In this example,
there was no effect on I-activity at vibration amplitudes
of 80 and 400 microns (B,C), a transient reduction (*) at
1300 microns (D), and complete termination at 2000 microns
(E).

distinction was made between 1°MSE and 2°MSE. The response of the MSE
and tension receptors to vibration was determined at 200 Hz and
various amplitudes. External intercostal MSE could be selectively
stimulated at vibration amplitudes under 100 um; at 90 um, 60% (29/48,
9 cats) of the MSE increased their firing rate. Twenty-one percent of
the internal intercostal MSE could be selectively stimulated at amplitudes under 40 um; at 90 um, 64% (42/66, 13 cats) of the MSE and 13%
(6/46) of the tension receptors were stimulated. As the vibration
amplitudes were increased, the activity of MSE and tension receptors
increased.

The response of dorsal and ventral respiratory group I-neurons and
phrenic activity to intercostal muscle vibration (200 Hz) was determined at various amplitudes (fig. 2). Animals in these experiments
were paralyzed with Flaxedil to prevent resting and reflex tone in the
muscles. Vibration of external (26 neurons in 10 cats) or internal
(29 neurons in 10 cats) intercostal muscles at amplitudes small enough
to selectively stimulate MSE (90 and 40 um, respectively) had no apparent effect on inspiratory activity. The only effect of vibration
on inspiratory activity was a transient reduction or complete termination at larger amplitudes (generally above 800 um), which we attribute
to the increased activity of tendon organs. These results suggest
that intercostal MSE effects on medullary respiratory activity are
minimal or absent.

Supported by USPHS Grant HL-17715 and a USF Division of Sponsored
Research Grant.

REFERENCES
1. Bolser, D.C., Lindsey, B.G. and Shannon, R. (1983). Inhibition of
medullary inspiratory drive by intercostal muscle tendon organs. Soc.
Neurosci. Abst., 9, 1160
2. Shannon, R. (1980). Intercostal and abdominal muscle afferent
influence on medullary dorsal respiratory group neurons. Respir.
Physiol., 39, 73-94
3. Shannon, R. and Freeman, D.L. (1981). Nucleus retroambigualis
respiratory neurons: responses to intercostal and abdominal muscle
afferents. Respir. Physiol., 45, 357-375
4. Bolser, D.C., Lindsey, B.G. and Shannon, R. (1984). Evidence that
intercostal muscle spindle endings do not reflexly modulate medullary
inspiratory drive. Fed. Proc., 43, 433

53

Effects of Short Term Exposure to Hypoxia on Breathing Pattern in Rabbit Pups

T. TRIPPENBACH, G. KELLY and R. AFFLECK

INTRODUCTION

Despite an increase in pulmonary stretch receptor activity caused by low oxygen content in the tracheobronchial tree (1) intact vagus nerves are not essential for the hypoxia-related increase in respiratory frequency found in adult animals (2,3,4). The role of vagal feedback in the ventilatory response to hypoxia in newborns is unknown. To define the contribution of vagal input to changes in the breathing pattern caused by hypoxia during the early postnatal period, we investigated the effects of hypoxia in intact and vagotomized rabbit pups.

METHODS

Rabbit pups during the second week of life were anaesthetized (i.m.) with a mixture of Ketamine (50 mg/kg) and acepromazine (3 mg/kg). Surgical preparation consisted of tracheostomy, cannulation of the left carotid artery, separation of both vagus nerves and the C_5 root of the phrenic nerve, and exposure of the T_6 segment of the right external intercostal muscles. Fine bipolar copper electrodes were

301

implanted into this muscle in the vicinity of the midaxillary line. Using bipolar silver electrode efferent phrenic activity was recorded from the cut C_5 root under mineral oil. Both intercostal muscle EMG and phrenic activity were rectified and integrated by a moving-average processor. Blood pressure was measured from a carotid catheter. Tidal volume was obtained by integration of the flow signal from a pneumotachograph. 'Integrated' phrenic activity (PHR) Integrated external intercostal EMG (INT), blood pressure (BP), and tidal volume (V_T) were recorded on a 6 channel pen recorder. Warmed and humidified hypoxic gas mixture (12% O_2 in N_2) was applied to the vicinity of the tracheal opening for 3-5 min. Microsamples of blood were taken for blood gas measurements during room-air breathing and 2 min after the onset of hypoxia before and after vagotomy.

RESULTS

When animals were exposed to hypoxia we observed almost an immediate increase in V_T, PHR and INT and a decrease in blood pressure. All the measurements were done when there were no further changes in these parameters after approximately 1 min from the beginning of the hypoxic run. Despite a decrease in inspiratory time (T_I) during hypoxia in most intact rabbits the group mean T_I was not significantly changed and expiratory time (T_E) was not affected either. However, the increase in V_T, PHR and INT was significant. PaO_2 of 50.5 ± 14.9 mmHg and $PaCO_2$ of 56.4 ± 8.8 mmHg recorded during control conditions decreased to 34.5 ± 5.4 mmHg, and 43.0 ± 8.3 mmHg respectively while breathing 12% O_2.

Vagotomy did not affect alveolar ventilation (PaO_2 = 56.6 ± 13.2 mmHg and $PaCO_2$ = 51.1 ± 8.6 mmHg). During hypoxic exposure these values dropped to 34.2 ± 11.5 mmHg and 41.5 ± 8.5 mmHg for PaO_2 respectively. In response to hypoxia, decrease in T_I became significant while T_E was not affected. As in intact animals V_T, PHR and INT increased. These changes in the respiratory variables led to a negative relationship between V_T vs T_I, PHR vs T_I and INT vs T_I in all vagotomized rabbit pups breathing hypoxic mixture. Group mean increases in PHR and INT were similar in both intact and vagotomized rabbits. However in six out of nine intact, and in three out of seven vagotomized pups, the increase in PHR was more pronounced than that in INT.

DISCUSSION

In this study we were investigating only the primary ventilatory response to hypoxia in newborns, i.e. ventilatory excitation. Records of phrenic activity allowed estimation of the effects of hypoxia on central nervous system output without the added influence of the mechanical characteristics of the respiratory system that might influence tidal volume. The results of this study show that vagal input is not essential for the early hypoxia-related changes in the pattern of breathing during the early postnatal period. Thus, like in adults, the decrease in T_I provoked by hypoxia in newborns is not a consequence of increased vagal stretch receptor activity at these conditions (1) but is due to a direct action of peripheral chemoreceptor activity on the mechanisms controlling respiratory frequency and depth of breathing. In fact, T_I shortening observed in response to hypoxia was significant only in rabbit pups with vagi cut.

The effect of vagotomy on hypoxia related T_I decrease is difficult to explain. It is possible that inconsistent effects of hypoxia on T_I in intact animals were related to input from the vagal irritant receptors stimulated by the chest distortions (5) that are augmented during increased inspiratory efforts. Stimulation of these receptors could rely on peripheral effects of hypoxia i.e., increase in airway resistance (6). The irritant receptors could also be responsible for greater changes in PHR and V_T in intact than vagotomized rabbit pups.

This study suggests that the contribution of intercostal muscle in increasing tidal volume during hypoxia in rabbit pups is less than that of the diaphragm. This different contribution of diaphragm and

inspiratory intercostal muscles in work of breathing was especially
seen during hyperpnea caused by prolonged airway occlusion at FRC in
rabbit pups (7).

REFERENCES

1.   Fisher, J.T., Sant'Ambrogio, F.B. and Sant'Ambrogio, G. (1983).
Stimulation of tracheal slowly adapting stretch receptors by
hypercapnia and hypoxia. Respir. Physiol. 53, 325-339

2.   Favier, R., Kepenekian, G., Desplanches, D. and Flandrois, R.
(1982). Effects of chronic lung denervation on breathing pattern and
respiratory gas exchanges during hypoxia, hypercapnia and exercise.
Respir. Physiol. 47, 107-119

3.   Gautier, H. (1975). Effects of hypoxia or hypercapnia on
ventilatory pattern of chronic cats before and after vagotomy. Bull.
Physiol. Pathol. Resp. 11, 84-90

4.   Ledlie, J.F., Kelsen, S.G., Cherniack, N.S. and Fishman, A.P.
(1981). Effects of hypercapnia and hypoxia on phrenic nerve activity
and respiratory timing. J. Appl. Physiol.: Respirat. Environ.
Exercise Physiol. 51, 732-738

5.   Martin, R.J., Nearman, H.S., Katona, P.G. and Klaus, M.H.
(1977).   The effect of low continuous positive airway pressure on
the reflex control of respiration in the preterm infant. J.
Pediatrics. 90, 976-981

6.   Nadel, J.A. and Widdicombe, J.G. (1962). Effect of changes in
blood gas tensions and carotid sinus pressure on tracheal volume and
total lung resistance to airflow. J. Physiol. 163, 13-33

7.   Trippenbach, T., Affleck, R. and Kelly, G. (1984). Respiratory
effects of progressive asphyxia in rabbit pups and adult rabbits. J.
Appl. Physiol.: Respirat. Environ. Exercise Physiol. 56, 940-947

54

An Effector Role of the Ventral Surface of the Medulla in Respiration

R. L. MARTIN-BODY, J. D. SINCLAIR and N. J. DAWSON

INTRODUCTION

Numerous studies have revealed that major cardiorespiratory disturbances result from experimental perturbations at the ventral surface of the medulla (VSM). Those interested in the role of the VSM in cardiovascular regulation agree that the VSM is part of an effector system. In contrast, those interested in respiratory control have considered the VSM to constitute a chemosensitive mechanism and therefore part of an afferent system (1). However, in the rat we have been unable to demonstrate a significant ventilatory response to superperfusion of the VSM with acid CSF when the temperature of the perfusate is carefully controlled (2). And experiments on rat brain slices have demonstrated similar numbers of neurones activated, inhibited or unresponsive to high PCO2/normal bicarbonate solutions (3).

Further, electrocoagulation of the VSM in the cat leads to a reduction in the ventilatory response to hypoxia (4), although the afferent terminations of the peripheral chemoreceptors are not known to be located near the VSM (5). An efferent pathway whereby changes in VSM pH modulate the drive from the carotid and aortic bodies has been suggested (6) but not regarded as important.

In this paper we present results which we cannot interpret on the rigid concept of the VSM constituting only an afferent pathway in respiratory control.

METHODS AND RESULTS

Experiments were performed on 33 male rats anaesthetised with urethane (675 mg kg-1) and chloralose (67.5 mg kg-1).

In the first series of experiments we repeated our previous

attempts to demonstrate a ventilatory response to perfusion of the VSM with CSF of low pH (2). One significant technical change was used: the dura was left intact except for two small punctures for the perfusion. Despite this procedure, instituted to minimise trauma to the VSM, we were unable to obtain a satisfactory ventilatory response to changes of perfusate pH from 7.52 to 7.21 when the temperature of the fluid was maintained at 37 C.

In a second series of experiments we applied the excitatory neurotoxin, kainic acid (KA) to the VSM. The kainic acid was made as a 1% solution in CSF with the fluorescent dye rhodamine as the tracer. KA applied over the intermediate and caudal areas of the VSM produced respiratory arrest, usually after an intense stimulation of ventilation. Muscle pinch or NaCN (400 ug kg-1) were ineffective in reinitiating ventilation. Further localisation with cottonwool pledgelets could not be achieved because apnoea did not always ensue. Clearly, drug effects on a critical mass of tissue are important but spread of a drug is difficult to control in small animals like the rat.

While the ventilatory response to kainic acid was developing the responses to tracheal hypercapnic and hypoxic gas mixtures was tested. The responses were abolished when KA covered the intermediate and caudal areas of the VSM and reduced after more discrete applications. If apnoea did not occur the responses to hypoxia and hypercapnia returned to control levels over about 2 hours despite the return of ventilation to baseline some time previously. Sodium glutamate, a less potent excitatory neurotoxin, produced the same pattern of response, and it was graded with respect to the concentration of the drug (46.9 mM to 4.69M).

The response to carbachol was abolished after application of KA to the VSM and remained abolished when the ventilatory response to CO2 was again perceptible.

DISCUSSION

Experiments in our laboratory have repeatedly failed to demonstrate a significant ventilatory response to decreases in the pH of CSF perfusing the ventral surface of the medulla. This is despite care in the method of approach to the problem and contrasts with the apparent sensitivity of the cat ventral medullary surface.

However, there is no doubt that respiratory responses can be elicited from this region in the rat. We chose to apply kainic acid to the VSM because it is reported to destroy cell bodies while leaving axons of passage intact (7). It was expected that the ventilatory response to hypercapnia would be abolished in animals with carotid bodies physiologically silenced by high oxygen as the carrier. The results demonstrated a dramatic excitatory response to KA but unexpectedly the ventilatory responses to both hypercapnia and hypoxia were reduced or abolished. And with wide application of KA apnoea occurred.

It has been argued that the magnitude of a ventilatory response depends critically on the input/output relationship for the system under consideration (8). Thus, the reduction or abolition of the chemoreflex responses could be understandable since KA had changed the basic level of ventilation. However, we repeatedly observed a return to control levels of ventilation which was not accompanied by a return of the chemoreflex responses.

It appeared therefore that the effects of KA on the hypoxic response could be due to diffusion of the drug into the nucleus ambiguus where afferent terminals and cell bodies of carotid sinus nerves axons have been located (5). In the rat we have been unable to confirm afferent terminations within the nucleus ambiguus in tracing studies using wheat germ agglutin conjugated horseradish peroxidase; nevertheless numerous cell bodies and extensive dentritic fields were noted (G.D. Housley, unpublished observations). Since the fluorescent dye with which the KA was mixed was never found deeper than 200u it appears unlikely that the effect is exerted on cells of the nucleus ambiguus, the closest cell being found 350 u from the surface.

We therefore believe that the most plausible explanation of our experimental results is that the VSM forms part of an effector pathway to the motoneurones of respiration.

Finally, we cannot support the hypothesis that the cholinergic cells of the VSM mediate the hypercapnic responses since KA abolished the carbachol response while leaving an appreciable hypercapnic response. This confirms our earlier studies in which analysis of the hypercapnic response in the presence of varying doses of carbachol on the VSM revealed an additive response (9).

REFERENCES

1. Schlaefke, M. E. (1981). Central chemosensitivity: a respiratory drive. Rev. Physiol. Biochem. Pharmacol., 90, 171-244

2. Malcolm, J. L., Sarelius, I. H. and Sinclair, J. D. (1980). The respiratory role of the ventral surface of the medulla studied in the anaesthetised rat. J. Physiol., 307, 503-15

3. Fukuda, Y. (1983). Difference between actions of high PCO2 and low [HCO3-] on neurons in the rat medullary chemosensitive areas in vitro. Pflugers Arch., 398, 324-30

4.   Schlaefke, M. E. (1976).   Central chemosensitivity: neurophysiology and contribution to regulation. In: Loeschcke, H. H. (ed). Acid Base Homeostasis of the Brain Extracellular Fluid and the Respiratory Control System.   pp. 66-80.   (Stuttgart: Thieme Publishers)

5. Davies, R. O. and Kalia, M. (1981). Carotid sinus nerve projections to the brain stem in the cat. Brain Res. Bulletin, 6, 531-41

6. Trzebski, A., Majcherczyk, S., Szulczyk, P. and Chruscielewki, L. (1976). Direct nervous mechanisms as the possible pathways of interaction of the central and peripheral chemosensitive areas. In: Loeschcke, H. H. (ed). Acid Base Homeostasis of the Brain Extracellular Fluid and the Respiratory Control System. pp. 130-45

7. McGreer, P. L. and McGreer, E. G. (1982). Kainic acid: the neurotoxic breakthrough. CRC Crit. Rev. Toxicol., 10, 1-26

8. Millhorn, D.E., Eldridge, F.1. and Waldrop, T.G. (1982). Effects of medullary ((s) cooling on respiratory response to chemoreceptor inputs. Resp. Physiol., 49, 23-39.

9. Hill, P. McN., Hughes, E. W. and Sinclair, J. D. (1983). The relationship between respiratory stimulation by the superficial medullary cholinergic mechanism and by inspired carbon dioxide in anaesthetised rats. J. Physiol. 334, 42P

55

Differing Responses of Medullary Respiratory Neurons to Peripheral and Central Chemoreceptor Stimuli

W. M. ST. JOHN and A. L. BIANCHI

In previous studies, we found that stimuli acting upon the peripheral or central chemoreceptors do not produce equivalent alterations in the activities of brainstem respiratory neurons (1,2). Compared to values in hyperoxic normocapnia, the discharge frequencies of most neurons increased in hyperoxic hypercapnia. Hyperoxic hypercapnia causes ventilatory changes primarily by actions upon the central chemoreceptors (3). Responses to stimulation of the peripheral, carotid chemoreceptors by pharmacological agents or isocapnic hypoxia differed from responses to hypercapnia in that the discharge frequency of many neurons declined. Neurons showing such a decline in discharge frequency during stimulation of the peripheral chemoreceptor were more prevalent in the ventral than the dorsal medullary respiratory nucleus. The dorsal nucleus is composed almost entirely of bulbospinal neurons, having efferent axons in the spinal cord; many of these bulbospinal axons impinge upon phrenic motoneurons (4,5). Bulbospinal neurons constitute only a portion of the ventral nucleus; the rest have axons that project in the recurrent laryngeal nerve (laryngeal units) or to other brainstem sites (not antidromically activated, propriobulbar) (4,5). Many bulbospinal neurons of the ventral nucleus have axons in the thoracic spinal cord (5).

We hypothesized that neurons whose discharge frequencies rose in hypercapnia but remained unchanged or declined in isocapnic hypoxia would be more prevalent among the laryngeal than the bulbospinal population. This hypothesis was based upon the difference in the

309

anatomical distribution of bulbospinal and laryngeal neurons, as discussed above, and also upon the finding of Bartlett (6) that vocal cord movements may differ in hypercapnia and hypoxia.

Data obtained in a recent study (7) did not support the hypothesis that excitatory afferents from the carotid chemoreceptors are preferentially distributed upon bulbospinal neurons, compared to the laryngeal population. In that study, medullary respiratory neurons were designated as 'bulbospinal' or 'laryngeal' if antidromic action potentials were recorded following stimulation, respectively, of the spinal cord and the recurrent laryngeal nerve. These neuronal activities were recorded during equivalent changes in peak integrated phrenic activity achieved in hyperoxic hypercapnia and normocapnic hypoxia. While the discharge frequencies of neurons typically rose in hypercapnia, many bulbospinal, laryngeal and not antidromically activated units exhibited declines in discharge frequencies in hypoxia. Results of this study imply that excitatory afferents from the central chemoreceptors are widely distributed among medullary respiratory neurons. In contrast, there is more discrete and unequal distribution of excitatory peripheral chemoreceptor afferents. Thus, in hypoxia, activity will increase only in those neurons which receive sufficient excitatory peripheral chemoreceptor afferents to overcome the direct depression of neuronal activity by brainstem hypoxia. A further implication of this study is that there must be a significant integration of bulbospinal activity within the phrenic motoneuron pool. Such integration is required since, as opposed to the differing responses of many bulbospinal neurons of the dorsal medullary respiratory nucleus to hypercapnia and hypoxia, phrenic motoneurons exhibit identical changes in activity in hypercapnia and hypoxia (8).

In a companion study to that described above, we evaluated mechanisms underlying changes in discharge frequencies of medullary respiratory neurons in hypercapnia and hypoxia (9). This evaluation was made by determining variations in the antidromic latencies of bulbospinal and laryngeal respiratory neurons throughout the respiratory cycle. Such variations in antidromic latencies reflect changes in membrane potential and, thus, the relative depolarization and hyperpolarization of the neuron under examination (10).

At normocapnic hyperoxia, antidromic latencies of bulbospinal and laryngeal neurons fell to minima during periods of spontaneous neuronal activity with maxima occurring between neuronal bursts. In hypercapnia or hypoxia, these minima were not altered, whereas maximum latencies typically rose for neurons whose discharge frequencies increased. However, the increased discharge frequencies most strongly correlated with increases in the difference between maximum and minimum latencies. No such correlation was evident for neurons whose discharge frequencies declined in hypoxia and/or hypercapnia. We conclude that the overall change of membrane potential during the respiratory cycle, rather than the absolute level of depolarization, primarily defines the discharge frequency of medullary respiratory neurons.

ACKNOWLEDGMENTS

These studies were supported by Grant 20574 from the National Heart, Lung and Blood Institute, National Institutes of Health (U.S.A.) and by grants from INSERM (82-6002) and the Ministère des Relations Extérieures (FRANCE).

REFERENCES

1. St. John, W.M. and Wang, S.C. (1977). Response of medullary respiratory neurons to hypercapnia and isocapnic hypoxia. J. Appl. Physiol., 43, 812-822

2. St. John, W.M. (1981). Respiratory neuron responses to hypercapnia and carotid chemoreceptor stimulation. J. Appl. Physiol., 51, 816-822

3. Sorensen, S.C. (1971). The chemical control of ventilation. Acta Physiol. Scand., 83 (suppl. 361), 1-72

4. Bianchi, A.L. (1971). Localisation et étude des neurones respiratoires bulbaires. Mise en jeu antidromique par stimulation spinale ou vagale. J. Physiol. (Paris), 63, 5-40

5. Cohen, M.I. (1979). Neurogenesis of respiratory rhythm in the mammal. Physiol. Rev., 59, 1105-1173

6. Bartlett, D. (1980). Effects of vagal afferents on laryngeal responses to hypercapnia and hypoxia. Respir. Physiol., 42, 189-198

7. St. John, W.M. and Bianchi, A.L. (1985). Responses of bulbospinal and laryngeal neurons to hypercapnia and hypoxia. J. Appl. Physiol.

(submitted)

8.   St. John, W.M. and Bartlett, D. (1979). Comparison of phrenic motoneuron responses to hypercapnia and isocapnic hypoxia. J. Appl. Physiol., 46, 1096-1102

9.   Bianchi, A.L. and St. John, W.M. (1985). Excitation and inhibition of medullary respiratory neurons in hypercapnia and hypoxia. J. Appl. Physiol. (submitted)

10. Lipski, J. (1981). Antidromic activation of neurons as an analytic tool in the study of the central nervous system. J. Neurosci. Methods, 4, 1-32

Section 5
Neuropharmacology of Respiratory Centers

56

Chemical Transmission and Central Respiratory Mechanisms

M. DENAVIT-SAUBIÉ, J. CHAMPAGNAT and M. P. MORIN-SURUN

Rhythmic respiratory activity can be recorded in decerebrated vagotomized animal, showing that this activity is generated in a rather restricted part of the brain stem. This brain stem area contains different types of respiratory neurons which are tightly interconnected in a network which is able to generate spontaneously a rhythmic activity. In the intact animal this network is submitted to several controls of peripheral as well as central origin. These controls are necessary to adapt the central generator to external conditions. The main control coming from the periphery is constituted by the vagal reflex whose circuitry is enclosed in the brain stem respiratory network (9). Other controls are of central origin and may come from the hypothalamus, either directly to the nucleus tractus solitarius (21) or indirectly through the parabrachial nuclei (4) to the nucleus ambiguus (19). Thus the respiratory neurons of the brain stem area have to integrate the peripheral and central messages to their rhythmic firing.

The strategic importance of this brain stem area has been demonstrated as early as 1842 by Flourens who called it "le noeud vital". With the new anatomical, pharmacological, immunohistochemical and electrophysiological techniques a bulk of new data is concentrating on this area, helping in understanding its functioning.

It is now possible to identify and to give the quantitative distribution of endogenous substances in different parts of the brain. It is even possible to localize by immunohistochemistry these substances in the neuron (L. Leger et al., this volume) and to localize their binding sites (N. Sales et al., this volume).

It appears from these neurochemical data that brain stem structures involved in the generation of respiratory rhythm, such as the dorsal or ventral respiratory group (1) contain a rich diversity of chemical substances. Some of these chemical substances act as classical neurotransmitters within a very short period. Such fast-switching transmission constitutes the basic process whereby the potential circuitry can be rapidly and precisely activated. This type of neurotransmitter can be involved in some specific events occurring in a precise part of the respiratory cycle. However the time course of a respiratory pattern of discharge suggests the involvement of neuroactive substances acting with a slower time course. Recently such endogenous substances have been described. On the one hand, slow muscarinic excitatory effects (10) have been found on respiratory neurons (15). They may be due to inactivation of an M current (2), which is suggested to play a role in rhythmogenesis of respiratory neurons (Richter et al. this volume) On the other hand a growing number of families of neuropeptides which function as chemical messengers in the central nervous system are described (8). Furthermore it has been shown that several neuropeptides and classical neurotransmitters could be co-localized in the same neuron (11,18). Several findings (12) suggest that   peptides amplify the effect caused by the classical transmitter on the postsynaptic cell. While classical neurotransmitters cause rapid and short lasting effects, neuropeptides are involved in responses characterized by slow onset and long duration.

A very rich palette of chemical substances can be released in the brain stem respiratory area. It may be assumed that only substances which are found in neurons with their somata and terminals included in the respiratory network, may be involved in the rhythmogenesis. Other substances which are found in terminals located in the respiratory network, while their immunoreactive somata are found outside

this area, are more probably involved in the control of the rhythmogenesis.

The aim of our study is to find out the action or the interaction
of these chemical messengers,in the elaboration or control of the
respiratory rhythmogenesis. These messengers are essential since
their impairment is able to suppress ventilation.

Different technics of investigations have been used. Iontophoretic
applications of agonists or antagonists of neurotransmitters are
useful to determine whether identified respiratory neurons possess
the receptor corresponding to the test substance, how the cell
reacts to this substance and what type of interaction may be effective. In medullar slices, application of neurotransmitters in the
liquid of perfusion enables to get information about the sensitivity
of the nucleus tractus solitarius cells to the chemical messenger.
Furthermore with this technic it is possible to get information
about the modification of membrane properties submitted to this
substance. Finally systemic administrations of the drug may give
information about its global effect on ventilation. At any rate it
is necessary to permanently correlated the data obtained with these
different technics in order to replace the informations obtained at
the elementary level in the diagram of the respiratory function.

1. Possible involvement of classical neurotransmitters

In the respiratory cycle the off switch mechanism occurs abruptly
to stop inspiration. Furthermore glycine sensitive receptive sites
have been shown to be dense at the level of bulbar respiratory
centers (17).

Using iontophoretic application on respiratory neurons of
threshold doses of strychnine, a specific antagonist of glycine-like
transmitters, it has been possible to demonstrate that such classical
neurotransmitter was involved in the off switch mechanism (fig. 1).
Thus in a precise part of the respiratory cycle, early expiratory
neurons may release a glycine-like neurotransmitter which is able to
transfer faithfully a coded information to an inspiratory neuron
and stop its discharge.

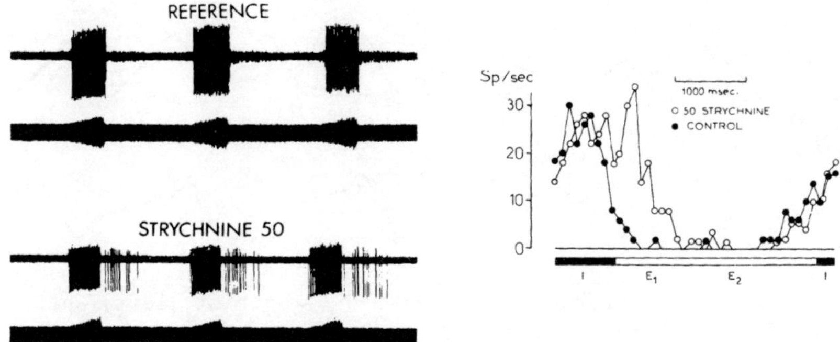

Figure 1   Effect of iontophoretic application of strychnine on
            inspiratory neuron. <u>Left</u> : samples of film taken before
            and 30 sec. after strychnine application (50 nA). Upper
            trace : spike recording, lower trace : phrenic nerve
            recording. <u>Right</u> : changes in discharge frequency during
            the respiratory cycle presented in cycle-triggered histo-
            grams using the beginning of inspiration to trigger the
            histogram. Discharge frequency in ordinate (Sp/sec) and
            respiratory cycle in abscissae ; I (black bar) : inspir-
            ation ; E_1 and E_2 : expiration. Control : (black circles) ;
            strychnine application : (open circles). Note that under
            strychnine, spikes occurred at the beginning of expiration.

2. Possible involvement of neuropeptide

In the respiratory area opioid peptides are located in somata as
well as in terminals (20). Therefore they may be involved in the
respiratory rhythmogenesis. These opioid neuropeptides act on different receptors. Using highly selective μ and δ agonists, two types of
receptors have been shown to be present, distinct and functionally
active in respiratory related neurons (14). However when injected
systemically the ventilatory effects of each agonist are quite different. The agonist δ reduces more particularly the respiratory rate
while the agonist μ reduces more particularly the tidal volume (13),
(figure 2).

The iontophoretic study does not give information about the
density of the receptors. Such information can be obtained either
pharmacologically by testing the sensitivity of brain slices at
different levels or anatomically by autoradiography. Using the latter
technic it appears that a higher density of μ receptors can be found
in medullar structure while δ receptors were more particularly located
in the pneumotaxic center (N. Sales et al., this volume). This would
suggest that the δ receptors located in the pneumotaxic center are

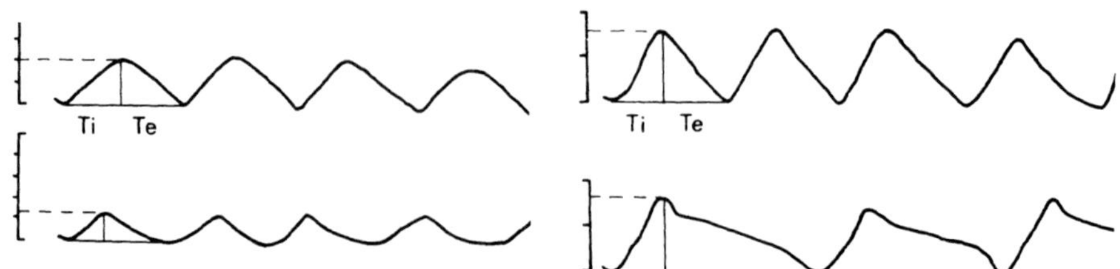

Figure 2   Example of the respiratory effects produced by intraperitoneal administration of TRIMU-4 (μ agonist, left) and
DSLET (δ agonist, right). Upper line : reference period ;
lower line : 10 minutes after injection.

particularly involved in the reduction of respiratory frequency
observed under opiate : a result which may be confirmed by the
study of Florez et al. (this volume).

3. Possible interaction between neuropeptides and neurotransmitters

Interaction between opioid peptides and excitatory action induced
by Glutamate or Substance P (SP) have been tested. A high concentration of opioid immunoreactive neurons are located in the respiratory
area, and interact with the respiratory rhythmogenesis (6). Furthermore the presence of SP in vagal and glossopharyngeal afferents
neurons and in nucleus tractus solitarius nerve terminals is well
documented (5,7). This neuropeptide may be involved in the afferent
control from the lung. Its action has been tested _in vivo_ and _in
vitro_. In anaesthetized animals, as well as in slices preparation
neurons were recorded in the area in which a field potential could
be evoked by electrical stimulation of vagal nerve or tractus
solitarius. The pattern of discharge of non-respiratory as well as
respiratory neurons identified in the anaesthetized animal was
reversibly increased by iontophoretic application of SP (16). Intracellular recordings in slice preparation have demonstrated that
excitations by SP (10^{-7}-10^{-9}M) were associated with a depolarization
and an increase of the neuronal input resistance (3).

In vivo and _in vitro_ application of glutamate (10^{-3}-10^{-5}M)
induces also an increase of firing discharge associated with a
depolarization. Simultaneous applications of enkephalin, SP and

glutamate, were used on respiratory neurons to look for a possible
interaction. Enkephalin was shown to reduce the effectiveness of
glutamate while simultaneously, the effect of SP was unaffected
(16). Thus opioid neuropeptides are able to selectively interact
with some excitations and not others. This is of functional importance since enkephalins were also found to reduce the periodic
excitations which are responsible for neuronal respiratory modulation.

These examples have been selected to illustrate the main aspects
of the chemical transmission which may be involved in the control
and generation of respiratory rhythm. This type of investigation on
neurochemical transmission of identified respiratory neurons must be
based on previous multidisciplinary studies giving general information about the localization of endogenous chemical substances and
about the appropriate pharmacological tools available to test the
presence of receptors.

The particularly high concentration of endogenous substances which
have been found in brain stem where they are co-localized in
several cases, may correspond to the characteristics required by
this system. On the one hand a high degree of fiability is
necessary to support this vital function. On the other hand a high
degree of adaptation is necessary. The powerful neurotransmission
is ensured by amino acids while opioid peptides may be more
particularly involved in adaptation processus.

REFERENCES

1. Bianchi, A.L. (1971). Localisation et étude des neurones respiratoires bulbaires. Mise en jeu antidromique par stimulation spinale
ou vagale. J. Physiol. (Paris), 63, 5-40.

2. Brown, D.A. (1983). Slow cholinergic excitation - a mechanism for
increasing neuronal excitability. Trends in Neurosciences, 6 , 302-
307.

3. Champagnat, J., Shen, K.F., Siggins, G.R., Koda, L. and Denavit-
Saubié, M. (1984). Effects of neurotansmitters and synaptic process-
ing of neurovegetative afferents in the brain stem of the rat.
J. Autonom. Nerv. Syst. (in press).

4. Conrad, L.C.A. (1976). Efferents from medial forebrain and hypothalamus in the rat. II. An autoradiographic study of the anterior
hypothalamus. J. Comp. Neurol., 169, 221-261.

5. Cuello, A.C. and Kanazowa, I. (1978). The distribution of substance P immunoreactive fibers in the rat central nervous system. J. Comp. Neurol., 178, 129-156.

6. Denavit-Saubié, M., Champagnat, J. and Zieglgansberger, W. (1978). Effects of opiates and methionine enkephalin on pontine and bulbar respiratory neurones of the rat. Brain Res., 155, 55-67.

7. Gillis, R.A., Helke, C.J., Hamilton, B.V., Norman, W.P. and Jacobowitz, D. (1980). Evidence that substance P is a neurotransmitter of baro- and chemoreceptor afferents in the nucleus of tractus solitarius. Brain Res., 181, 476-481.

8. Iversen, L.L. (1983). Neuropeptides - What next ? Trends in Neurosciences, 6 (n°8), 293-294.

9. Kalia, M. and Mesulam, M.N. (1980). Brain stem projections of sensory and motor components of the vagus complex in the cat. II. Laryngeal, trachebronchial, pulmonary, cardiac and gastrointestinal branches. J. Comp. Neurol., 193, 467-508.

10. Krnjevic, K. (1975). Acetylcholine receptors in vertebrate CNS. In : Iversen, L.L., Iversen, S.D. and Snyder, S.H. (eds). Handbook of Psychopharmacology, vol. 6, pp. 97-126 (Plenum Press, New York, London).

11. Lunberg, J.M. and Hökfelt, T. (1983). Coexistence of peptides and classical neurotransmitters. Trends in Neurosciences, 6 (n°8) 325-333.

12. Mitchell, R. and Fleetwood-Walker, S. (1981). Substance P, but not TRH, modulates the 5-HT autoreceptor in ventral lumbar spinal cord. Eur. J. Pharmacol., 76, 119-120.

13. Morin-Surun, M.P., Boudinot, E., Gacel, G., Champagnat, J., Roques, B.P. and Denavit-Saubié, M. (1984). Different aspects of and opiate agonists on respiration. Eur. J. Pharmacol., 98, 235-240.

14. Morin-Surun, M.P., Gacel, G., Champagnat, J., Denavit-Saubié, M. and Roques, B.P. (1984). Pharmacological identification of and opiate receptors on bulbar respiratory neurons. Eur. J. Pharmacol., 98, 241-247.

15. Morin-Surun, M.P., Champagnat, J., Denavit-Saubié, M. and Moyanova, S. (1984). The effects of acetylcholine on bulbar respiratory related neurons. Consequences of anaesthesia by pentobarbital. Naunyn Schmied. Arch. Pharmacol., 325, 205-208.

16. Morin-Surun, M.P., Jordan, D., Champagnat, J., Spyer, K.M. and Denavit-Saubié, M. (1984). Excitatory effects of iontophoretically applied substance P on neurons in the nucleus tractus solitarius of the cat : lack of interaction with opiates and opioids. Brain Res., 307, 388-392.

17. Palacios, J.M., Wamsley, J.K., Zarbin, M.A. and Kuhar, M.J. (1981). GABA and glycine receptors in rat brain : autoradiographic localization. In : Mandel, P. and Defeudis, F.V. (eds). Amino acid Transmitters (Raven Press, New York).

18. Potter, D.D., Fushpan, E.J. and Landis, S.C. (1981). Multiple-transmitter status and "Dale is principle". Neuroscience Commentaires, 1, 1-9.

19. Riche, D., Denavit-Saubié, M. and Champagnat, J. (1979). Pontine afferents to the medullary respiratory system : anatomofunctional correlation. Neurosci. Lett., 12, 151-155.

20. Simantov, R., Kuhar, M.J., Uhl, G.R. and Snyder, S.H. (1977). Opioid peptide enkephalin : immunohistochemical mapping in rat central nervous system. Proc. Natl. Acad. Sci., U.S.A., 74, 2167.

21. Swanson, L.W., Sawchenko, P.E., Wiegand, S.J. and Price, J.C. (1980). Separate neurons in the paraventricular nucleus project to the median eminence and to the medulla or spinal cord. J. Comp. Neurol., 198, 190-195.

57

Morphofunctional Studies on the Role of Somatostatin-Enkephalin Interactions in the Regulation of Central Respiratory Mechanisms

K. FUXE, A. HÄRFSTRAND, M. KALIA, L. F. AGNATI
and L. TERENIUS

Over a number of years we have been interested in the role of catecholamines (CA) and neuropeptides in the regulation of central respiratory mechanisms (1, 2, 3, 4, 5, 6, 7) All these studies have been performed in the α-chloralose anaesthetized rat. The drugs and the neuropeptides were injected i.c. These results together with others (8) indicated the existence of central dopamine (DA) mechanisms involved in respiratory regulation and that α2-adrenergic receptors probably linked to adrenergic nerve terminals can exert an inhibitory control of respiratory frequency. However, no major control of respiratory activity seems to be exerted by dopaminergic or adrenergic nerve terminal networks. In line with these functional data there is a relatively sparse innervation of the respiratory subnuclei of the nucleus tractus solitarius (NTS) by CA nerve terminals (ventral subnucleus and ventrolateral subnucleus) (9). However, the adrenergic comodulator neuropeptide Y (NPY) has been shown to reduce ventilation in α-chloralose anaesthetized rats (4), probably by increasing the coupling of the α2-adrenergic recognition sites to their biological effector (10, 11). These results underline that adrenergic nerve terminal networks in the dorsal respiratory cell groups may play a role as inhibitory regulators of respiratory frequency. In line with this view avian pancreatic polypeptide can potentiate clonidine induced bradypnea (5). In our functional analysis the most interesting results have been obtained following i.c. injections of somatostatin into the α-chloralose anaesthetized male rat. (6, 7, 12). Thus, it was found that somatostatin and somatostatin analogues can produce apnea upon i.c. administration, indicating a possible role of somatostatinergic mechanisms in apneic syndromes. It was suggested that this somatostatinergic mechanism may at least in part be located within the ventrolateral and ventral subnuclei of the NTS, since they were shown to contain a substantial numbers of somatostatin immunoreactive nerve terminals and cell bodies. These two subnuclei receive afferents from lung stretch receptors and play a role in the respiratory off-switch mechanism in the medulla oblongata (13, 14).

In the present paper we will mainly report on morphological and functional results obtained in the analysis of the role of somatostatin and enkephalin interactions in central respiratory mechanisms. We have previously demonstrated (12) that substantial numbers of enkephalin immunoreactive nerve terminals exist within the respiratory subnuclei of the NTS. In the present study we can demonstrate that central and peripheral administration of the opiate receptor antagonist naloxone can prevent somatostatin induced apnea. The existence of large numbers of enkephalin immunoreactive nerve cell bodies of different types has also been demonstrated in the cardiovascular and respiratory subnuclei of the NTS underlining i.a. the importance of enkephalinergic mechanisms within the dorsal respiratory group of the medulla oblongata (15, 16).

METHODOLOGY

Male specific pathogen free Sprague-Dawley rats have been used as in previous studies (4, 12, 17). In the morphological analysis the indirect immunoperoxidase method of Sternberger (18) was used for the demonstration of enkephalin-like immunoreactivity in nerve cell bodies and terminals of the dorsal and ventral respiratory groups of the medulla oblongata. The enkephalin antibody was raised in rabbits against metenkephalin conjugated to bovine serum albumin For further details, see Kalia et al. (17). The respiratory effects of i.c. application of somatostatin and mμ and delta opiate receptor agonists were studied in α-chloralose (100 mg/kg i.v.) anaesthetized Sprague-Dawley rats. Intraoesophageal pressure was used to record respiratory excursions (inspiration and expiration) and respiratory frequency. Heart rate and arterial blood pressure were recorded by means of a catheter in the common carotide artery using a Statham P23 DC pressure transducer connected to a Grass 7 polygraph. For further details, see 4, 17.

THE DISTRIBUTION OF SOMATOSTATIN AND ENKEPHALIN IMMUNOREACTIVE NERVE CELL BODIES AND NERVE TERMINAL NETWORKS WITHIN THE SUBNUCLEI OF THE NUCLEI OF THE TRACTUS SOLITARIUS

We have previously demonstrated (12, 17) that there exist somatostatin immunoreactive nerve terminals of a moderate density in the ventral and ventrolateral subnuclei of the NTS, which are associated predominantly with pulmonary afferents and in the interstitial subnucleus of the NTS, which is primarily associated with the laryngeal afferents. In addition, somatostatin immunoreactive nerve cell bodies exist within the ventral and ventrolateral subnuclei (12). In contrast, all the respiratory subnuclei including also the intermediate nucleus of the NTS which is associated with tracheal afferents showed only scattered enkephalin immunoreactive nerve terminals. It is noteworthy, however, that the ventral parasolitary region contained dense collections of enkephalin immunoreactive nerve terminals, since this region is adjacent to the respiratory neurons within the ventral and ventrolateral subnuclei and receive dendrites from these neurons

(17). The important role of enkephalinergic mechanisms within the cardiovascular and respiratory subnuclei of the NTS has been further underlined in the present study by the demonstration of substantial numbers of enkephalin immunoreactive nerve cell bodies within practically all the subnuclei of the NTS (see figs. 1 and 2).

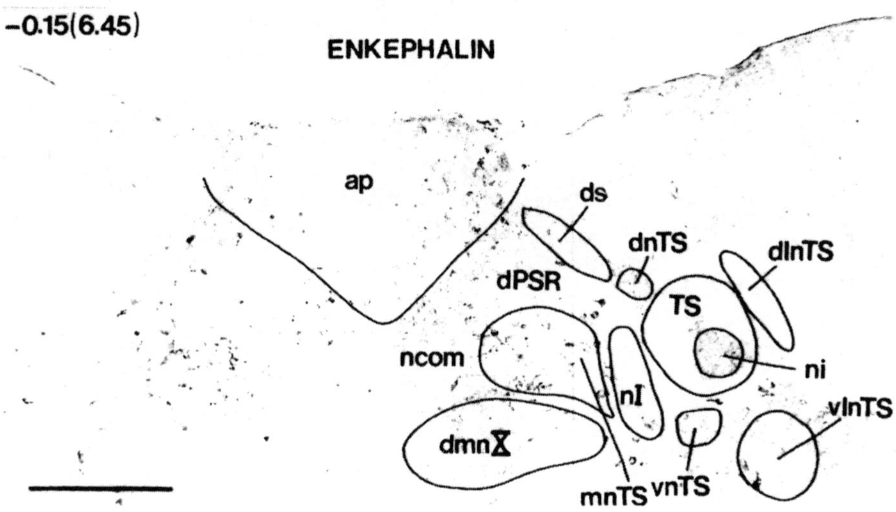

FIGURE 1   Distribution of enkephalin immunoreactive cell bodies is shown in a coronal section of the dorsal medulla of the male rat. Colchicine treatment (120µg/10µl; 18-24 h earlier). The rostrocaudal level is indicated in relation to the obex. In parenthesis also the level of de Groot is shown (17). The Horsley-Clarke stereotaxic coronal plane has been used. Bar= 250 µm. Abbreviations used: TS= tractus solitarius; ni= interstitial nucleus of nTS; nI= intermediate nucleus of nTS; dnTS= dorsal subnucleus of the nTS; vlnTS= ventrolateral subnucleus of the nTS; vnTS= ventral subnucleus of the nTS; mnTS= medial subnucleus of the nTS; dmnX= dorsal motor nucleus of the vagus; ap= area postrema.

Enkephalin immunoreactive nerve cell bodies are observed in the medial, intermediate, interstitial, ventral, and ventrolateral subnuclei of the NTS. They are usually small to medium sized and oval shaped but within the ventral and ventrolateral subnucleus also certain large strongly immunoreactive nerve cell bodies exist (fig. 2). These results open up the possibility that enkephalin immuno

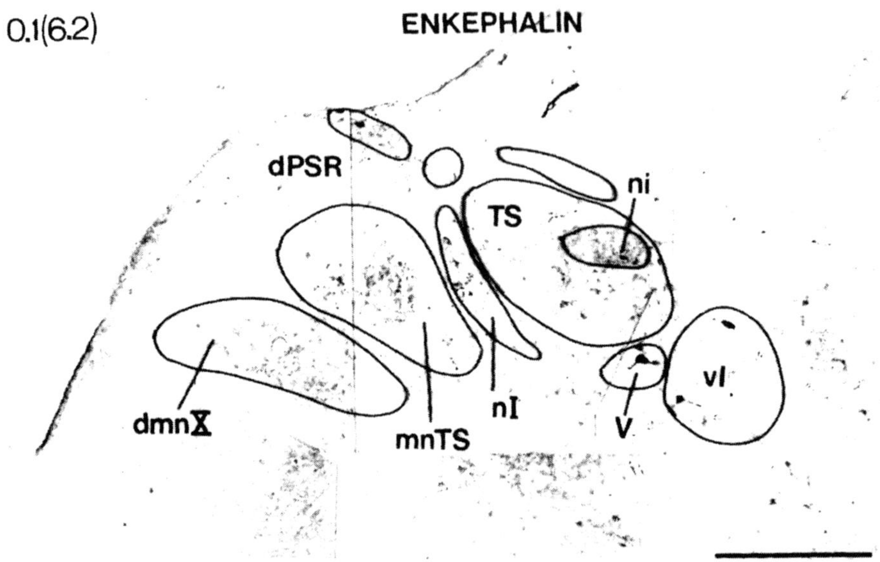

FIGURE 2   Distribution of enkephalin immunoreactive cell bodies is
shown in a coronal section of the dorsal medulla of the
male rat at a level (0.1 mm in front of obex) rostral to
that in fig. 1. For further details, see text to fig. 1.

reactive neurons in the ventral and ventrolateral subnuclei of the
NTS can represent not only respiratory interneurons but also bulbospinal projection neurons controlling respiratory motor neurons of
the spinal cord. The results also indicate that enkephalin neurons
within the respiratory nuclei can be innervated by somatostatin
immunoreactive nerve terminals and vice versa. It is also conceivable
that they both may control activity within the excitatory (Rα) and
inhibitory (Rβ) respiratory neurons of this region, exerting i.a. an
inhibitory action on the Rα neurons, which drive inspiration. It must
be noted however that also the ventral respiratory groups within the
ventrolateral reticular formation of the medulla oblongata contain
substantial numbers of enkephalin immunoreactive nerve cell bodies at
various rostrocaudal levels (see fig. 3). Enkephalin immunoreactive
nerve cell bodies are medium sized to large (nucleus giganto cellularis reticularis ventralis). The medium sized enkephalin immuno-

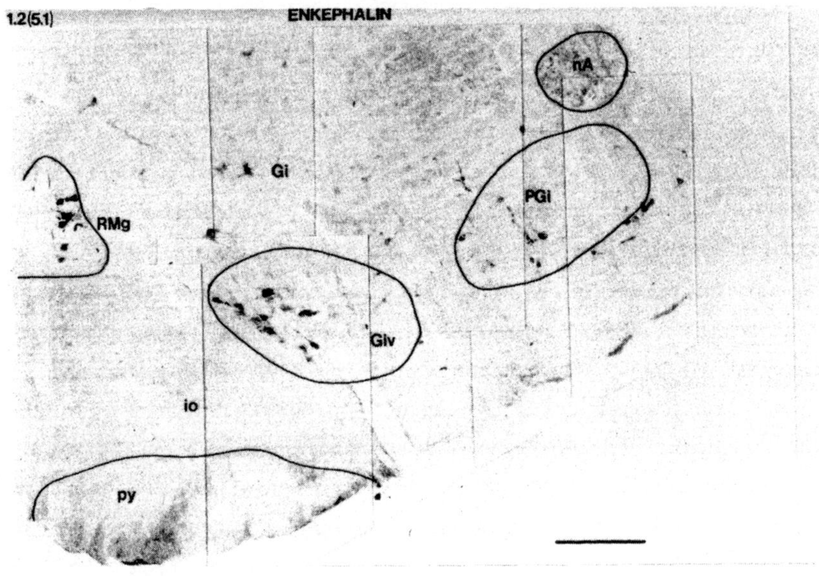

FIGURE 3   Distribution of enkephalin immunoreactive cell bodies in
the ventrolateral medulla of the male rat at a level +1.2
mm rostral to the obex. For further details, see text to
fig. 1. Giv= nuc. gigantocellularis, pars ventralis; PGi=
nuc. paragigantocellularis reticularis; Py= pyramid; RMg=
nuc. raphe magnus; Gi= nuc. gigantocellularis reticularis;
io= inferior olive; nA= nuc. ambiguus. For further details,
see text to fig. 1.

reactive nerve cell bodies are found partly within the nucleus reticularis lateralis and within the nucleus paragiganto cellularis reticularis as well as within adjacent regions of the reticular formation. A substantial enkephalin innervation is observed within the
nucleus ambiguus, (fig. 3), nucleus paraambigualis and within the
nucleus retroambiguus. Thus, there exist morphological evidence that
enkephalinergic mechanisms can regulate central respiratory mechanisms also by an action in the ventrolateral medulla where mμ opiate
receptor activation may lead to depressant actions on respiration
(lateral reticular nucleus) (16). Also when metenkephalin is applied
to the ventral brain surface of the cat it induces a depression of
ventilation (19).

SOMATOSTATIN INDUCED APNEA: EFFECTS OF BILATERAL VAGOTOMY AND LOW DECEREBRATION

An i.c. injection of somatostatin in a dose of 6 nmol was found to produce within 6-10 min. an expiratory apnea, which was preceeded by a lowering of respiratory frequency and by a gradual decline of arterial blood pressure and heart rate. Somatostatin induced apnea could not be prevented by bilateral vagotomy or by low cerebration (caudal to the inferior collicle). (12). Following these experimental procedures the expiratory apnea was associated with a change in the set point functional residual capacity probably related to a change in the set point between central inspiratory-expiratory activity. The results indicate that the apnea is not induced by a modulation of release of transmitters from the pulmonary stretch afferents or by modulation of activity within descending fibers from the higher centers.

Instead the somatostatin receptors involved are probably located within the medulla oblongata itself and at least partly within the respiratory subnuclei of the NTS. It is presently unclear to which extent the somatostatin receptors involved are related to somatostatin immunoreactive vagal sensory neurons or to somatostatin interneurons (12).

SOMATOSTATIN INDUCED APNEA: PREVENTION BY CENTRAL ADMINISTRATION OF A SOMATOSTATIN ANTISERUM

The somatostatin antiserum used in these experiments have been raised in rabbits against synthetic somatostatin conjugated to keyhole limpet hemocyanine (20). The characteristics of this somatostatin antiserum has previously been described (12, 17). It was found that pretreatment with this antiserum given i.c. in a volyme of 10 µl and diluted up to 1:8000 could completely prevent the development of somatostatin induced apnea (6 nmol/10 µl). The antiserum was administered 20 min prior to the somatostatin injection. It was also demonstrated that this prevention by the somatostatin antiserum of the somatostatin induced apnea was not produced by the presence of the antibodies within the cerebrospinal fluid, since repeated rinsing of the fourth ventricle did not antagonize the protective effects of the somatostatin antiserum. Thus, it is assumed that the somatostatin antiserum acts by penetration through the floor of the fourth ventricle, so that it can reach the respiratory subnuclei of the NTS. Normal rabbit serum (10 µl) did not exhibit any protective action against somatostatin induced apnea when given i.c. These results indicate that the somatostatin induced apnea is produced by somatostatin or a somatostatin related peptide, which can bind to the somatostatin antiserum.

SOMATOSTATIN INDUCED APNEA: INTERACTION WITH HYPOXIA AND HYPERCAPNEA

We have recently studied the effects of hypoxia and hypercapnea on the production of somatostatin induced apnea during rebreathing experiments. (21). Hypoxia and hypercapnea caused a shortening of the latency of somatostatin induced apnea from around 7 minutes to around 3 minutes. Thus, the sensitivity of the medullary respiratory centers to the apneic action of somatostatin can be altered by the chemical environment as well as by changes in the activity in the chemoreceptor afferent inflow from the carotide body. The effects are probably not caused by a low pH, since saline (pH 6.0) did not produce such a reduction in latency. Thus, in fact an interplay of chemical and neuronal signals could represent an importand integrative mechanism of respiratory regulation (22). A sudden infant death apneic syndrome is usually associated with hypoxia and hypercapnea which could further sensitize the respiratory centers to an error in the mechanisms controlling somatostatin transmission. The location of such an interaction could be medullary respiratory neurons involved in the control of inspiration in the respiratory cycle such as the Rα and Rβ-cells. (14).

SOMATOSTATIN INDUCED APNEA: PREVENTION BY CENTRAL AND PERIPHERAL ADMINISTRATION OF THE OPIATE RECEPTOR BLOCKING AGENT NALOXONE

As seen in Table 1 we have recently been able to demonstrate in the α-chloralose anaesthetized rat that naloxone given i.c. (306 nmol or i.v. (15 µmol/kg) can significantly increase the survival of the rats after i.c. injections of somatostatin (6 nmol in 10 µl). (23). Instead i.c. administration of of the α2-receptor blocking agent RX781094 (24) could not prevent the development of somatostatin induced apnea. Nor did this pretreatment affect the latency to the onset of somatostatin induced apnea. Also various doses of Substance P and CCK-8 in the nanomolar range given i.c. do not exert an protective action against somatostatin induced apnea. These results indicate that opiate receptor activity participate or play a permissive role in the development of somatostatin induced apnea. In view of the morphologcial analysis reported above it seems possible that interactions between the somatostatinergic and enkephalinergic mechanisms in the regulation of respiratory rhythm take place in the respiratory subnuclei of the NTS.

EFFECTS OF CENTRAL AND PERIPHERAL ADMINISTRATION OF NALOXONE ON THE CENTRALLY INDUCED SOMATOSTATIN APNEA

Experimental groups	Basal values			% Peak change of basal value after pretreatment			Latency of apnea (Proportion of surviving rats)	% change of basal in min value 60' after SRIF inj.		
INTRACISTERNAL PREVENTION	RR (breaths/min)	ABP (mmHg)	HR (beats/min)	RR	ABP	HR		RR	ABP	HR
CSF 10µl + SRIF 6 nmol ic	71±5	102±8	349±13	-7±2	-3±3	-1±1	15±2 (3/10) ⎤ p<0.02	-	-	-
Naloxone 306nmol + SRIF 6 nmol ic	79±3	115±4	395±10	-3±1	0±2	-2±3	- (10/11) ⎦	-14±5	5±3	-1±5
INTRAVENOUS PREVENTION										
Saline 0.3 ml iv + SRIF 6 nmol ic	76±6	99±3	369±14	-9±2	-5±3	-1±2	11±1 (1/9) ⎤ p<0.02	-	-	-
Naloxone 15µmol/kg + SRIF 6 nmol	73±4	114±6	395±19	-9±4	-1±3	-4±2	- (7/9) ⎦	-19±4	-2±4	-6±2

The mµ type of opiate receptors is preferentially sensitive to the blocking activity of naloxone (25, 26). However, in the present experiments a fairly high dose of naloxone had to be used (15 µmol/kg) to counteract the somatostatin induced apnea, indicating that the delta-subtype of opiate receptors may have to be blocked in order to obtain a protective action. In agreement with this view it has been found in preliminary experiments that morphinoceptin, a selective mμ opiate agonist , (CRB, Cambridge, England) given i.c. in a dose of 19 nmol, does not aggravate somatostatin induced apnea. Nor does treatment with this dose by itself produce signs of apneic episodes. Instead an i.c. injection of the delta opiate receptor agonist, (D-Ser^2)-Leu-Enkephalin-Thr (13 nmol) into the α-chloralose anaesthetized male rat leads to a marked reduction of respiratory frequency and to the development of apnea within 5-15 minutes. The present pharmacological evidence therefore indicates that naloxone mainly exert its protective activity by blocking the delta-type of opiate receptor within respiratory centers of the medulla oblongata. In line with these results it has also been suggested that the mμ type of opiate receptors produce exitatory effects on respiration within the NTS. (16). It has also previously been shown that exertiness treatment with an enkephalin analogue (27) in the neonatal rat can produce irregular breathing or apnea, an action which is blocked by naloxone. It has also recently been reported that naloxone decreases the duration of primary apnea in neonatal asphyxia (28). Taken together the present results give rise to the hypothesis that the somatostatin and/or delta-type of opiate receptors represent major receptor systems for the suppression of inspiratory neuronal activity within the medulla oblongata, which under pathological conditions may led to the development of apneic syndromes.

ACKNOWLEDGEMENT

This work has been supported by a Grant (04X-715) from the Swedish Medical Research Council. We are grateful to Mrs. Siv Nilsson, Miss Barbro Tinner and Miss Catharina Nilsson for excellent technical assistance and to Mrs. Anne Edgren and Mrs. Ulla-Britt Wedin for excellent secreterial assistance.

REFERENCES

1.   Bolme, P. and Fuxe, K. (1973). Pharmacological studies on a possible role of central noradrenaline neurons in respiratory control. J. Pharm. Pharmacol., 25, 351-352.

2.   Bolme, P, Corrodi, H., Fuxe, K., Hökfelt, T., Lidbrink, P. and Goldstein, M. (1974). Possible involvement of central adrenaline neurons in vasomotor and respiratory control. Studies with clonidine and its interactions with piperoxane and yohimbine. Eur. J. Pharmacol., 28, 89-94.

3.   Bolme, P., Fuxe, K., Hökfelt, T. and Goldstein, M. (1977). Studies on the role of dopamine in cardiovascular and respiratory control: Central versus peripheral mechanisms. Adv. Biochem. Psychopharmacol., 16, 281-290.

4.   Fuxe, K., Agnati, L.F., Härfstrand, A., Zini, I., Tatemoto, K., Merlo Pich, E., Hökfelt, T., Mutt, V. and Terenius, L. (1983). Central administration of neuropeptide Y induces hypotension, bradypnea and EEG synchronization in the rat. Acta Physiol. Scand., 118, 189-192.

5.   Fuxe, K., Agnati, L.F., Härfstrand, A., Lundberg, J.M., Hökfelt, T., Calza, L., Kimmel, J. and Bernardi, P. (1982b). Intracisternal administration of avian pancreatic polypeptide lowers respiration rate and enhances the clonidine induced reduction of respiration rate in α-chloralose anesthetized rats: Possible interactions with an α_2-adrenergic receptor. Acta Physiol. Scand., 115, 381-384.

6.   Fuxe, K., Agnati, L.F., Ganten, D., Andersson, K., Calza, L., Vincent, M., Sassard, J., Yukimura, T., Eneroth, P., Goldstein, M., Hökfelt, T., Rosell, S., Härfstrand, A., Vale, W., Brown, M. and Rivier, J. (1982c). Central and peripheral hormones and peptides: Focus on the involvement of noradrenaline and adrenline neurons, opiod peptides, substance P and somatostatin in central cardiovascular regulation of the rat. In: Rascher, W., Clough, D. and Ganten, D. (eds). Hypertensive Mechanisms. The Spontaneously Hypertensive Rat as a Model of Study Human Hypertension. pp. 417-441. (Schattauer Verlag, Stuttgart and Berlin).

7.   Fuxe, K., Agnati, L.F., Härfstrand, A., Mutt, V., Andersson, K., Hökfelt, T., Vale, W., Brown, M. and Rivier, J. (1982a). Cardiovascular and respiratory actions of somatostatin peptides follwing intracisternal injection into the α-chloralose anesthetized rat. Neurosci. Lett., Suppl. 10, 189.

8.   Hedner, J. (1983). Neuropharmacological Aspects of Central respiratory regulation. An experimental stydy in the rat. Acta Physiol. Scand., Suppl. 524.

9.   Kalia, M., Fuxe, K. and Goldstein, M. (1985). Rat medulla oblongata: II. Dopaminergic, noradrenergic and adrenergic neurons, nerve fibers and nerve terminals in the caudal medulla. J. Comp. Neurol., in press.

10. Agnati, L.F., Fuxe, K., Benfenati, F., Battistini, N., Härfstrand, A., Tatemoto, K., Hökfelt, T. and Mutt, V. (1983). Neuropeptide Y in vitro selectively increases the number of α-adrenergic binding sites in membranes of the medulla oblongata of the rat. Acta Physiol. Scand., 118, 293-295.

11. Fuxe, K., Agnati, L.F., Härfstrand, A., Martire, M., Goldstein, M., Grimaldi, R., Bernardi, P., Zini, I., Tatemoto, K. and Mutt, V. (1984). Evidence for a modulation by neuropeptide Y of the α-2 adrenergic transmission line in central adrenaline synapses. New possibilities for treatment of hypertensive diosorders. Clin. Exp. Hypertension, in press.

12. Kalia, M., Fuxe, K., Agnati, L.F., Hökfelt, T. and Härfstrand, A. (1984). Somatostatin produces apnea and is localized in medullary respiratory nuclei: a possible role in apneic syndromes. Brain Res., 296, 339-344.

13. von Euler, C. (1983). Brain-stem mechanisms for the generation and control of the breathing pattern. In: The Respiratory System. Handbook of Physiology, in press.

14. Euler, C. von and Lagercrantz, H. (1979). Central Nervous Control Mechanisms in Breathing, Pergamon Press, Oxford.

15. Denavit-Saubie, M., Champagnat, J. and Zieglgansberger, W. (1978). Effects of opiates and methionine-enkephalin on pontine and bulbar respiratory neurons of the rat. Brain Res., 155, 55-67.

16. Hassen, A.H., Feuerstein, G., Pfeiffer, A. and Faden, A.I. (1982). versus μ receptors: cardiovascular and respiratory effects of opiate agonists microinjected into nucleus tractus solitarius of cats. Regulatory Peptides, 4, 299-309.

17. Kalia, M., Fuxe, K., Hökfelt, T., Johansson, O., Lang, R., Ganten, D., Cuello, C. and Terenius, L. (1984). Distribution of neuropeptide immunoreactive nerve terminals within the subnuclei of the nucleus of the tractus solitarius of the rat. J. comp. Neurol., 222, 409-444.

18. Sternberger, L.A. (1979). Immunocytochemistry, Wiley, New York.

19. Flórez, J. and Mediavilla, A. (1977). Respiratory effects of met-enkephalin applied to the ventral surface of the brain stem. Brain Res., 138, 585-590.

20. Schultzberg, M., Hökfelt, T., Nilsson, G., Terenius, L., Rehfeld, J.F., Brown, M., Elde, R., Goldstein, M. and Said, S. (1980). Distribution of peptide- and catecholamine-containing neurons in the gastrointestinal tract of rat and guinea-pig: immunohistochemical studies with antisera to substance P, vasoactive intestinal polypeptide, enkephalins, somatostatin, gastrin/cholecystokinin, neurotensin and dopamine beta-hydroxylase. Neurosci., 5, 689-744.

21. Härfstrand, A., Kalia, M., Fuxe, K., Kaiser, L. and Agnati, L.F. (1984). Somatostatin induced apnea: interaction with hypoxia and hypercapnea. Neurosci. Lett., in press.

22. Mitchell, R.A. and Herbert, D.A. (1974). The effect of carbon dioxide on the membrane potential of medullary respiratory neurons. 75, 345-349.

23. Härfstrand, A., Fuxe, K., Kalia, M. and Agnati, L.F. (1984). Somatostatin induced apnea: Prevention by central and peripheral administration of the opiate receptor blocking agent naloxone. Acta Physiol Scand., in press.

24. Doxey, J.C., Roach, A.G. and Smith, C.F.C. (1983). Studies on RX781094: a selective, potent and specific antagonist of α_2-adrenoceptors. Br. J. Pharmac., 78, 489-505.

25. Martin, W.E., Eades, C.G., Thompson, J.A., Hupper, R.E. and Gilbert, P.E. (1976). The effects of morphine and nalorphine-like drugs in non-dependent and morphine dependent chronic spinal dog. _J. Pharmacol. Exp. Ther._, _197_, 517-532.

26. Miller, R.J. (1982). Multiple opiate receptors for multiple opioid peptides. _Med. Biol._, _60_, 1-6.

27. Hedner, T., Hedner, J., Bergman, B., Jonason, J. and Lundberg, D. (1981b). Transient apnea after an enkephalin analogue in the preterm rabbit. **Biol. Neonate,** _39_, 290-294.

28. Chernick, V., Madansky, D.L. and Lawson, E.E. (1980). Naloxone decreases the duration of primary apnea with neonatal asphyxia. _Pediat Res._, _14_, 357-359.

58

Dissociative Respiratory Activities of Opioids and Pentobarbital at the Medullary and Pontine Levels

J. FLÓREZ, M. A. HURLÉ and A. MEDIAVILLA

INTRODUCTION

The overall respiratory effect of centrally acting drugs is clearly the consequence of their action at each of the nuclei and systems involved in the genesis of breathing. Because each nucleus may display a different sensitivity towards a particular drug, the ultimate effect will depend on the affinity, the type of receptor interaction, and the dose.

Opiates and barbiturates are among the most conspicuous respiratory depressant drugs existing in therapy. Since their action on respiration has been carefully examined both in humans and animals, a functional dissection of the restricted activity at the nuclei related to respiration seems particularly appropriate. Furthermore, both types of drugs are now considered as the exogenous equivalents of endogenous neuroregulators mediating the activity of the opioid and the GABA system. These two systems are widely distributed in the brain stem nuclei related to respiration. Consequently, the more we restrict the selective action of the drugs to a particular component of the respiratory circuitry, the more knowledge we can obtain on the possible function of these endogenous systems in the respiratory regulation.

As a first approximation, we have studied the effects induced by pentobarbital and by two opioid peptides restrictively applied to the ventral medullary surface and to the dorsorostral surface of the pons, the μ-agonist D-Ala^2-Me-Phe^4-Met(O)ol^5-enkephalin (FK-33824) and the δ-agonist D-Ala^2-D-Leu^5-enkephalin (DADLE).

METHODS

The experiments were carried out on lightly anesthetized cats. The ventral medullary surface was exposed from the lower pons down to C_1 and 4 to 5 mm laterally from the midline, as previously described (5). For approaching the dorsal pontine surface, the cerebellum was suctioned and the bony tentorium was removed, the floor of the fourth ventricle being exposed in a horizontal plane. In some animals decerebration was added at the midcollicular level. When the vagi were to be transected, they were isolated at the midcervical level. Body temperature was maintained at 37±0.5 ºC.

Respiration was monitored by pneumotachography and integration to yield the full respiratory wave. End-tidal CO_2 was recorded with a CO_2 analyzer. Respiratory function was assessed by measuring the spontaneous resting respiratory parameters under continuous 100 % O_2 breathing, as well as the responses to challenges with 5 % CO_2 in O_2. Blood pressure and heart rate were recorded through appropriate transducers.

For application of drugs in the medullary ventral surface, two plastic chambers 10 mm high of a kidney shape, fixed to a holder, were lowered bilaterally. The covered area included those described as M, S and L (8). In the dorsorostral pontine surface, a single oval chamber was applied; the longitudinal and transverse axis were 8 and 5 mm, respectively, the anterior border remaining just behind the inferior colliculi. The lower rims of the chambers were siliconed and rested under light pressure on the semected areas, to prevent any leakage of the content. Each dose of the sodium pentobarbital and the opioid peptides was diluted in a volume of 80 µl of saline. Each cat received only one drug and a single dose. When appropriate, iv naloxone (0.1 mg/kg) was injected immediately after the chambers were withdrawn from the surfaces.

RESULTS

In the <u>pontine rostrodorsal surface</u>, pentobarbital depressed respiration dose-dependently (fig. 1). The dose of 4 mg was inactive, but 8 and 16 mg induced a slowly progressing reduction of frequency down to 50 % of the control value, which peaked between 15 and 45 min; tidal volume either was not modified or it was moderately increased. End-ti-

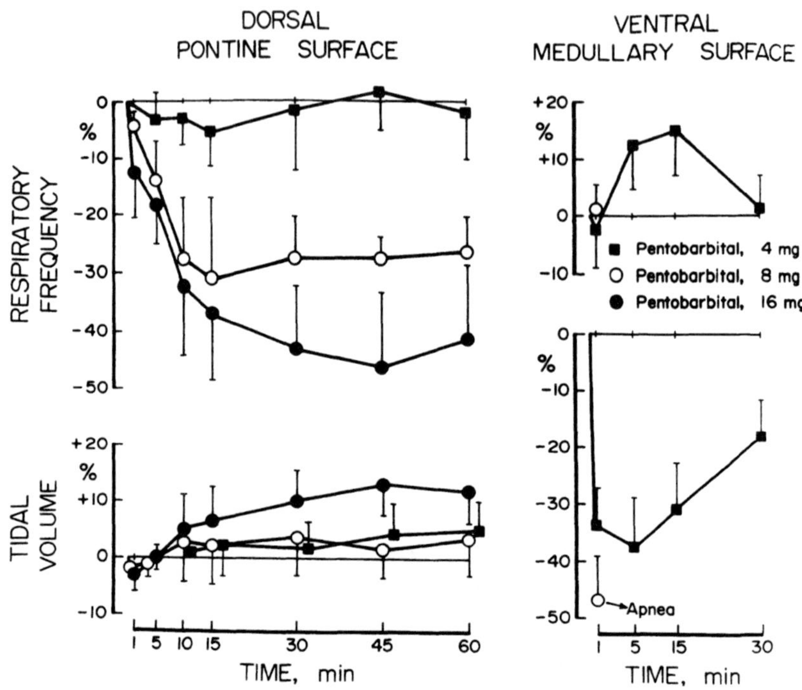

FIGURE 1 Time courses of the respiratory effects induced by increasing
doses of pentobarbital, applied to the pontine dorsal surface and to the medullary ventral surface in cats. Symbols represent mean values (4 to 6 animals)±SEM.

dal CO_2 values followed the course of the frequency. The action of
both opioids (0.7 and 7.0 nmol) was qualitatively similar to that of
pentobarbital (fig. 2): they also depressed dose-dependently the respiratory frequency, but the intensity of the depression was higher. In
fact, in the course of the experiment some animals developed apnea
which was consistently reached by a progressive slowing of the respiratory rhythm. Tidal volume remained unchanged during the first hour,
but as the breathing became very slow it was increased. Naloxone reversed fully and consistently the opioid-induced respiratory depression, but not that induced by pentobarbital. Similar effects were observed in decerebrate animals.

In the <u>ventral medullary surface</u>, the lowest dose of pentobarbital
induced a rapid and severe reduction of tidal volume which almost
reached its peak during the first minute (fig. 1); the frequency was

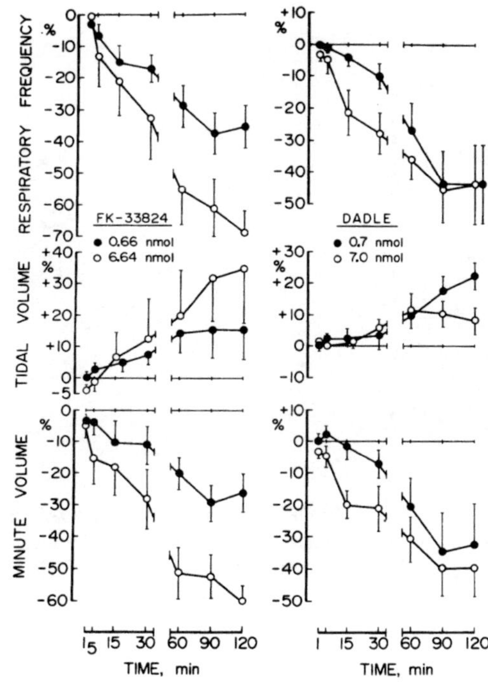

FIGURE 2 Time courses of the respiratory effects induced by DADLE and FK-33824 applied to the dorsorostral surface of the pons.

not modified initially, but subsequently it was slightly elevated. Higher doses induced apnea in a few minutes, accounted for by a rapidly progressing decline in the respiratory amplitude. The response to the opioid peptides was again similar to that of pentobarbital (fig. 3). A decline in tidal volume was immediately induced, which reached 50–75 % of the maximal depression by the first minute. Subsequently, it progressed and persisted during the two hr of the experiment. After a latency of about 5 min, frequency was consistently increased, it reached a peak at 30 min and then returned towards the baseline. End-tidal CO_2 was increased during the first hour. Similar effects were observed in vagotomized cats. The opioid effects were reversed by naloxone.

CO_2 responsiveness was tested in the cats subjected to the application of opioids. The response was consistently depressed, but the component varied depending on the level of application. In the medullary surface, the tidal volume response was sharply reduced, whereas in the pontine surface the sluggishness of the frequency response accounted for most of the depression.

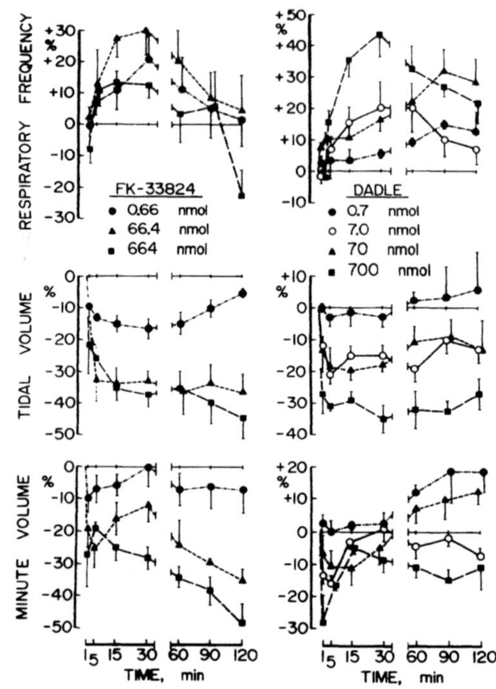

FIGURE 3 Time courses of the respiratory effects induced by increasing doses of DADLE and FK-33824, applied bilaterally to the ventral medullary surface.

DISCUSSION

Two major observations become apparent in the present study: 1) dissimilar patterns of respiratory response were induced by delivering drugs to the medullary and pontine surfaces; 2) although pentobarbital and opioids induced qualitatively similar effects at both levels of the brain stem, the sensitivity of the centers to each type of drug differed considerably. If we assume that the drugs applied to a given area in the surface preferentially disturbed the function of the respiration-related nuclei existing at that level, it appears that, in the medulla, they depressed more specifically the mechanisms responsible for the generation of tidal volume and its response to CO_2, whereas in the pons they selectively blunted some of the mechanisms which regulate the respiratory rhythmicity.

In the pons the drugs depressed selectively the respiratory frequency by increasing both inspiratory and expiratory times. The area of application was close to the nucleus parabrachialis medialis; it is therefore likely that the drugs reached the nucleus by tissue diffusion, since there is no vascularization in the surface, and directly depres-

sed the activity of a center which is known to play a prominent role
in the regulation of the medullary rhythmogenesis. The latency observed after application supports the existence of a diffusion process.
The pontine mechanisms showed a higher sensitivity to the opioid depressant action than the medullary structures, because doses which in
the medulla scarcely modified the respiration, in the pons induced a
substantial depression (compare 0.7 nml of DADLE in both surfaces).
Furthermore, in terms of overall effect, the opioids in the pons induced a more severe depression than did similar doses in the medulla
oblongata. These results support earlier observations obtained in unitary neuronal recordings (4). On the other hand, the sensitivity to
pentobarbital was higher in the medulla than in the pons, because doses that readily caused apnea when applied to the medulla they induced
moderate depression when applied to the pontine surface.

In contrast to previous experiments (5,7), in this study the ventral
medullary surface included all the chemosensitive areas, thereby faci-
litating a massive interaction of the drugs with several medullary
structures related to respiration, either at the level of the ventral
surface (8) or after the rapid penetration from the pial surface into
the medullary parenchyma, along blood vessels and nerve fibers (1). In
these experiments, the tidal volume and its response to CO_2 were rapidly depressed without occurring any change in frequency. Subsequently,
the reduction of tidal volume persisted while frequency was first increased and then declined toward the baseline. Similar results have
been described after applying the opiate fentanyl (7), and pentobarbital (5,9) to the intermediate area in lightly anesthetized cats.

Several mechanisms appear responsible for the persistent depression
of the tidal volume and its response to CO_2. a) The drugs may depress
the sensitivity of the CO_2 central chemoreceptors, or their input into
the central inspiratory generator located in the medulla which is highly responsive to the stimulation by CO_2 (H^+) and is independent of the
vagal afferencies. A reduction in central CO_2 sensitivity should be accompanied by a reduction in the threshold of the inspiratory off-switch
mechanism, which is known to be raised by chemical drive factors (3);
as a consequence, the inspiratory activity would progress to a lower
intensity and would produce a smaller tidal volume. b) The drugs could
directly depress the neurons constituting the central inspiratory gene-

rator, and reduce their output on the inspiratory bulbospinal premotor
neurons. In this regard it has been shown that the opioids, applied mi-
croiontophoretically to respiration related neurons, changed the modulation of discharges (2). c) Attention should also be paid to the nucleus paragigantocellularis, located near the ventral surface, because
it contains enkephalinergic neurons and exerts a strong influence on
breathing mechanisms.

How these effects, induced by selective application of drugs, correlate to those observed after systemic injection? A common feature in
the respiratory action of opiates is the reduction of frequency and the
depression of CO_2 responsiveness. As for pentobarbital, the response to
iv injection of small doses is characterized by a reduction in tidal
volume; higher doses may also depress frequency (9). These descriptions
are consistent with the the present data, because the pontine structures related to the rhythmogenesis showed higher sensitivity to the opioids than to pentobarbital, whereas the medullary nuclei related to the
genesis of amplitude were more sensitive to pentobarbital than to the
opioids. Moreover, the opioids applied to the medullary surface blunted
severely the CO_2 responsiveness.

Our results do not support the contention that μ- and δ-opioids can
induce a different pattern of respiratory depression (6). The μ-agonist
was slightly more potent than the δ-agonist when applied to the medullary surface but no clear difference in potency was observed in the
pontine surface. Furthermore, no differences between the opioids were
recognized in the respiratory patterns generated at each level of the
brainstem. The differences observed by others can be explained by differences in doses and potency of the compounds, rather than by the sub-
type of receptor affected by the drug. However, a δ-agonist more selective than DADLE must be studied.

In conclusion, drugs restrictively applied to selective areas of the
brain stem can induce differential patterns of respiratory activity.
Each drug, by interacting with specific receptors, will initiate its
own molecular mechanisms in different respiratory neuronal populations.
But the respiratory output so originated will differ, depending on the
group of neurons selectively affected and the function they subserve.

ACKNOWLEDGMENTS

This investigation was supported by a grant from FIS (Spain). We thank

Ms. Corina Oceja for her excellent technical assistance.

REFERENCES

1. Borison, H.L., Borison, R. and McCarthy, L.E. (1980). Brain stem penetration by horseradish peroxidase from the cerebrospinal fluid spaces in the cat. Exp. Neurol., 69, 271-289

2. Denavit-Saubié, M., Champagnat, J. and Zieglgänsberger, W. (1978). Effects of opiates and methionine-enkephalin on pontine and bulbar respiratory neurones of the cat. Brain Res., 155, 55-67

3. Euler, C. von (1983). On the central pattern generator for the basic breathing rhythmicity. J. Appl. Physiol., 55, 1647-1659

4. Hassen, A.H., St. John, W.M. and Wang, S.C. (1976). Selective respiratory depressant action of morphine compared to meperidine in the cat. Eur. J. Pharmacol., 39, 61-70

5. Hurlé, M.A., Mediavilla, A. and Flórez, J. (1982). Morphine, pentobarbital and naloxone in the ventral medullary chemosensitive areas: differential respiratory and cardiovascular effects. J. Pharmacol. Exp. Ther., 220, 642-647

6. Morin-Surun, M.P., Boudinot, E., Gacel, G., Champagnat, J., Roques, B.P. and Denavit-Saubié, M. (1984). Different effects of μ and δ opiate agonists on respiration. Eur. J. Pharmacol., 98, 235-240

7. Pokorski, M., Grieb, P. and Wideman, J. (1981). Opiate system influences central respiratory chemosensors. Brain Res., 211, 221-226

8. Schlaefke, M.E. (1981). Central chemosensitivity: A respiratory drive. Rev. Physiol. Biochem. Pharmacol., 90, 171-244

9. Yamada, K.A., Moerschbaecher, J.M., Hamosh, P. and Gillis, R.A. (1983). Pentobarbital causes cardiorespiratory depression by interacting with a GABAergic system at the ventral surface of the medulla. J. Pharmacol. Exp. Ther., 226, 349-355

59

Anatomical Organization of the Aminergic and Peptidergic Systems in the Respiratory Areas in the Brain Stem of the Cat

L. LEGER, Y. CHARNAY, J.-A. CHAYVIALLE, F. DRAY, J. ROSSIER, L. WIKLUND, M.-P. DUBOIS, and M. JOUVET

The aim of the present communication is to describe the distribution of the monoaminergic (MA) and peptidergic (substance P (SP), enkephalins (ENK) and somatostatin (SRIF)) neurones in the 3 respiratory areas of the brainstem : the pontine area including nuclei parabrachialis medialis (PBM) and Kölliker-Fuse (KF), and the 2 bulbar areas corresponding to the ventrolateral part of the nucleus tractus solitarius (NTS) and ambiguus nucleus (NA). This desciption mainly concerns the cat, animal of choice for neurophysiologists.

In the cat pons, catecholaminergic (CA) neurones are distributed over the dorsolateral region around the mesencephalic root of the trigeminal nerve and superior cerebellar peduncle (A4, A6 and A7 CA groups in the rat) (Fig. 5). The respiratory area of the nuclei PBM and KF contains about 1,200 of these cells (or 13 % of the total CA cell population) according to a quantitative estimation from fluorescence histochemical preparations (11) (Fig. 1). The same preparations show the presence of some serotonin (5HT) cell bodies, mixed with the CA cells (Fig. 5). The number of the 5HT cell bodies would not exceed 40. In its dorsalmost part, nucleus PBM contains a rich plexus of CA fibres, probably corresponding to ascending axons (4). In its ventral part and in KF the CA and 5HT varicosities are numerous among the MA and non MA cell bodies and correspond to axon terminals (Fig. 2). The CA varicosities would belong to axons taking their origin in the neighbouring areas (7) and the 5HT varicosities to axons originating in nucleus raphe dorsalis (10).

With the use of immunohistochemistry (indirect immunofluorescence method) we were able to visualize dense networks of SP-, ENK- and SRIF-immunoreactive fibres in PBM and KF. Some of these fibres closely surround cell bodies. These immunoreactive networks are in continuity with fibres occupying the other subdivisions in dorsolateral pons and an area associated with the exiting roots of the facial nerve. The origin of these fibres is not known at present. After an injection of colchicine in the mesencephalic reticular formation into the area of the ascending pontine CA fibres, a few SP- and a large number of ENK-immunoreactive cell bodies were visualized in PBM and KF (Fig. 4). With the use of adjacent 8 μm-thick sections, one section being treated with an anti-ENK antiserum, the other with an anti-dopamine-β-hydroxylase antiserum, it was possible to show that most if not all CA cells contain an ENK-immunoreactive material (6) (Figs. 3-4). Further, the presence in these cell bodies of an immunoreactivity to synenkephalin, a fragment of the bovine adrenal medulla proenkephalin A, suggests that they would be able to synthetize the ENKs (3).

In medulla, NTS contains the CA cell groups equivalent to A2 (noradrenergic cells) and C2 (adrenergic cells) in the rat. The CA cell bodies are gathered in the caudal two thirds of the nucleus, in the medial, intermediary, ventrolateral and commissural subdivisions (1,11) (Fig. 5). These cell bodies are surrounded by a dense plexus of CA fibres, the origin of which would be partly located in the CA Al group in ventrolateral bulbar reticular formation (2). Our observations show that NTS possesses a rich 5HT innervation, especially dense in lateral and ventrolateral subdivisions (Fig. 6). In addition to the MA innervation, rich plexuses of SP-, ENK- and SRIF-immunoreactive fibres are visualized in NTS (Figs. 7-9). The lateral and commissural regions are the most densely supplied for the 3 peptides (see also ref. 8). The ENK- and especially the SP-immunoreactive fibres form a very dense horizontal band in the dorsal part of commissural subdivision. After an injection of colchicine in the bulbar reticular formation or cisterna magna, scattered SP-immunoreactive cell bodies are apparent in the medial, lateral parvocellular and commissural subdivisions (Fig. 7). ENK- immunoreactive cell bodies are visualized in the same areas plus intermediary and ventrolateral subdivisions (Fig. 8). SRIF-immunoreactive cells were seen in parvocellular and commissural nuclei (see also ref. 9).

The third respiratory area in brainstem, namely NA area, contains a small number of CA cell bodies and some rare 5HT cell bodies (11) (Fig. 5). These cells are located in reticular formation adjacent to NA. The nucleus itself is richly supplied with CA fibres (5). Our immunohistochemical observations show that in and outside NA a rather dense 5HT fibre network is present. In NA, these fibres closely surround the large cell bodies. SP-immunoreactive axons are very few in NA itself. ENK- and SRIF-immunoreactive axons form medium-dense plexuses (Fig. 10). When colchicine is injected in the vicinity of NA, a group of ENK-immunoreactive cell bodies is visualized in the caudal pole of NA and SRIF-immunoreactive cells are located in the reticular formation just dorsal and medial to NA (Fig. 10).

In conclusion, it can be said that the distribution of SP-, ENK- and SRIF-immunoreactive neurones in pons and medulla is in favor of a role of these neuropeptides in the control of the respiratory cycle. This is particularly true for the ENKs, known for their depressant effect on the respiratory neurones.

References

1. Blessing W. W. et al., Brain Research, 192 (1980) 69-75.
2. Blessing W. W. et al., Cell Tissue Res., 220 (1981) 27-40.
3. Charnay Y. et al., Ann. Endocrinol., 1984, in press.
4. Chu N. S. and Bloom F. E., Brain Research, 66 (1974) 1-21.
5. Jones B. E. and Friedman L., J. comp. Neurol., 215 (1983)382-96.
6. Léger L. et al., Neuroscience, 9 (1983) 525-546.
7. Levitt P. and Moore R. Y., J. comp. Neurol., 186 (1979) 505-528.
8. Maley B. and Elde R., Neuroscience, 7 (1982) 2469-2490.
9. Maley B. et al., J. comp. Neurol., 217 (1983) 405-417.
10. Sakai K. et al., Brain Research, 119 (1977) 21-41.
11. Wiklund L. et al., J. comp. Neurol., 203 (1981) 613-647.

Abbreviations for the figures :

AP : area postrema ; BC : brachium cunjunctivum ; CM : cisterna magna ; Ic : nucleus intercalatus ; KF : Kölliker-Fuse nucleus ; LC : nucleus locus coeruleus ; LLV : ventral nucleus of the lateral lemniscus ; LRN : lateral reticular nucleus ; M : medial subdivision of solitary tract nucleus ; NA : nucleus ambiguus ; PBM : nucleus parabrachialis medialis ; PC : parvocellular subdivision of solitary tract nucleus; TS : solitary tract ; VL : ventrolateral subdivision of solitary tract nucleus ; X : dorsal motor nucleus of the vagus ; XII : hypoglossal nucleus.

Fig. 1- Catecholamine cell bodies in PBM and KF as visualized on a frontal section treated with an antiserum to dopamine-β-hydroxylase. Their distribution would be identical on a fluorescence histochemical preparation. The perikarya display a medium-size and oval to fusiform shape. Bar : 80 μm.

Fig. 2- Serotonin fibres and varicosities in PBM as seen on a preparation reacted with an anti-serotonin antibody. Note the ubiquitous presence of the varicosities in PBM. Bar : 50 μm.

Fig. 3-4- Homologous fields taken from 2 adjacent 8 μm-thick frontal sections across KF after reaction with an anti-dopamine-β-hydroxylase antiserum (Fig. 3) and anti-enkephalin antiserum (Fig. 4). Most profiles visible on one section have a positive counterpart on the other section. Bar : 50 μm.

Fig. 5- Schematic representation of the distribution of catecholamine (circles) and serotonin (black dots)- containing perikarya as visualized with fluorescence histochemistry and immunohistochemistry at the level of pneumotaxic center (PBM + KF) in pons and nuclei of solitary tract (TS) and ambiguus (NA) in medulla oblongata in the cat. Stereotaxic levels are indicated according to Horsley-Clarke coordinates. Each circle or dot represents 1-5 cell bodies.

Fig. 6- Photomontage showing NTS on a frontal section after reaction with an antiserum against 5HT. Bar : 50 μm.

Fig. 7- Photomontage showing the distribution of SP-immunoreactive fibres in NTS and adjacent areas. Midline is to the left. Note the dense plexuses in dorsal and lateral parts of NTS. Ventrolateral subdivision is only poorly supplied. Arrows indicate immunoreactive cell bodies at the border of ventrolateral subdivision and in nucleus intercalatus. Bar : 100 μm.

Fig. 8- This photomontage shows the distribution of the ENK-immunoreactive fibres in the area of NTS. Midline is to the right. The lateral subdivision is the most densely supplied with ENK-fibres. Inset : two ENK-immunoreactive cell bodies in ventrolateral subdivision. Bars : 70 μm .

Fig. 9- Distribution of SRIF-immunoreactive fibres and cell bodies in the area of NTS. Cell bodies are indicated by arrows. Bar : 70 um.

Fig. 10- SRIF-immunoreactive fibres and cell bodies in the area of NA. The immunoreactive fibres surround the large cell bodies of NA. The immunoreative cell bodies are located dorsal to NA. Bar : 60 μm.

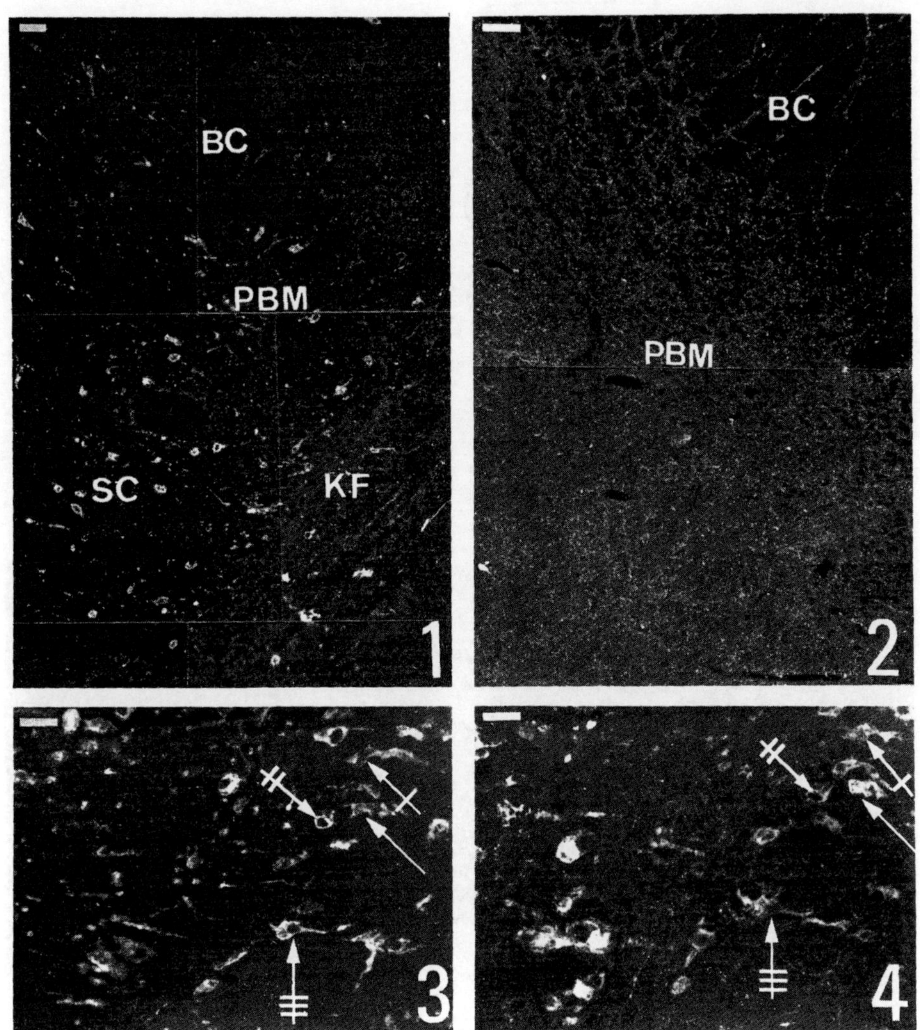
BC
PBM
SC
KF
1
BC
PBM
2
3
4

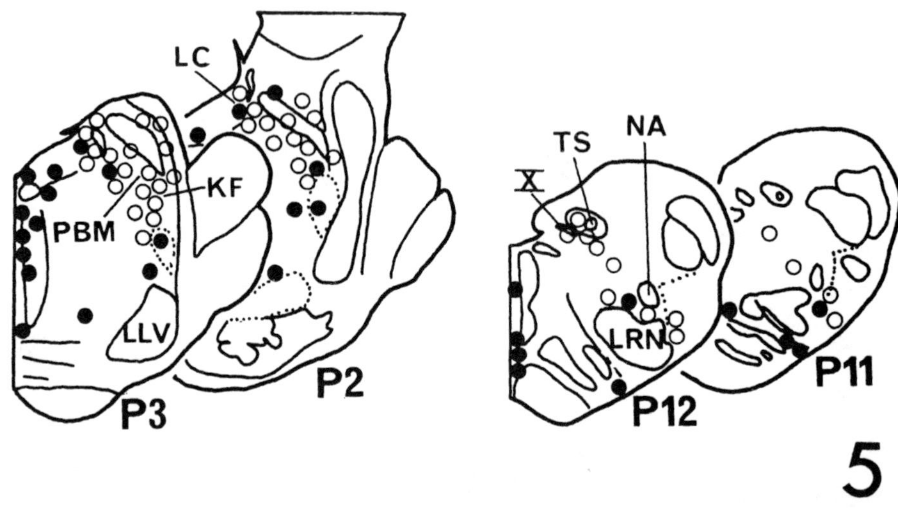
LC
KF
PBM
LLV
P3
P2
TS
NA
X
LRN
P12
P11
5

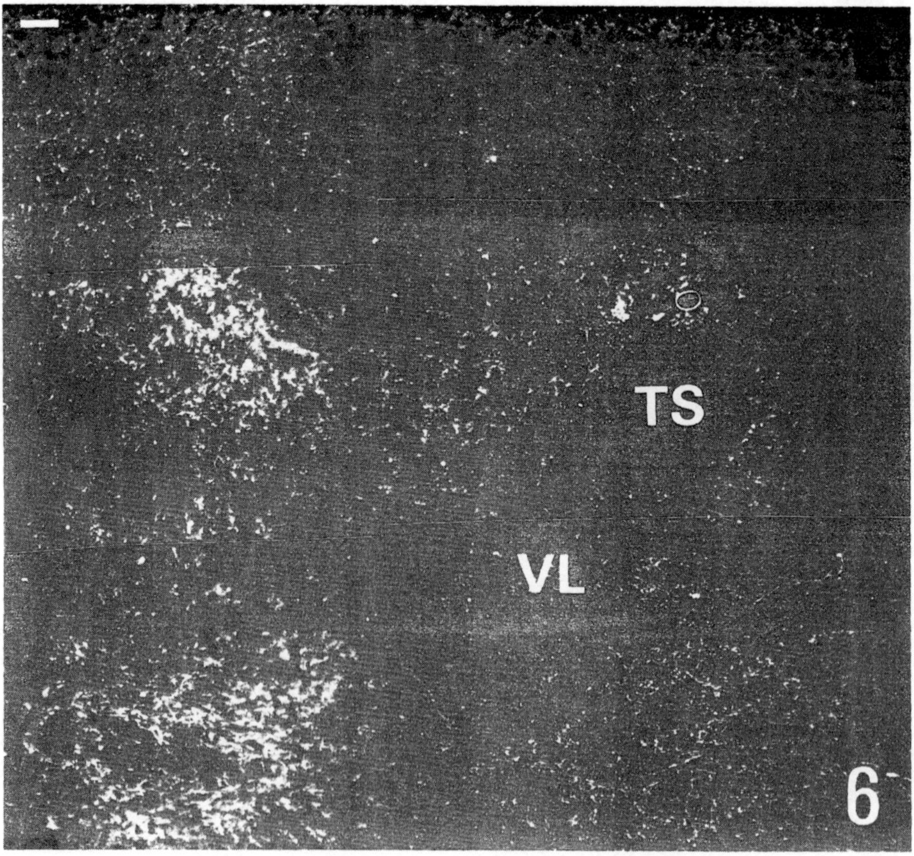

TS
VL
6

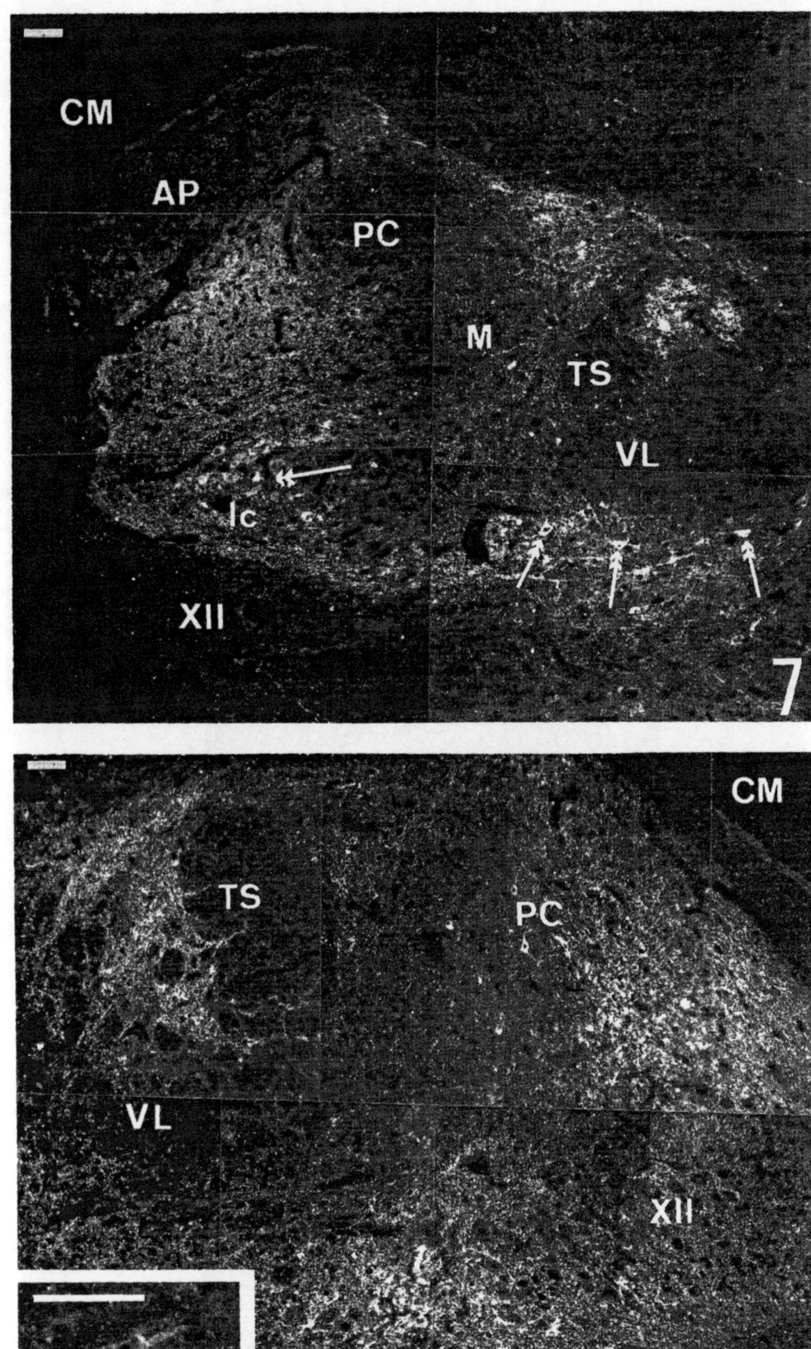
CM
AP
PC
M
TS
VL
Ic
XII
7
TS
CM
PC
VL
XII
8

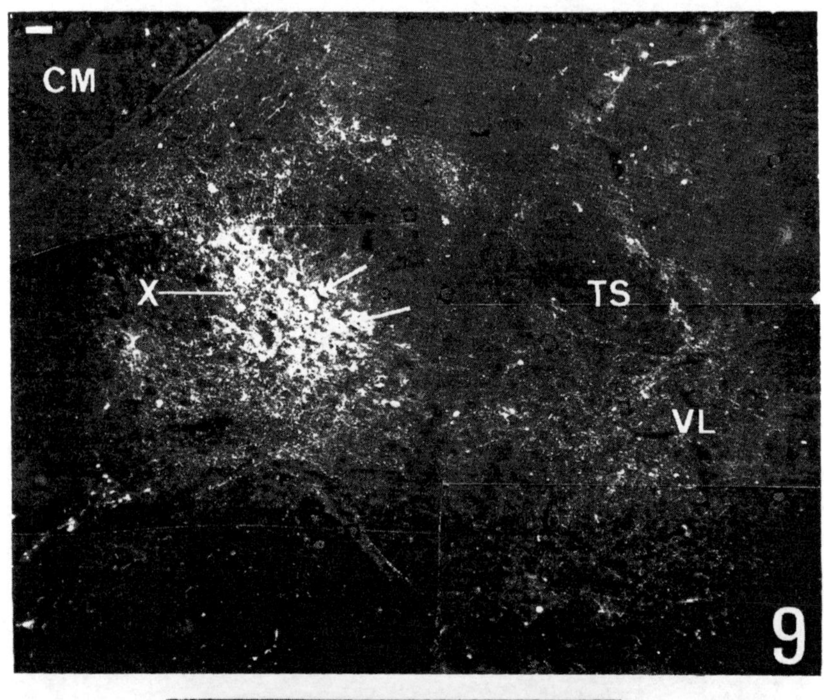

CM
X
TS
VL
9

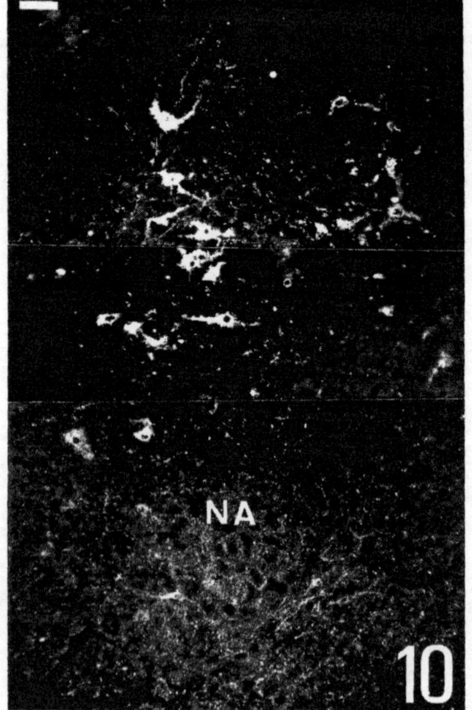

NA
10

60

Specific Probes to Investigate Involvement of μ and δ Opioid Receptors in Respiration : Autoradiographic Studies

N. SALES, J. M. ZAJAC and B. P. ROQUES

INTRODUCTION

Respiratory depression caused by opiate drugs is a well-
known and deletereous effect in which more than one sub-
type of opioid receptor seems to be involved. The biological
relevance of autoradiographic studies aimed at the localization of such sub-types of binding sites remains crucially
dependent on the use of selective radiolabelled ligands,
ideally able to interact with only one sub-type, whatever
the concentration of the others. This situation being rare-
ly satisfied, it is essential to compute the cross-reacti-
vity of a given ligand from its K_D for each binding sub-
type. Based on computer calculations, one can estimate that
specific autoradiographic determinations of binding sites
requires ligands which cross-react less than 5 % with other
sites. (^3H)-DTLET, a ligand satisfying this requirement was
designed (1) and used for visualization of δ binding sites
in cat respiratory centers ; μ binding sites were evidenced
after binding with (^3H)-DAGO (2).

We first determined the best conditions allowing the distinction of μ and δ opioid sites in brain cryostat sections.
In these conditions, both μ and δ sites were found in cat
medulla and pons. This finding is coherent with the presence of endogenous opiates in enkephalin like immuno-

Fig. Autoradiograms from adjacent sections of cat medulla (1a and 1b) and pons (2a and 2b). Cryostat 10 µm sections were incubated in the presence of 3.5 nM (^{3}H)-DAGO (1a and 2a) for µ sites or of 3.5 nM (^{3}H)-DTLET (1b and 2b) for δ sites. Non-specific binding with 1 µM levorphanol added (not shown) was low and homogeneously distributed.
(Dtn : dorsal tegmental nucleus, IOn : inferior olivary nucleus, PBn : parabrachial nuclei, STn : solitary tract nucleus, X-n : vagal nerve nucleus, XIIn : hypoglossal nerve nucleus)

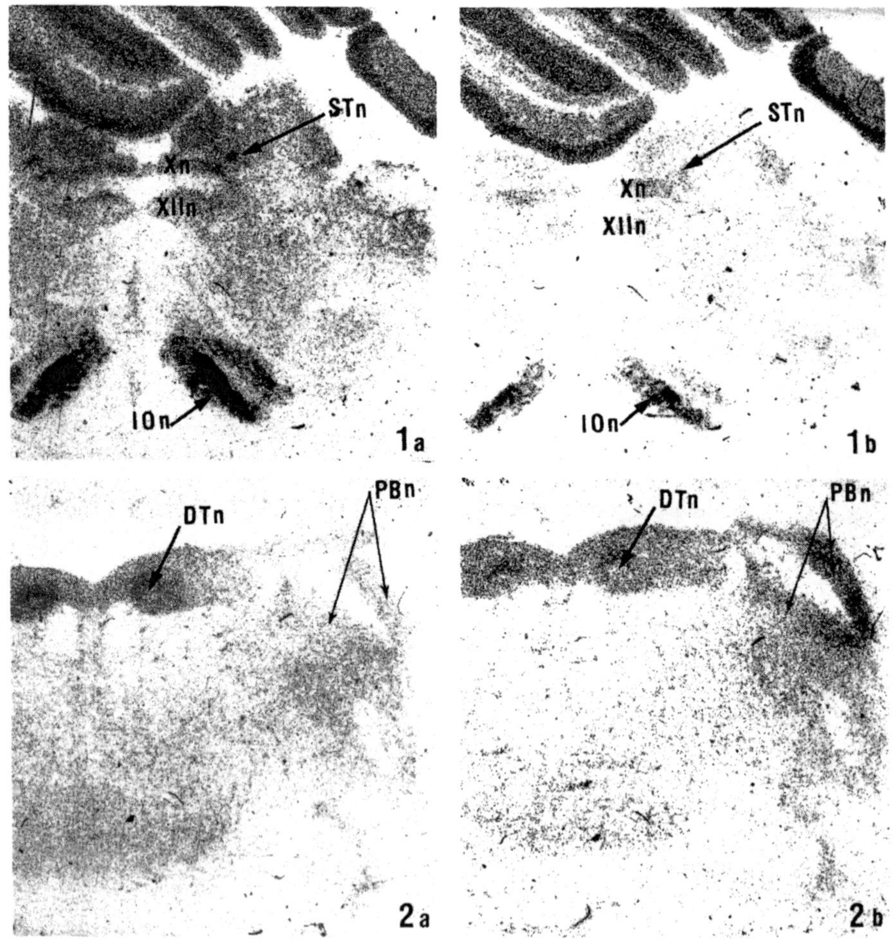

reactive fibers and cells located within the same structures
in several species including the rat (3, 4) and the cat
(see L. Léger, same issue).

The use of highly discriminant ligands points to a clearent
anatomical separation of μ sites situated in the lateral
part of the nucleus of the solitary tract and of δ sites
more abundant in the parabrachial region. (Fig.)

Whereas the cat is the most commonly used species in electrophysiological experiments, previous autoradiographic studies were usually performed in the rat with less specific
ligands (5, 6, 7, 8, 9). They are likely to predominantly
reveal μ sites which they label in the same medullary and
pontine nuclei. On the other hand, it must be noted that,
using the same specific ligands as us, QUIRION et al. (10)
observed almost exclusively μ sites both in the nucleus of
the solitary tract and in the parabrachial nuclei of the
rat. A possible explanation for this discrepancy with our
results would be a species difference. Such a difference
has been shown to exist for other structures such as the
Central grey (6).

In vivo studies concerning the effects of μ and δ agonists
on several respiratory parameters have been performed in
the rat (11, 12, 13, 14, 15) and in the cat (16). While a
general agreement seems to exist concerning the predominant
role played by the μ receptors of the nucleus of the solitary tract, the involvement of δ receptors (possibly mediating complementary through different effects) remains unclear and still controversial. The uncovery in the cat of a
pontine structure enriched with δ receptors would offer
an alternative explanation for the diversity of the effects
observed after the general administration of δ and μ types
agonists.

Références :
1. Zajac, J.M. et al. (1983). Biochem. Biophys. Res. Comm.
111, 390-397.
2. Handa, B.K. et al. (1981). Eur. J. Pharmacol., 70, 531-540.

3. Frederickson et al. (1977). _Life Sci._, <u>21</u>, 23-42.

4. Lewis, M.E., Khachaturian H., Watson S.J. (1982), _Life Sci._, <u>31</u>, 1347-1350.

5. Atweh, S.F. , Kuhar, M.J. (1977a) _Brain Res._, <u>124</u>, 53-67.

6. Atweh, S.F. , Kuhar, M.J. (1977b) _Brain Res._, <u>129</u>, 1-12.

7. Herkenham M., Pert, C. (1980) _Proc. Natl.Acad. Sci._, <u>77</u>(9), 5532-5536.

8. Goodman, R.R., Snyder, S.H., Kuhar, M.J., Scott, Y.III.W. (1980), _Proc. Natl. Acad. Sci._, <u>77</u>, (10), 6239-6243.

9. Herkenham, M., Pert, C.B. (1982), _J. Neurosci._, <u>2</u>(8), 1129-1149.

10. Quirion, R., Zajac, J.M., Morgat, J.L., Roques, B.P. (1983), _Life Sci._, <u>33</u> (suppl.I), 227-230.

11. Holaday J.W. (1982), _Peptides_, <u>3</u>, 1023-1029.

12. Hassen, A.H. et al. (1982), _Peptides_, <u>3</u>, 1031-1037.

13. Pazos,A. et al. (1983), _Eur. J. Pharmacol._, <u>87</u>, 309-314.

14. Pazos,A. et al. (1984), _Eur. J. Pharmacol._, <u>99</u>, 15-21.

15. Morin-Surun, M.P. et al. (1984), _Eur. J. Pharmacol._, <u>98</u>, 235-240.

16. Morin-Surun, M.P. et al. (1984), _Eur. J. Pharmacol._, <u>98</u>, 241-247.

61

Involvement of Cholecystokinin in the Respiratory Control

M. A. HURLÉ, M. P. MORIN-SURUN, A. S. FOUTZ, J. CHAMPAGNAT and M. DENAVIT-SAUBIÉ

INTRODUCTION

The active octapeptide terminal of cholecystokinin (CCK-8) is widely distributed in the central nervous system including several brain stem respiratory related nuclei (1). The identification of CCK-8 binding sites (4) and CCK-8 immunoreactive nerve fibers and terminals (3) in the Nucleus Tractus Solitarius (NTS) and in the pontine pneumotaxic center, suggests an involvement of this peptide in the central control of breathing. However, targets other than the respiratory centers cannot be excluded. In fact, binding sites for CCK-8 and immunoreactive nerve terminals have also been demonstrated in several central and peripheral structures engaged in the modulation of respiratory pattern (3,4,5). This is the case for the cortex, the hypothalamus and the vagus nerve.

The purpose of the present study was to determine the possible role of CCK-8 in the control of respiratory function and the mechanisms whereby the CCK-8 exerts its effects. In an attempt to localize the different targets of CCK-8, the neuropeptide was applied either locally or systemically in cats : locally, by intracerebral pressure pulses or iontophoretic applications in the NTS ; systemically, by intravenous injections (i.v.) in different types of preparations.

RESULTS

1) Effects of CCK-8 in the Nucleus Tractus Solitarius (NTS)

Iontophoretic application of CCK-8 to NTS respiratory neurons exerts
a predominantly inhibitory effect on spike discharge (2). This effect
was not produced by iontophoretic application of the inactive desulfated form of CCK-8 (CCK-8-NS). On the other hand, the effects of two
substances which have been reported to antagonize the effect of CCK-8
were tested.Proglumide was tested on 9 neurons and antagonized CCK-8
effect in 3 cases. The CCK-27-32 was tested on 7 neurons ; it
suppressed in 5 cases the effect of CCK-8. Furthermore, in some cases
iontophoretical applications of these molecules produced a depression
of firing.

In an attempt to investigate the consequences on ventilation of the
application of CCK-8 on a group of neurons located in the respiratory
area, we applied CCK-8 by microinjections. Pressure pulses were used
to eject CCK-8 into the NTS. Location of the pipette was verified by
recording field potentials evoked by vagus nerve stimulation and by
post-mortem histology. These microinjections produced depression of
the phrenic nerve discharge which could be : a reduction of burst
amplitude, a reduction of burst frequency or a reduction of both parameters of the integrated phrenic nerve discharge.

2) Effects on respiratory and cardiovascular parameters

Effects of intravenous injection (i.v.) CCK-8 (10 µg/kg) were
studied in 3 different groups of preparations : cats chronically
implanted, decerebrated cats, vagotomized and non-vagotomized. Respiration was continuously monitored by using a Fleisch pneumotachograph.
Cardiovascular parameters were analyzed by recording the blood
pressure (BP) and electrocardiogram (EEG).

In all preparations a dose of 10 µg/kg of CCK-8 induced an immediate
drop in blood pressure and cardiac rate. Blood pressure returned to
control value in the course of the first two minutes while heart rate
returned to control level within 5 to 10 minutes. During these first
 minutes, alterations of respiratory parameters occurred, associated
with the drop in blood pressure and heart rate. After those initial

rapid shifts, when cardiovascular parameters had recovered, respiratory parameters became stable. Thus the respiratory modifications occurring 5-30 minutes after the injection were analyzed.

In chronically implanted cats, i.v. injections of CCK-8 induced an important increase of respiratory frequency and a decrease of tidal volume, resulting in no significant change in minute volume. To determine the role of suprapontine regions in these respiratory effects of CCK-8, the effects of CCK-8 on chronically implanted cats were compared with its action in decerebrated cats. The global effect of CCK-8 administration in decerebrated cats was a respiratory depression. This action was characterized only by depression of tidal volume while no significant change of respiratory frequency was observed (Fig. 1).

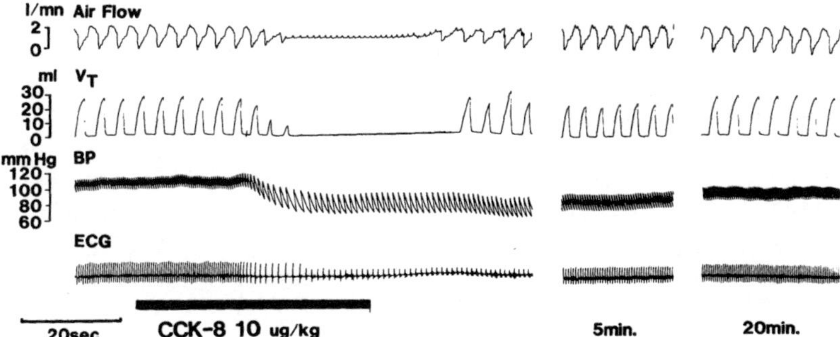

Figure 1 : Effect of intravenous CCK-8 administration (10 µg/kg) on heart rate (EEG), blood pressure (BP), tidal volume (VT) and air flow simultaneously recorded before, during and after (5 min., 20 min.) the injection.

Furthermore, the vagal component of CCK-8 administration on respiration was determined by comparison between decerebrated cats with intact vagi and those with cut vagi. In vagotomized, decerebrated cats, a stimulation of respiration was observed due only to an increase in tidal volume while the respiratory frequency did not change. In all preparations, respiratory parameters gradually returned to normal range within varying lengths of time (10 min to 30 Min).

Furthemore the i.v. injection of the non active desulfated form of CCK-8 induced no modifications of cardiovascular or respiratory parameters.

DISCUSSION

In conclusion, respiratory modifications induced by systemic injection of this neuropeptide are the resultant of several effects : an effect on the bulbar network which generates respiration and an effect on the central and peripheral control of this pattern generator. It appears from microiontophoresis and microinjection experiments that the bulbar respiratory network is depressed by CCK-8, and from systemic injection that this network is further depressed by the action of CCK-8 on vagal afferents. In addition, CCK-8 injected systemically exerts a stimulatory effect on respiration through other structures also involved in the control of breathing.

REFERENCES

1. Emson, P.C. (1979) Peptides as neurotransmitter candidates in the mammalian CNS. Prog. Neurobiol., 13, 61-116.

2. Morin, M.P., De Marchi, P., Champagnat, J., Vanderhaeghen, J.J., Rossier, J. and Denavit-Saubié, M. (1983) Inhibitory effect of cholecystokinin octapeptide on neurons in the nucleus tractus solitarius. Brain Res., 265, 333-338.

3. Vanderhaeghen, J.J., Lotstra, F., Demey, J. and Gilles, C. (1980) Immunohistochemical localization of cholecystokinin and gastrin-like peptides in the brain and hypophysis of the rat. Proc. Natl. Acad. Sci., U.S.A., 77, 1190-1194.

4. Zarbin, M.A., Innis, R.B., Wamsley, J.K., Snyder, S.H. and Kuhar, M.J. (1983) Autoradiographic localization of cholecystokinin receptors in roden brain. J. Neurosci., 3, 877-906.

5. Zarbin, M.A., Wamsley, J.K., Innis, R.B. and Kuhar, M.J. (1981) Cholecystokinin receptors : presence and axonal flow in the rat vagus nerve. Life Sci., 29, 697-705.

The authors wish to thank E. Boudinot and H. Hryn for their assistance. This work was supported by Grants from the C.N.R.S., D.R.E.T. and Fondation pour la Recherche Médicale.

62

Localization of the Respiratory Effects of TRH in the Rat Medulla

T. J. McCOWN, A. C. TOWLE, J. HEDNER, R. A. MUELLER and G. R. BREESE

INTRODUCTION

It has previously been shown that thyrotropin releasing hormone (TRH) induces a shortening of both inspiratory and expiratory (T_I and T_E, respectively) phases of the respiratory cycle [1,2]. The shortening of T_E, is later in onset [3]. Together with a significant decrease in tidal volume (V_T), these changes result in an increase in inspiratory drive (V_T/T_I) while respiratory timing is significantly decreased (T_I/T_{TOT}) (see Table 1 [3]). Since we have observed many TRH immuno-reactive nerve terminals throughout the nucleus tractus solitarius (NTS) and nucleus raphe obscurus (RO), we sought to examine the relationship between TRH localization and the respiratory response to microinjection of TRH in these areas of the medulla.

METHODS

Rats were prepared for TRH immunocytochemistry by acrolein fixation and immunoreactivity was demonstrated using antiserum raised against a TRH-BSA conjugate formed with bis-diazotized benzidine. Generally, TRH reactive cells were found in lateral and medial portions of the ventral medulla. Fibers were widely distributed throughout the medulla and appeared concentrated in the NTS and the midline raphe structures.

For the respiratory studies, rats were implanted stereotaxically with 26-guage stainless steel cannulae which terminated 1.5mm dorsal

359

to the RO (-4.8 interaural, 0.0 lateral, 0.5 vertical, according to the atlas of Paxinos and Watson, 1982 [4]). After five days of recovery, the rats were placed in a closed chamber plethysmograph, breathing 0.7% halothane in O_2. Microinjection of 100ng of TRH (0.5μl total volume over 3 minutes through a 32-guage injection cannulae) into the area of the RO caused significant decreases in the T_I and the respiratory duty cycle relative to vehicle control injections (see Fig. 1). Concommitantly, no changes were observed in the respiratory frequency, minute volume, $PaCO_2$, blood pressure or heart rate for a 15-minute post-treatment period. Conversely, when TRH (100ng) was microinjected into the central canal, lateral NTS or ambiguous nucleus, no changes were found. Similar injections into the nucleus raphe dorsalis or medial septal nuclei were also without effect on respiration.

TABLE 1

	0	1	5	15	45 min
V_T (ml/100g)	0.46±0.02	0.49±0.01	0.42±0.03	0.35±0.02[**]	0.32±0.01[***]
T_I(s)	0.24±0.01	0.19±0.02[*]	0.16±0.02[***]	0.11±0.01[***]	0.11±0.02[***]
T_E(s)	0.33±0.02	0.28±0.02	0.29±0.05	0.19±0.02[**]	0.19±0.04[**]
T_{TOT}(s)	0.57±0.03	0.46±0.04	0.45±0.06	0.30±0.03[***]	0.30±0.06[***]
V_T/T_I (ml/s·100g)	1.92±0.06	2.65±0.20	2.78±0.49	3.19±0.37[*]	3.15±0.48[*]
T_I/T_{TOT}	0.52±0.02	0.39±0.03[*]	0.39±0.05[*]	0.33±0.04[**]	0.35±0.06[*]

Effects of TRH (5μg i.c.v.) on tidal volume, respiratory time parameters, inspiratory drive and respiratory duty cycle at different time intervals after administration. Shown are means ± s.e.m. Statistics by one way analysis of variance followed by t-test.

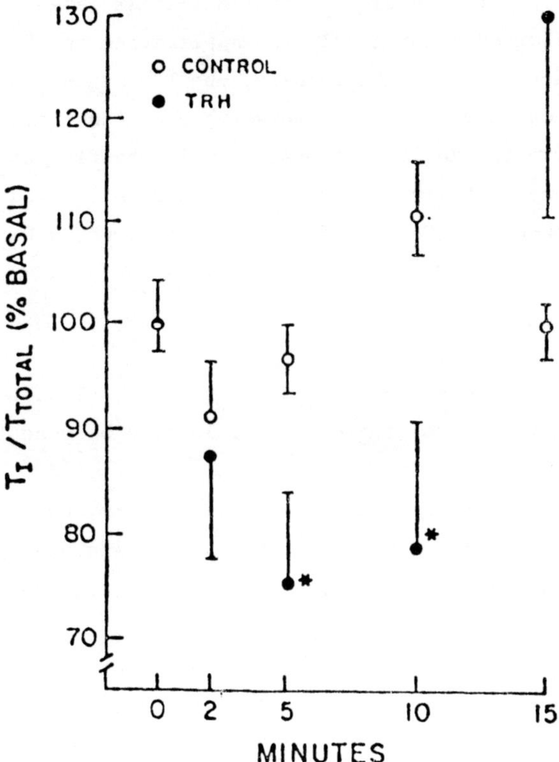

FIGURE 1   TRH or saline was administered at zero time.   Each value
represents the mean $\pm$ SEM of 7-9 observations with verified
cannula placements.   *p<.05

RESULTS

Since the respiratory response to i.c.v. TRH was increased in rats
given neonatal pargyline and 5,7-dihydroxytryptamine, [5] we examined
the response to TRH in this group as well (see Fig. 3).   Preliminary
results suggest that the decrease in T_I/T_{TOT} also is more pronounced
at lower doses (50ng) in the adult rats treated neonatally with the
neurotoxin.

These findings indicate that TRH's action on T_I appear to be
localized to an area around the RO in the medulla where TRH reactive
fibers are present, while TRH's effects on T_E most probably occur
elsewhere in the brain.   Possibly, areas rostral to the level of the
colliculi are essential to the T_E shortening response as Yamamoto
et al [6] did not find a tachypneic breathing pattern in unanesthetized

rabbit pups after brain section at the collicular level. However, the response was elicited after local TRH application to the brain stem surface of intact animals. Conversely, the TRH receptors in the mid-line of the caudal medulla, which subserve the effect on T_I, appear to be supersensitive to TRH after neonatal administration of pargyline and 5,7-dihydroxytryptamine, a treatment which selectively destroys serotonergic neurons. (Supported by HL-31424 from the NHLBI and NS-17509).

REFERENCES

1. Mueller, R. A. Breese, G. R. and Kilts, C. D. (1981). Stimula-tion of respiration by thyrotropin releasing hormone (TRH). Pharmacologist, 23, 139

2. Hedner, J., Hedner, T., Wessberg, P., Lundberg, D., and Jonason, J. (1983). Effects of TRH and TRH analogues on the central regulation of breathing in the rat. Acta Physiol. Scand., 117, 427-437

3. Hedner, J. (1984). Neuropharmacological aspects of central respiratory regulation. An experimental study in the rat. Acta Physiol. Scand. Suppl., 524, 1-109

4. Paxinos, C. and Watson, C. (1982). The Rat Brain in Stereotaxic Coordinates. (New York: Raven Press)

5. Mueller, R. A., Towle, A. C. and Breese, G. R. (1984). Supersensitivity to the respiratory stimulatory effect of TRH in 5,7-dihydroxytryptamine-treated rats. Brain Res., 298, 370-373

6. Yamamoto, Y., Lagercrantz, H. and von Euler, C. (1981). Acta Physiol. Scand., 113, 541-543

63

Dopaminergic and Serotonergic Interaction with Central Respiratory Regulation

D. LUNDBERG, R. A. MUELLER and J. HEDNER

Over the years the majority of information on the central regulation
of respiration has been provided by neurophysiological studies.
Lately, however, the area has been explored also by means of pharmacological tools developed in other fields of CNS research such as behavioral and cardiovascular pharmacology. Increased knowledge of the
neuropharmacology of respiratory control would certainly help in
finding specific respiratory stimulating drugs and CNS-active drugs
lacking respiratory depressant effects. The studies have so far been
worth-while and have recently been reviewed (1).

The aim of the present paper is to review some of the studies concerning the involvement of central serotonergic and dopaminergic neuronal systems in central respiratory control.

DOPAMINE
It has been known for a while that dopamine-containing cells in the
carotid body may be involved in peripheral chemoreceptor mechanisms
(2). Dopaminergic nerve terminals have also been demonstrated in
brain stem areas intimately related to respiratory control like the
lateral reticular nucleus, the nuclei of the solitary tract and the
nucleus locus coeruleus, for refs see (1). Thus, it would appear
plausible that changes in dopaminergic activity in those regions

could affect respiration. Bolme et al (3) noted that i.v. apomorphine, a dopamine receptor agonist, reduced respiratory rate in
chloralose-anesthetized rats. However, experiments by Lundberg et al
(4) in intact rats anesthetized with halothane have demonstrated respiratory stimulation after i.v. or i.p. apomorphine. The ventilatory
response in the latter study consisted of an increase in respiratory
rate and minute ventilation accompanied by a certain decrease in
tidal volume. Apomorphine also increased the sensitivity to CO_2 exposure. The apomorphine response was abolished by haloperidol. On the
other hand the effect was potentiated in rats pretreated at birth
with 6-hydroxy-dopamine to destroy CNS dopamine-containing neurons
(5). Although vagotomy abolished the apomorphine-induced increase in
respiratory rate, tidal volume was then increased, so minute ventilation remained elevated (6). Glossopharyngectomy did not alter the
response to apomorphine. These findings would suggest that CNS dopaminergic neurons participate in ventilatory control. Recent experiments (7) in halothane-anesthetized rats treated with i.c.v. apomorphine confirm this assumption. It was then demonstrated that apomorphine, thus administered centrally, affected respiration in a
biphasic manner; an initial shortlasting decrease in respiratory rate
and increase in tidal volume followed by a sharp and prolonged increase in frequency and a decrease in tidal volume. Systemic haloperidol, a dopamine receptor blocker, inhibited the apomorphine
response. Injected alone i.c.v. haloperidol produced a response which
resembled the initial response of i.c.v. apomorphine, i.e. a decrease
in rate and a slight increase in tidal volume. The presumed presynaptic dopaminergic agonist 3-(-3-hydroxyphenyl)-N-n-propylpyridine
(3-PPP) produced a similar response. These results would fit the view
that there is a certain tonic postsynaptic dopaminergic stimulation
probably related to respiratory timing that increases respiratory
rate.

There is an evident disparity between the respiratory effects of
dopamine agonists in different experiments beeing sometimes despres-
sant and sometimes stimulatory. One apparent reason is that the apomorphine response seems to be affected by the anesthetic used. Thus,
Hedner (8) has recently found that whereas the increase in respira-

tory rate after i.c.v. apomorphine was easily demonstrated in halo-
thane- or urethane-anesthetized rats it was not obtainable in animals
anesthetized with pentobarbital or enflurane which both clearly
tended to decrease respiratory rate _per se_. Thus, the effect of the
dopamine agonists appears to vary with the basal state of CNS-activa-
tion.

Aminophylline is used nowdays as a respiratory stimulant in certain
clinical situations such as neonatal apnoea. In both rat (9) and cat
(10) the stimulatory effect of systemic aminophylline is blunted or
abolished by haloperidol or α-methyltyrosine. Acute sections of the
vagus and glossopharyngeal nerves did not affect the aminophylline
response (9). Rats with deficient CNS dopamine-containing neurons
after neonatal treatment with desmethylimipramine and 6-hydroxydopa-
mine did not respond to aminophylline whereas neonatal destruction of
CNS serotonin-containing neurons with pargyline plus 5,7-dihydroxy-
tryptamine (5, 7,-DHT) induced a supersensitivity to aminophylline
(11). These results demonstrate that the respiratory response to
aminophylline is dependent on both CNS dopamine and serotonin neuronal systems.

SEROTONIN
It appears that brain stem nuclei whose neurons are tied to respiratory activity recieve an appreciable number of serotonin-containing
nerve terminals (for refs see 1) Hence, as with dopamine, the anatomical basis for serotonergic mechanisms in central ventilatory regu-
lations seems to be fairly well established.

Already 1972 Florez et al (12) reported that inhibition of serotonin
synthesis induced by p-chlorophenylalanin (PCPA) pretreatment increased respiratory frequency, minute volume and CO_2 sensitivity
in cats. Later the same group found that the serotonin precursors
1-tryptophan and 5-hydroxytryptophan (5-HTP) depressed respiratory
rate and minute volume (13). Subsequently Lambert et al (14) have
shown that small amounts of serotonin administered into the cerebral
ventricles of urethane-anesthetized rats decreased respiratory rate

and depth. Independently Mueller et al (15) showed in halothane-
anesthetized rats that serotonergic agonists or antagonists inter-
fered with both resting and CO_2 stimulated ventilation. Serotonin
agonists depressed tidal volume and minute volume in a dose dependent
way and induced a respiratory acidosis, whereas PCPA or methysergide,
a serotonin receptor blocker, alone produced an increase in respiration and methysergide blocked the effects of the serotonergic
agonists. It has also been found (15, 16) that i.c.v. administration
of the serotonin agonist 5-methoxy-N,N-dimethyltryptamine (5-MDMT) to
rats treated neonatally with intracisternal 5,7-DHT caused a greater
decrease in minute ventilation than when given to normal rats. The
depressant effects of serotonin agonists given systemically were
observed even after glossopharyngectomy or vagotomy (16). Olsson et
al (17) have recently reported that serotonergic mechanisms are involved in respiratory regulation also in awake rats. These findings
taken together strongly favours the view that CNS serotonin receptors
are involved in the control of basal and CO_2 stimulated respiration.

A peculiar finding in the rat after higher doses of serotonin
agonists is that as the tidal volume progressivly declines and the
CO_2 response is obtunded, the respiratory pressure tracing re-
sembles the apneustic pattern seen in aminals treated by vagotomy.
Furthermore, this serotonin effect is unchanged by denervation of the
vagus (16). This finding together with the fact that methysergide is
supposed to inhibit only excitatory responses (18) indicates that a
serotonin mediated excitatory response would somehow decrease minute
ventilation, resembling that of afferent vagal stretch or J-receptor
impulses whose effects are to decrease minute ventilation.

Occasionally stimulation of respiratory activity after administration
of serotonin agonists has also been reported (19, 20). These results
could be due to that serotonin agonists exert different effects in
certain regions of the brain. Thus depending upon the region of the
brain where the drug effect is more profound an increase or a decrease in respiration can be produced.

Several nuclei in the rat medulla oblongata have recently been found
to contain both serotonin and substance P or serotonin and thyro-
tropin-releasing hormone (TRH), for refs see (1). The fact that substance P and TRH, both of which are respiratory stimulants (1), co-
excist with serotonin in neurons of nuclei associated with respiratory regulation indicates the presence of new and complex neuropharmacological interactions and mechanisms in the already compli-
cated field of ventilatory control.

In summary both dopaminergic and serotonergic neuronal systems
located in the brain stem area appear to be included in respiratory
control. There seems to be a reciprocal interaction in that dopaminergic activation is mainly stimulatory and serotonergic activity
inhibitory.

REFERENCES

1. Mueller, R.A., Lundberg, D.B.A., Breese, G.R., Hedner, J.,
Hedner, T, and Jonason, J. (1982). The neuropharmacology of respiratory control. Pharm. Rev., 34, 255-285
2. Sampson, S.R. (1972). Mechanism of efferent inhibition of carotid
chemoreceptors in the cat. Brain Res., 45, 266-270
3. Bolme, P., Fuxe, K., Hökfelt, T., and Goldstein, M. (1977).
Studies on the role of dopamine in cardiovascular and respiratory
control: Central versus peripheral mechanisms. Adv. Biochem. Phycho-
pharmacol., 16, 281-290
4. Lundberg, D., Breese, G.R., and Mueller, R.A. (1979). Dopaminergic interaction with the respiratory control system in the rat.
Eur. J. Pharmacol., 54, 153-159
5. Lundberg, D.B., Mueller, R.A., and Breese, G.R. (1979). Dopaminergic and serotonergic modulation of respiratory drive. E. Usdin,
I.J. Kopin, and J. Barchas (Eds). In: Catecholamines: Basic and
clinical frontiers. (Pergamon Press, New York), pp 526-528
6. Lundberg, D., Mueller, R.A., and Breese, G.R. (1982). Effects of
vagotomy and glossopharyngectomy on respiratory response to dopamine-
agonists. Acta Phys. Scand., 114, 81-89

7. Hedner, J., Hedner, T., Jonason, J., and Lundberg, D. (1982). Evidence for a dopamine interaction with the central respiratory control system in the rat. Eur. J. Pharmacol., 81, 603-615

8. Hedner, J. (1983). Neuropharmacological aspects of central respiratory regulation. An experimental study in the rat. Acta Physiol. Scand., suppl 524

9. Lundberg, D.B., Breese, G.R., and Mueller, R.A. (1981). Aminophylline may stimulate respiration in rats by activation of dopaminergic receptors. J. Pharmacol. Exp. Ther., 217, 215-211

10. Eldrigde, F.L., Millhorn, D.E., and Waldrop, T.G. (1981). Mechanism of hyperpnoe induced by aminophylline. Physiologist, 24, 102

11. Mueller, R.A., Lundberg, D.B., and Breese, G.R. (1918). Alteration of aminophylline-induced respiratory stimulation by perturbation of biogenic amine systems. J. Pharmacol. Exp. Ther., 218, 593-599

12. Florez, J., Delgado, G., and Armijo, J.A. (1972). Adrenergic and serotonergic mechanisms in morphine-induced respiratory depression. Psychopharmacologica, 24, 258-274

13. Armijo, J.A. and Florez, J. (1974). The influence of increased brain 5-hydroxytryptamine upon respiratory activity of cats. Neuropharmacology, 13, 977-986

14. Lambert, G.A., Friedman, E., Buchweitz, E, and Gershon, S. (1978). Involvement of 5-hydroxytryptamine in the central control of respiration, blood pressure and heart rate in the anesthetized rat. Neuropharmacology, 17, 807-813

15. Mueller, R.A., Lundberg, D., and Breese, G.R. (1980). Evidence that respiratory depression by serotonin agonists may be exerted in the central nervous system. Pharmacol. Biochem. Behav., 13, 247-255

16. Lundberg, D., Mueller, R.A., and Breese, G.R. (1980). An evaluation of the mechanism by which serotonergic activation depresses respiration. J. Pharmacol. Exp. Ther., 212, 397-404

17. Olson, E.B.Jr., Dempsey, A.J., and McCrimmon, D.R. (1979). Serotonin and the control of ventilation is awake rats. J. Clin. Invest., 64, 689-693

18. Aghajanian, G.K. and Wang, R.Y. (1978). Physiology and pharmacology of central serotonergic neurons. M.A. Lipton, A. DiMascio, and K.F. Killam (Eds) In: Psychopharmacology. A generation of Progress. (Raven Press, New York), pp 171-183

19. Armijo, J.A., Mediavilla, A., and Florez, J. (1979). Inhibition of the activity of the respiratory and vasomotor centers by centrally administered 5-hydroxytryptamine in cats. <u>Rev. Esp. Fisiol.</u>, <u>36</u>, 219-227

20. Millhorn, D.E., Eldridge, F.L., and Waldrop, T.G. (1980). Prolonged stimulation of respiration by a new central neural mechanism. <u>Respir. Physiol.</u>, <u>41</u>, 87-103

64

Respiratory – Vagal Interactions in the Nucleus Ambiguus of the Cat

D. JORDAN, M. P. GILBEY, D. W. RICHTER, K. M. SPYER
and L. M. WOOD

There is now substantial evidence in the literature for interactions between respiratory and cardiovascular control systems. Anrep and his colleagues in 1936 [1] showed that the respiratory related changes in heart rate, known as sinus arrhythmia, were dependent both on centrally generated respiratory drive and sensory input related to lung inflation. It has since been suggested [2] that this phenomenon, which is primarily of vagal origin, is a consequence of a respiratory variability in the effectiveness of the baroreceptor input to excite cardiac vagal motoneurones (CVMs). Indeed, there is a marked respiratory related component in the discharge of CVMs of many species [3,4,5,6,7] and the powerful excitatory input from arterial baroreceptors appears to excite cardiac vagal efferent fibres only if timed to coincide with expiration [5].

The site of interaction of respiration and cardiac control has been a matter of speculation and the subject of investigations in our laboratory over the past few years. We have demonstrated that the baroreceptor afferent terminals in the nucleus tractus solitarius are not themselves affected presynaptically by central respiratory activity, though some slowly adapting lung stretch afferents

do show such modulation [8,9]. Further, by studying
subliminal inputs to CVMs, McAllen and Spyer [10] suggested
that the baroreceptor input is able, at least in part, to
reach these neurones in all phases of the respiratory
cycle. Taken together with the recordings from cardiac
vagal efferent fibres [5] these observations would suggest
that at least part of the respiratory modulation of the
baroreceptor-cardiac reflex activity is acting post-
synaptically at the level of the CVMs themselves.

The present studies have concentrated on the neural
mechanisms acting at the level of the CVM which might
account for these observations. Recordings have been made
from CVMs, which in the cat are located in the nucleus
ambiguus. These neurones were identified on the basis of
properties previously described in detail [6,10,11]. In
summary, they were shown to project to the heart by their
antidromic response to electrical stimulation of the
cardiac branches of the vagus. Measurements of antidromic
response latency and conduction distance confirmed that
they all had axons in the range of B fibres. In open chest,
artificially ventilated cats CVMs have either a low tonic
firing rate or are silent. However, when tonically firing,
or firing in response to ionophoretic application of D,L-
homocysteic acid (DLH), all were shown to receive an
excitatory input from arterial baroreceptors, as demonstrated
by R-wave triggered histograms of ongoing activity.
Similarly, phrenic-triggered histograms illustrated the
profound respiratory modulation of their discharge.

NEUROPHARMACOLOGICAL EXPERIMENTS
These experiments were performed to gain information
regarding the possible neurotransmitters acting on CVMs.
So far we have concentrated on two possible neurotransmitters; first, we have studied the possibility that
cholinergic mechanisms may act in the nucleus ambiguus
since there is evidence in the literature that subclinical
doses of atropine sulphate can have a central

parasympathomimetic action [12]. Secondly, microinjections
of GABA and its antagonists into nucleus ambiguus have been
shown to have profound cardiac actions [13] and so we have
studied the possibility that GABA may act directly on CVMs.

The putative neurotransmitters and their antagonists
were applied by microionophoresis from multibarrelled glass
pipettes onto extracellularly recorded CVMs. All cells
fired in response to low doses (1-40 nA) of the excitant
amino acid DLH. Application of GABA (0-10 nA) reduced or,
in most instances, totally silenced such neurones (8/9)
(Fig. 1). The simultaneous application of the GABA
antagonist bicuculline methiodide (5-100 nA) abolished the
effect of GABA in all cases. When applied alone
bicuculline increased the firing rate of these neurones
which may indicate a tonic action for GABA at this site.

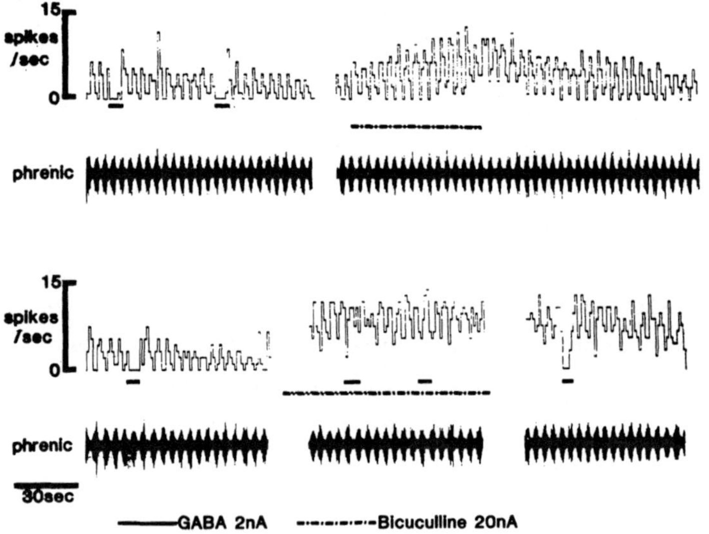

FIGURE 1. Effect of GABA and bicuculline on the discharge
           rate of a CVM. Each panel shows a ratemeter record
           of CVM activity and, below, phrenic nerve activity

When acetylcholine was applied (20-100 nA) it too
reduced the DLH-evoked discharge of 12/15 CVMs (Fig. 2).
In ten of these neurones simultaneous application of
atropine sulphate (4-50 nA) antagonised the effect of

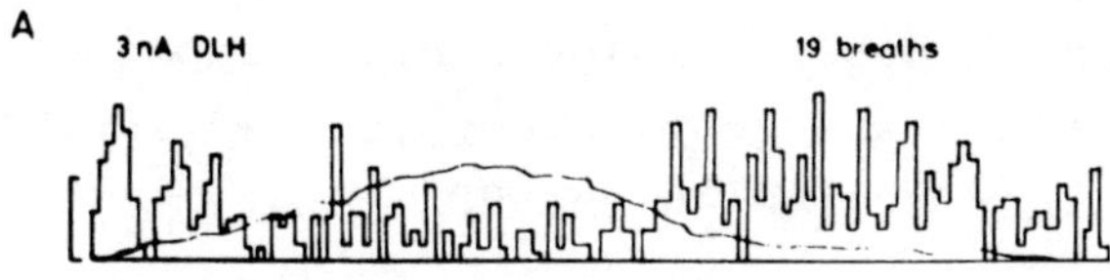

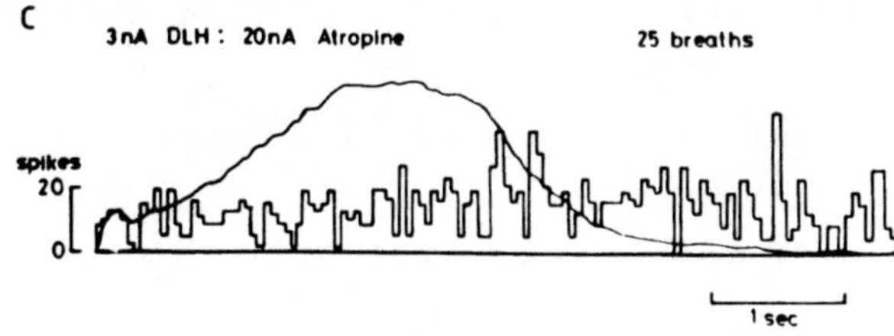

FIGURE 2. Effect of DL-homocysteic acid (DLH), acetylcholine
(Ach) and atropine on the discharge of a CVM.
Each panel shows the averaged integrated phrenic
nerve discharge superimposed upon a histogram of
CVM discharge triggered from the rise in phrenic
nerve activity. (Reproduced, with permission,
from [14]).

acetylcholine. Our results suggest that this inhibitory
cholinergic mechanism is tonically active since atropine
applied alone enhanced neuronal activity. As can be seen
from Fig. 2 this enhancement was particularly marked during
the phase of inspiration. This effect is not unlike that
which we previously noted when applying atropine to
expiratory neurones located in this same region [15].
These too were inhibited by application of acetylcholine.
The present observations would suggest that CVMs receive
a phasic inhibition during inspiration and that this input
involves a cholinergic mechanism.

We have previously shown that CVMs also receive an
inhibitory input from the hypothalamic defence area (HDA)

in the cat [16], a site which evokes tachycardia in
association with other cardiovascular responses. Atropine,
whilst abolishing the respiratory related inhibitory input
to CVMs was ineffective on the HDA input.

INTRACELLULAR EXPERIMENTS

Our ionophoretic experiments would suggest that the
respiratory modulation of CVMs is imposed by an inspiratory-
related inhibitory input superimposed upon a tonic
excitatory drive to these neurones. To study more directly
whether the pattern of activity of CVMs was due to synaptic
inputs related to central respiratory drive, intracellular
recordings have been made from identified CVMs in
anaesthetised, paralysed and artificially ventilated cats.
Phrenic nerve activity was used as a monitor of central
respiratory activity.

Inspection of the membrane potential of the neurones
showed that they received depolarising synaptic potentials
110-120 ms after the onset of each femoral blood pressure
wave which is consistent with a baroreceptor input [10].

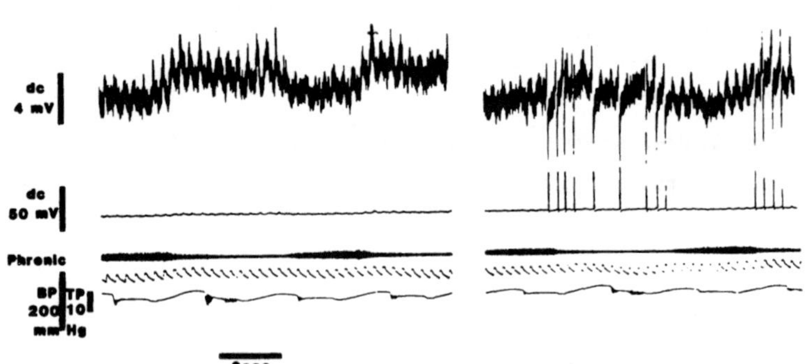

FIGURE 3. Intracellular record from a CVM showing
respiratory related changes in membrane potential.
Five minutes after penetration the cell stopped
firing action potentials (1st panel) but could be
made to fire by injecting low depolarising
currents (2nd panel). In each panel traces from
above: high and low gain records of membrane
potential, phrenic nerve activity, arterial
blood pressure and tracheal pressure.

These pulse-rhythmic waves of e.p.s.p.s were superimposed
on slower fluctuations in the membrane potential occurring
in phase with the central respiratory cycle (Fig. 3).
During inspiration maximal levels of membrane potential of
-50 — -60 mV were reached after which the cells depolarised
and discharged spikes during the phase of post-inspiration.
This spike discharge decreased during a slow repolarisation
in expiration. After a period of impalement, or after
injection of Cl$^-$, the cells stopped firing action
potentials. After prolonged passage of negative (Cl$^-$)
current (1 nA for 3-5 mins) the membrane hyperpolarisation
seen during inspiration reversed to a wave of
depolarisation which terminated abruptly at the onset of
post-inspiratory activity (Fig. 4). This would indicate
that the hyperpolarisation was due to the arrival of a wave
of chloride-dependent i.p.s.p.s during the phase of phrenic
activity. This is compatible with the changes in neurone
input resistance measured by injecting small pulses of
current. Input resistance was highest during post-
inspiration and fell to its minimum during inspiration.

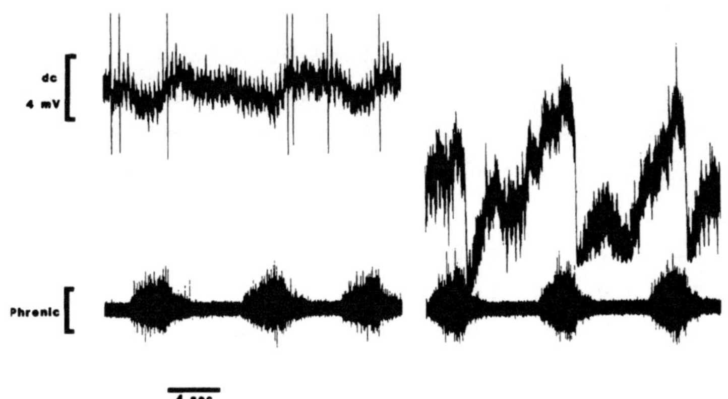

FIGURE 4. Intracellular record from a CVM showing
           respiratory related changes in membrane potential
           (1st panel). After passage of negative hyper-
           polarising current (1 nA for 3 min) hyper-
           polarising waves reversed (2nd panel). In each
           panel traces from above: high gain dc recording
           of membrane potential and phrenic nerve activity.

The present study has attempted to identify the synaptic mechanisms responsible for the discharge patterns seen in CVMs. We have identified two distinct inhibitory inputs to these neurones, one, related to HDA activity and a second related to inspiratory drive. This latter input is mediated by chloride dependent i.p.s.p.s and is antagonised by atropine sulphate. In terms of excitatory input, the neurones receive baroreceptor mediated e.p.s.p.s throughout the respiratory cycle and are at their most excitable during the phase of post-inspiration.

Our data with atropine confirms earlier suggestions of a central action [12], and a recent report has shown a similar effect on sinus arrhythmia in man [17]. The cholinergic input to CVMs appears to be specific for the respiratory-related modulation of their activity since the inhibitory input from the HDA was unaffected by application of atropine.

The powerful inspiratory-related inhibition of CVMs we have seen in this study may well be of significant physiological importance since it sculptures the membrane potential of CVMs and could provide the mechanisms required for respiratory modulation of vagally mediated responses. In addition, it would be predicted that by this mechanism the effectiveness of any reflex input may well be modified by the action of any other input which simultaneously alters respiratory drive.

1. Anrep, G.V., Pascual, W. and Rossler, R. (1936).
    Respiratory variations of the heart rate. I. The
    reflex mechanism of sinus arrhythmia. Proc. Roy. Soc.
    B, 119, 191-217.
2. Koepchen, H.P., Lux, H.D. and Wagner, P.H. (1961).
    Untersuchungen uber Zeitbedarf und Zentrale
    Verarbeitung des Pressoreceptorischen Herzreflexes.
    Pflügers Arch., 273, 413-430.

3. Jewett, D.L. (1964). Activity of single efferent
    fibres in the cervical vagus of the dog with special
    reference to possible cardioinhibitory fibres.
    J. Physiol., 175, 321-357.

4. Kunze, D.L. (1972). Reflex discharge patterns of
    cardiac vagal efferent fibres. J. Physiol., 222, 1-15.

5. Davidson, N.S., Goldner, S. and McCloskey, D.I. (1976).
    Respiratory modulation of baroreceptor and
    chemoreceptor reflexes affecting heart rate and
    cardiac vagal efferent nerve activity. J. Physiol.,
    259, 523-530.

6. McAllen, R.M. and Spyer, K.M. (1978). Two types of
    vagal preganglionic motoneurones projecting to the
    heart and lungs. J. Physiol., 282, 353-364.

7. Jordan, D., Khalid, M.E.M., Schneiderman, N. and
    Spyer, K.M. (1982). The location and properties of
    preganglionic vagal cardiomotor neurones in the rabbit.
    Pflügers Arch., 395, 244-250.

8. Jordan, D. and Spyer, K.M. (1979). Studies on the
    excitability of sinus nerve afferent terminals.
    J. Physiol., 297, 123-134.

9. Ballantyne, D., Donoghue, S., Jordan, D., Meesman, M.,
    Richter, D.W. and Spyer, K.M. (1981). Respiratory
    influences on vagal afferent endings in the medulla:
    a study on baroreceptor and lung stretch neurones.
    J. Physiol., 312, 28-29P.

10. McAllen, R.M. and Spyer, K.M. (1978). The baroreceptor
    input to cardiac vagal motoneurones. J. Physiol.,
    282, 365-374.

11. McAllen, R.M. and Spyer, K.M. (1976). The location of
    cardiac vagal preganglionic motoneurones in the
    medulla of the cat. J. Physiol., 258, 187-204.

12. Katona, P.G., Lipson, D. and Dauchot, P.J. (1977).
    Opposing central and peripheral effects of atropine
    on parasympathetic cardiac control. Am. J. Physiol.,
    232, H146-H151.

13. DiMicco, J.A., Gale, K., Hamilton, B. and Gillis, R.A. (1979). GABA receptor control of parasympathetic outflow to the heart: Characterization and brainstem localization. Science, <u>204</u>, 1106-1109.

14. Gilbey, M.P., Jordan, D., Richter, D.W. and Spyer, K.M. (1984). Synaptic mechanisms involved in the inspiratory modulation of vagal cardio-inhibitory neurones in the cat. J. Physiol., in press.

15. Jordan, D. and Spyer, K.M. (1981). Effects of acetylcholine on respiratory neurones in the nucleus ambiguus-retroambigualis complex of the cat. J. Physiol., <u>320</u>, 103-111.

16. Jordan, D., Khalid, M.E.M., Schneiderman, N. and Spyer, K.M. (1980). The inhibitory control of vagal cardiomotor neurones. J. Physiol., <u>301</u>, 54-55P.

17. Raczkowska, M., Eckberg, D.L. and Ebert, T.J. (1983). Muscarinic cholinergic receptors modulate vagal cardiac responses in man. J. Autonomic Nerv. Syst., <u>7</u>, 271-278.

This work was supported by grants from the British Heart Foundation, the Medical Research Council and Deutsche Forschungsgemeinschaft.

65

Muscarinic and Nicotinic Effects of ACh on the Activity of Respiratory Related Neurons in the Rabbit

G. BÖHMER, K. SCHMID, P. SCHMIDT and J. STEHLE

INTRODUCTION

Medullary respiratory related units (RRU) can be affected by iontophoretically applied ACh (1, 2, 3). Results of these studies were inconsistent, concerning the excitatory or inhibitory action of ACh. This
may be due to the fact that in the cat excitatory effects of ACh on
RRU are sensitive to anaesthesia (4). In this study both excitation
and inhibition of RRU were considered as muscarinic effects of ACh.
However, in the rat nicotinic excitation of RRU (5) and the presence
of both muscarinic and nicotinic receptors in the brain-stem was
demonstrated (6, 7).

Our study aimed to examine the action of ACh on the activity of RRU
in the rabbit. By application of receptor antagonists as well as of
receptor agonists the contribution of muscarinic and nicotinic recep-
tor-excitation to this action of ACh was examined.

METHODS

Experiments were performed on spontaneously breathing or immobilized
and artificially ventilated urethane-anaesthetized rabbits. In the
latter animals endexpiratory CO_2 was kept at 4-5 %. Body temperature
was kept at 37 $^{\circ}$C. Neuronal activity was recorded extracellularly.
Four-barreled glass microelectrodes (Clark) were used. Activity of RRU
was classified according to its phase-relation to phrenic nerve activity. Inspiratory neurons (I) and expiratory neurons (E) were discri-
minated from phasespanning EI and IE neurons.

The effects of ACh (1 M, pH 4-5), bethanechol (0.5 M, pH 5.5), DMPP

(0.1 M, pH 7.5–8.5), atropine (0.1 M, pH 5) and hexamethonium (0.1 M, pH 4) on neuronal activity of RRU were examined. Drugs were ejected by cationic currents of 10–200 nA. Neuronal activity was analyzed with digital methods. The frequency of neuronal action potantials was recorded continuously. During each test spike-density (f_i) and burst-duration (t_{bd}) were calculated periodically.

RESULTS

Spike-density of 70 % of all RRU examined was affected by ACh (Table 1 A). In E neurons f_i was increased or decreased in about the same incidence of cases, while in I neurons reduction of f_i preponderated over excitation. However, in most phase-spanning RRU affected, f_i was increased. In 67 % of the RRU examined t_{bd} was altered by ACh (Table 1 B). In most of these neurons t_{bd} was reduced. Reduction of t_{bd} was observed in all phase-spanning RRU affected. Increase of t_{bd} was

Table 1. Frequency of excitatory and inhibitory effects of ACh on f_i (A) and t_{bd} (B) of medullary RRU (+ = increase, – = decrease, 0 = unaffected).

A: f_i	+	–	0		B: t_{bd}	+	–	0
EI	6	2	2		EI	0	7	3
I	14	25	14		I	6	26	21
IE	19	4	5		IE	0	21	7
E	29	27	33		E	14	47	28
RRU	68	58	54		RRU	20	101	59

induced in 11 % of all RRU, only (I and E neurons), most of which were recorded from Ncl. XII. In a high incidence of cases t_{bd} of RRU recorded from NA was not altered. In more than 50 % of all RRU examined, ACh affected only f_i (29 %) or only t_{bd} (26 %). When f_i and t_{bd} were affected simultaneously, decrease of t_{bd} in most cases was accompanied by increase (24 %) or decrease of f_i (15 %).

In the studies using receptor-specific antagonists or agonists, effects of ACh which could be antagonized or mimicked, respectively, were classified as 'muscarinic' or 'nicotinic' effects (see Fig. 1).

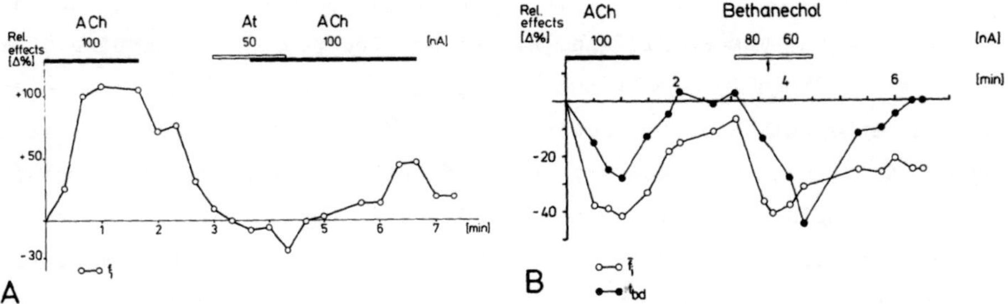

FIGURE 1 A: Antagonistic action of atropine on the effects of ACh on f_i of a tonic I neuron (muscarinic effect). B: Action of ACh and the muscarinic agonist bethanechol on f_i and t_{bd} of an I neuron (muscarinic effects).

Only excitatory effects of ACh on f_i were antagonized by both atropine and hexamethonium (17 RRU=55 %). The inibitory action of ACh on t_{bd} was sensitive to atropine and to hexamethonium (13 RRU=54 %). The few excitatory effects of ACh on t_{bd} were antagonized by atopine. Muscarinic receptors, sensitive to bethanechol, mainly mediated inhibitory effects of ACh on f_i and on t_{bd}. Increase of t_{bd} was mediated by muscarinic receptors. Nicotinic receptors, sensitive to DMPP, mediated increase or decrease of f_i, but only decrease of t_{bd}. Nicotinic receptors modulating t_{bd} were demonstrated in E neurons, only.

DISCUSSION

In the rabbit both excitatory and inhibitory effects of ACh on RRU were observed, which is in accordance with effects of ACh on RRU of decerebrated cats (4). There are, however, some differences between phase-types of RRU: in I neurons reduction of f_i preponderated over increase of f_i, while in E neurons f_i was increased or decreased in the same incidence of cases. In most RRU examined t_{bd} was reduced by ACh. The result that in more than 50 % of all RRU examined the effects on t_{bd} were independent from alterations of f_i or vice versa, suggests that both parameters can be affected by independent cholinergic mechanisms. We speculate that at least part of the effects on t_{bd} may be mediated by presynaptic action on phasic inhibitory afferents of RRU. A presynaptic influence of superior laryngeal nerve stimulation on lung stretch receptor afferents was demonstrated in the cat (8).

The effects of muscarinic and nicotinic antagonists on the action of

ACh suggest that the effects of ACh on t_{bd} can be mediated by excitation of both muscarinic and nicotinic receptors. Concerning the effects of ACh on f_i, only excitatory effects on RRU could be blocked by the antagonists used. However, the effects of receptor-specific agonists demonstrated the existence of both muscarinic and nicotinic receptors mediating decrease of f_i. Results suggest that f_i of inspiratory neurons predominantly is affected by excitation of muscarinic receptors, while f_i of expiratory neurons more frequently is affected by excitation of nicotinic receptors.

REFERENCES

1. Kirsten, E.B., Satayavivad, J., St. John, W.M. and Wang, S.C. (1978). Alteration of medullary respiratory discharge by iontophoretic application of putative neurotransmitters. Br. J. Pharmacol., 63, 275–281

2. Champagnat, J., Denavit-Saubie, M., Henry, J.L. and Leviel, V. (1979). Catecholaminergic depressant effects on bulbar respiratory mechanisms. Brain Res., 160, 57–68

3. Jordan, D. and Spyer, K.M. (1981). Effects of acetylcholine on respiratory neurons in the nucleus ambiguus-retroambigualis complex of the cat. J. Physiol. (London), 320, 103–111

4. Morin-Surun, M.P., Champagnat, J., Denavit-Saubie, M. and Moyanova, S. (1984). The effects of acetylcholine on bulbar respiratory related neurones. Consequences of anaesthesia by pentobarbital. Naunyn-Schmiedeberg's Arch. Pharmacol., 325, 205–208

5. Bradley, P.B. and Lucy, A.P. (1983). Cholinoceptive properties of respiratory neurones in the rat medulla. Neuropharmacol., 22, 853–858

6. Wamsley, J.K., Lewis, M.S., Young, W.S. and Kuhar, M.J. (1981). Autoradiographic localization of muscarinic cholinergic receptors in rat brainstem. J. Neurosci., 1, 176–191

7. Hunt, S. and Schmidt, J. (1978). Some observations on the binding patterns of alpha-bungarotoxin in the central nervous system of the rat. Brain Res., 157, 213–232

8. Ballantyne, D., Meesmann, M. and Richter, D.W. (1981). Presynaptic depolarization of lung stretch receptor afferents. Pflügers Arch., Suppl. 389, R 25

66

Electrophysiology of Intraocular Grafts from Respiratory Related Parts of the Medulla in the Rat

M. RUNOLD, H. LAGERCRANTZ, L. OLSON and C. V. EULER

The basic respiratory rhythm-generating neurons and the neurotrans-
mittors involved have not yet been identified. Two major groups, the
ventral and the dorsal respiratory groups, however, have each been
proposed as the "pace-setter". Different approaches have been applied
to study this question. In intact animal preparations it is difficult
to differentiate rhythmic activity of local origin in groups of neuron
from respiration-related activity produced elsewhere in the brain
stem. Brain-slice preparation from the medulla (1) might give rise to
artificial patterns of activity.

We have tried another approach. Parts of the medulla from fetal
rats of different gestational age were transplanted to the anterior
chamber of the eye in adult rats, through a small incision in the
cornea. The wound is healed rapidly and the vision is generally preserved. During the following weeks the graft became innervated and
vascularised from the iris and maturation of the graft occured. The
growth of the graft could be measured with a stereo-microscope. The
graft can be selectively reinnervated by differential denervation (2).
For electrophysiological experiments, the host animals are anaesthetized and the cornea partially removed. A perspex cylinder shaped to
fit the eye was put around the eye and sealed with agar. This formed
the recording chamber, which could be superfused with mock CSF.

Recordings of neural activity in the graft is then performed with tungsten electrodes.

Different manipulations such as changing pH or administration of transmittor substances in the superfusing fluid or electrical stimulation (direct in the graft or indirectly via different nerves innervating the iris) can easily be performed with this experimental setup. Another advantage with this approach is the possibility to study transplants at different developmental stages.

References:

1. Champagnat, J., Denavit-Saubie, M. & Siggins, G.R. (1983). Rhythmic neuronal activities in the nucleus of the tractus solitarius isolated in vitro. Brain Res. 280, 155-159.

2. Olson, L. Freedman, R. Seiger, A. & Hoffer, B. (1977). Electrophysiology and cytology of hippocampal formation transplants in the anterior chamber of the eye. 1. Intrinsic organization. Brain Res. 119, 87-106.

67

A Neuromodulator (Adenosine) involved in the Respiratory Hypoxic Response During Development

H. LAGERCRANTZ and M. RUNOLD

During hypoxia adenosine is released (1). It has been found to be
formed in the brain during increased neural activity (2) and inhibits
the firing rate of neurons in many parts of the CNS. The respiratory
effect was already reported in 1929 by Drury and Scent-Györgyi(3). A
similar effect has recently been found with an adenosine analogue
(e.g. 4,5,6). Most of the recent studies have been performed on
anaesthetized animals why we aimed to investigate the respiratory
effects of an adenosine analogue in unanaesthetized rabbit pups and
kitten.

Respiration was recorded by the barometric method (7). The animals
were placed in a plexiglas chamber; this and an identical reference
chamber were connected to a highly sensitive differential transducer.
The animal chamber was then flushed with humidified air. The in- and
outlet tubes were clamped during recording. All parameters were calculated as percent of control values. Immediately after the clamping
period the chamber was connected to a CO_2-analyzer. The CO_2-concen-
tration was never allowed to exceed 0.1 %. The temperature of the
chambers were kept constant at 28^0 within 0.2^0 by a thermoradiator.
A thin catheter was placed under local anaesthesia intraperitoneally.

The effect of a stable adenosine analogue L-phenyl-isopropyl-adeno-
sine (L-PIA) was studied in rabbit pups and kittens (1-8 days).

385

Ventilatory minute volume decreased maximally after about 15 minutes.
Both frequency and tidal volume decreased, though in the majority of
the animals the frequency drop was most prominent (Fig. 1).

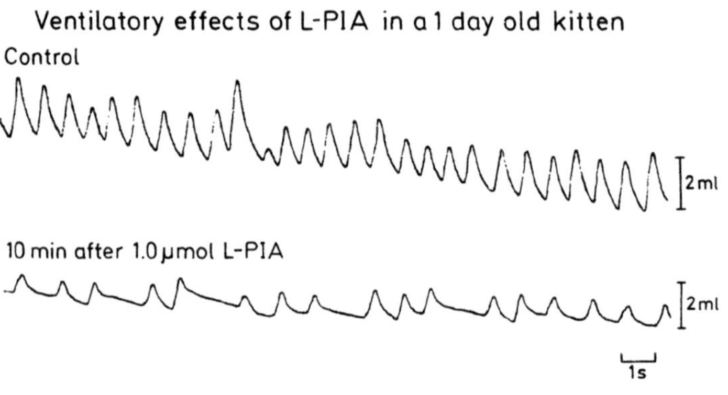

FIGURE 1.   A representative recording of respiration by the
             barometric method before and after administration
             of L-PIA i.p.

The effect lasted the whole period studied (e.g. one hour).

The youngest rabbit pups were more affected than the older ones by
1.0 umol/kg of L-PIA (Table 1).

Table 1.   Effect of L-PIA on minute ventilation in rabbit pups
           (% of control).

1 day:	$- 65.2 \pm 6.0$	(S.E., n = 4)
3 days:	$- 43.4 \pm 5.8$	(S.E., n = 4)
8 days:	$- 31.0 \pm 13.5$	(S.E., n = 3)

A dose-dependent respiratory depression was found in the kittens
after administration of adenosine in a concentration range from
0.01 - 1.0 umol/kg (Fig. 2).

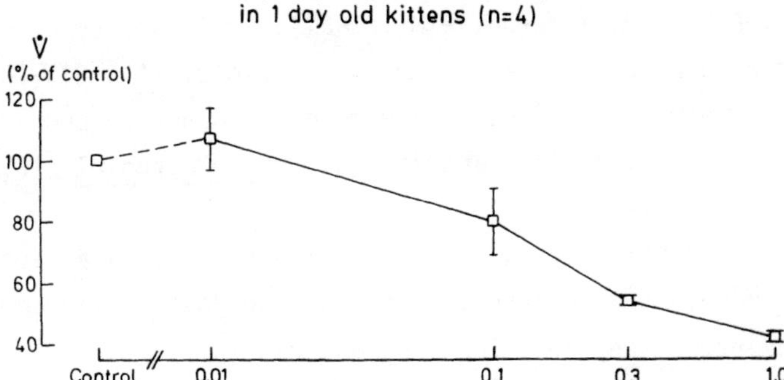

FIGURE 2.   Dose-response curve of L-PIA on minute ventilation in
kittens.

Theophylline (10 mg/kg) did block the ventilatory depressive effect
of L-PIA considerably when given before the L-PIA as well as it could
reverse the effect when given 40 minutes after the L-PIA. After administration of L-PIA the body temperature decreased 1-3°C, however
much slower than the decrease of ventilation why it is less likely
that the respiratory depression caused by L-PIA is due to decreased
metabolism. These findings confirm in unanaesthetized animals previous observations that on anaesthetized animals adenosine or its
analogue is a very potent respiratory depressive substance (5, 6)
The present study also shows that this effect is most pronounced in
the youngest animals. The level of adenosine released during hypoxia
has been found to reach about 2-10 μmol/l (8) why it is possible that
the hypoxic depression seen particularly in the neonate might be
caused by endogenously released adenosine. The peripheral chemoreceptors have been found to be stimulated by adenosine (9). This
effect might overcome the central depressive effect in older animals,
but not in neonates.

<u>References</u>

1.   Winn, H.R., Rubio, R. & Berne, R.M. (1981). Brain adenosine concentration during hypoxia in rats. <u>Am. J. Physiol.</u>, <u>241</u>, 235-242

2.   Phillis, J.W. & Wu, P.H. (1981). The role of adenosine and its nucleotides in central synaptic transmission. <u>Prog. Neurobiol.</u> <u>16</u>, 187-239.

3.   Drury, A.N. & Scent-Györgyi, A. (1929). The physiological activity of adenine compounds with special reference to their action upon the mammalian heart. <u>J. Physiol. (London)</u> <u>68</u>, 213-237.

4.   Hedner, T., Hedner, J., Wessberg, P. & Jonasson, J. (1982). Regulation of breathing in the rat; indications for a role of central adenosine mechanisms. <u>Neuroscience Letters</u>, <u>33</u>, 147-151

5.   Lagercrantz, H., Yamamoto, Y., Fredholm, B.B., Prabhakar, N.R. & Euler, C.v. (1984). Adenosine analogues depress ventilation in newborn neonates. Theophylline stimulation of respiration via adenosine receptors? <u>Pediatric Res.</u> <u>18</u>, 387-390

6.   Eldridge, F.L., Millhorn, D.E. & Kiley, J.P. (1984). Respiratory effects of a long-acting analogue of adenosine. <u>Brain Res.</u> (in press).

7.   Drorbaugh, J.E. & Fenn, W.O. (1955). A barometric method for measuring ventilation in newborn infants. <u>Pediatrics 16</u>, 81-87.

8.   Zetterström, T., Vernet, L., Ungerstedt, U., Tossman, U., Jonzon, B. & Fredholm, B.B. (1982). Purine levels in the intact brain. Studies with an implanted perfused hollow fibre. <u>Neurosci.Lett.</u> <u>29</u>, 111.

9.   Ribeiro, J.A. & Mc Queen, D.S. (1983). On the neuromuscular depression and carotid chemoreceptor activation caused by adenosine. In: Daly, J.W., Kuroda, Y., Phillis, J.W., Shimizu, H. & Ui, M.(eds). <u>Physiology and Pharmacology of Adenosine Derivatives</u>. Raven Press, New York.

68

Adenosine and Methylxanthines: Involvement in Regulation of Respiration

D. E. MILLHORN, F. L. ELDRIDGE and J. P. KILEY

It has long been recognized that various methylxanthines, caffeine, theophylline, and aminophylline (a complex of two molecules of theophylline and the diamine, ethylenediamine) stimulate breathing (1,2,3,4,5). Despite a number of studies showing this effect, little has been done to elucidate either the site of action or the responsible mechanism. Neither caffeine nor theophylline acts via vagal or carotid body reflexes (5,6,10). Because the methylxanthines cause central nervous system stimulation (e.g. nervousness, restlessness, insomnia, tremors, and other signs of central nervous system stimulation) (7), it has usually been assumed that the respiratory response is due to direct activation of brain stem respiratory neurons. It is, of course, axiomatic that the respiratory neurons are involved, but there is no experimental proof that methylxanthines act directly on these neurons. Alternatives include activation by the methylxanthines of non-respiratory neurons in the central nervous system that, in turn, stimulate respiratory neurons, dorsal root reflexes, changes of medullary blood flow that might change medullary pH and central chemoreceptor activity, and humoral release of some substance that causes respiratory stimulation.

In addition to a potentially direct action of theophylline on brainstem respiratory neurons, it has been suggested that various neurotransmitters are involved in the respiratory response. Pretreatment with anti-dopamine agents has been shown to attenuate or abolish the response to aminophylline (4,6,10).

It also has been shown that aminophylline improves respiratory muscle contractility (8) and Aubier and colleagues (9) have suggested that an increase of diaphragmatic contractility rather than an increase in the activity of the central respiratory controller is the mechanism by which aminophylline increases respiration.

The present studies were undertaken to attempt to elucidate the mechanism(s) by which the methylxanthines stimulate respiration. Experiments were performed on cats that either were anesthesized with chloralose and urethane or where decerebrate. In order to avoid central and peripheral chemical feedback, and to eliminate the possibility of changing muscular function, we used phrenic nerve activity to quantify changes in respiratory output in paralyzed, vagotomized and glomectomized cats whose end-tidal PCO_2 was kept constant during a given experiment by means of a servocontrolled ventilator. In some experiments medullary extracellular fluid (ECF) pH was measured from the ventral surface by means of a flat-tipped electrode.

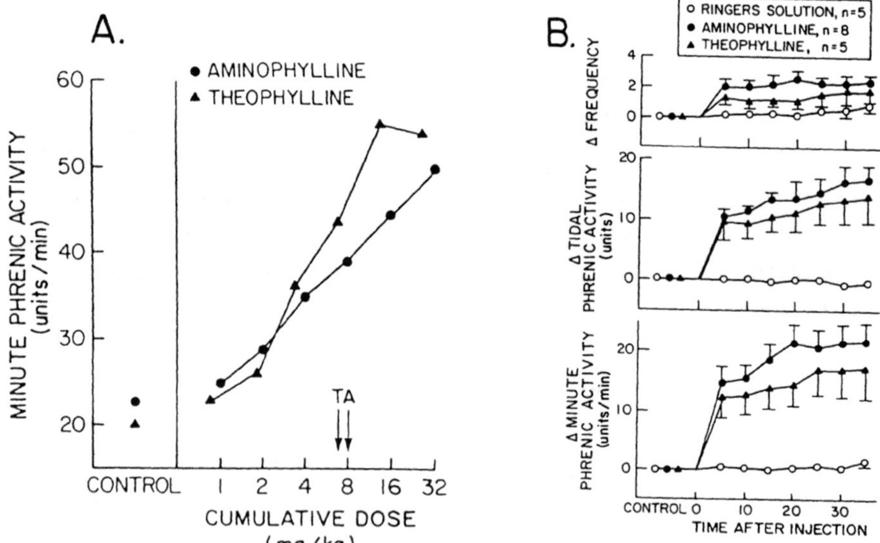

Fig. 1A. Respiratory response to cumulative doses of intravenously administered aminophylline and theophylline. Fig. 1B. The time course of phrenic responses to intravenously administered Ringer's solution, aminophylline and theophylline. Brackets indicate SEM. Reproduced from Respiration Physiology with permission.

Dose-response curves of minute phrenic activity (the product of tidal phrenic and respiratory frequency) for aminophylline and theophylline are shown in fig 1A. Both tidal activity and frequency participated in the increase of minute activity. The intravenous doses (denoted by arrows) of aminophylline (8mg/kg) and theophylline (6.8 mg/kg) used in most subsequent experiments were those which caused a moderately large respiratory response. Fig. 1B shows the time courses of the respiratory responses to intravenously administered Ringer's solution, aminophylline and theophylline. Baseline control values for minute phrenic activity, tidal phrenic activity and respiratory frequency were similar in all three groups. Ringer's solution had insignificant effects on these variables. Both aminophylline and theophylline caused a significant ($P<0.05$) increase in all variables. The greatest increase occurred during the initial 10 minutes after the injection. All variables continued to increase at a slower rate during the subsequent 25 minutes. The increased activity usually lasted for several hours. There were no significant differences in the responses to the two drugs.

The effect on the phrenic response to theophylline after transection of the spinal cord at various levels (L_{1-2}, T_{4-5}, $C_{7}-T_{1}$) and after decerebration was tested (10). There were no significant differences between the intact group and any of those with chordotomies, but the magnitude of the response in the decerebrate group was significantly larger. These findings show that neither dorsal root input nor suprapontine structures are required for the increase in respiration following intravenous theophylline.

The effect on respiration and medullary extracellular fluid pH of intravenously administered theophylline was measured in 8 cats (10). Respiration increased promptly after drug infusion, but medullary pH did not change significantly. This finding rules out the possibility that methylxanthines might stimulate respiration by increasing medullary vascular resistance, and causing a local increase in PCO_2 and H+ activity and chemoreceptor stimulation.

We also studied the effect of different monoamine antagonists on the respiratory response to intravenously administered theophylline

(10). Animals pretreated with the serotonin receptor antagonist, methysergide, had the same respiratory response as did the untreated animals. In contrast, haloperidol, a dopamine receptor antagonist, and alpha-methyltyrosine, an inhibitor of biosynthesis of dopamine, both caused a significant (P<0.05) reduction in the respiratory response to theophylline, to about one half that measured in the untreated cats. Thus, dopamine, but not serotonin, appears to be involved in the respiratory response to theophylline.

While theophylline may lead to an increase in resting ventilation due to improved muscle contractility (8,9), our study (10) which recorded neural output from the brain in paralyzed animals shows that there is a significant increase in central neural respiratory drive that originates in the medulla and pons.

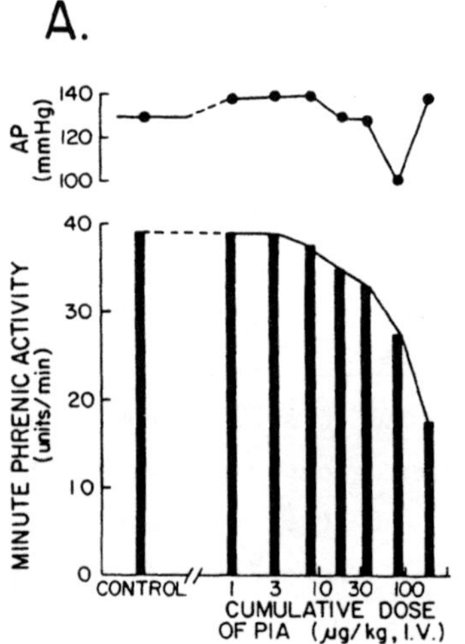

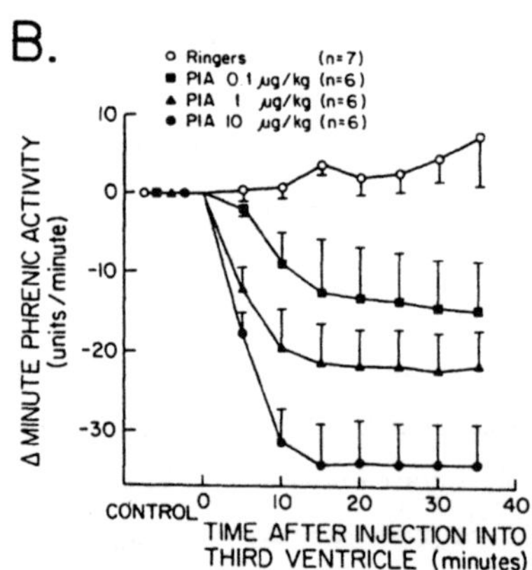

Fig 2A. Effect on phrenic activity of multiple intravenous doses of PIA in one cat. Measurements were made 30 minutes after each dose. Fig. 2B. Time courses of mean effects on phrenic activity of injections of control vehicle, and three different doses of PIA into the third cerebral ventricle. Reproduced from Brain Research with permission.

One possible explanation for the increased respiratory drive involves the purine nucleoside, adenosine, which is present in low concentration in most regions of the brain (11). Adenosine inhibits spontaneous and evoked firing of many neurons in the brain, an action most readily explained by its binding to specific receptors on the neuron's membrane (11). The inhibitory effect of adenosine on neuronal activity is reversed by methylxanthines (13). We recently performed studies to determine the effect of an analog of adenosine, L-2-phenylisopropyladenosine (PIA), on respiration (12). Whether given intravenously (fig. 2A) or into the third cerebral ventricle (fig. 2B), PIA caused a dose-related depression of respiration, involving both tidal activity and respiratory frequency. In a separate group of cats the effects on medullary ECF pH of intraventricular PIA was determined. PIA was associated with development of a metabolic acidosis in the medulla, but still led to marked depression of respiration. We conclude that the adenosine analog, PIA acts to inhibit directly neurons in the brain that are involved in the control of respiration.

We have recently performed experiments of two types to determine if PIA induced depression of respiration is reversed or prevented by theophylline (14). We found that the inhibitory effect of intraventricularly administered PIA was reversed by both systemically and intraventricularly administered theophylline. Furthermore, prior treatment with theophylline (13.6 mg/kg, IV) prevented the depression of respiration following systemic administration of PIA. We suggest that methylxanthines exert their excitatory effect on respiration by antagonizing the inhibitory effect of adenosine.

That an exogenously administered agonist affects neural respiratory activity does not, of course, prove that the endogenously generated adenosine is of physiological importance. Recent findings from our laboratory, however, suggests that this might be the case (15). We found that in the absence of afferent input from the peripheral chemoreceptors, hypoxia ($PaO_2 < 30$ torr) causes severe depression of respiration. Upon resumption of ventilation with 100% O_2, respiration increases but continues to show a significant degree of inhibition for more than an hour (fig. 3A). Because hypoxia causes an increase in medullary blood flow, it is possible that the depression is due to an alkalinity of the ECF and decreased chemoreceptor stimulation. To test this possibility we measured ECF pH in 3 cats and found that hypoxia caused a significant acid shift that lasted for the duration of the recovery period.

Since it has been shown that adenosine levels in the brain increase during severe hypoxia (16), we wondered if the long-lasting inhibition of respiration in glomectomized cats following exposure to hypoxia is mediated by endogenous adenosine. To examine this possibility, we pretreated a group of cats with theophylline (13.6 mg/kg IV) at least 20 minutes before exposure to hypoxia. The finding from one such experiment is shown in fig. 3B. As in the untreated animal (fig. 3A), hypoxia caused a depression of phrenic activity which, however, was not as severe as that found in the untreated animal. Unlike the untreated animal, phrenic activity returned to the control level within 15 minutes after return to hyperoxia. This is evidence that adenosine is released endogenously during hypoxia and causes depression of respiration.

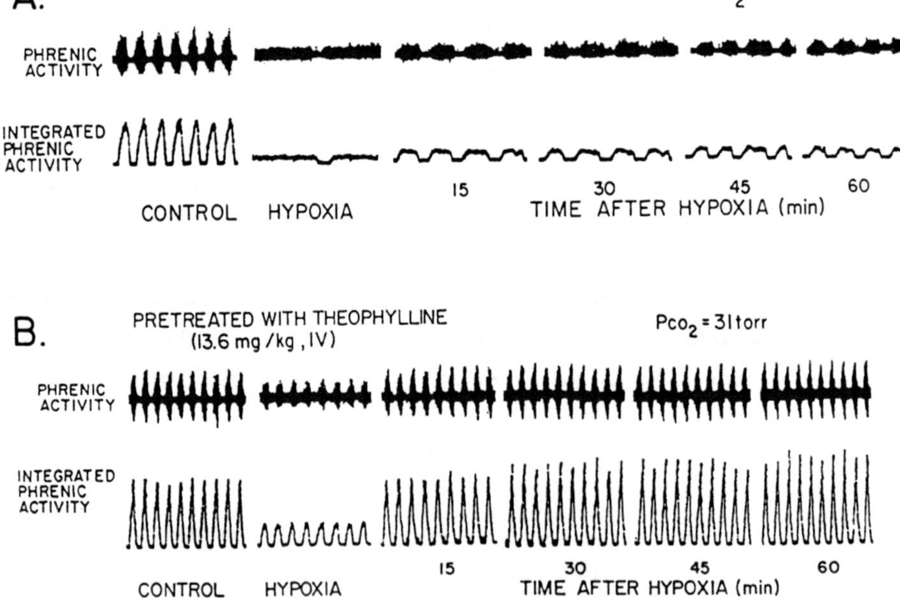

Fig. 3A. The effect on phrenic nerve activity of a brief exposure (10 minutes) to severe hypoxia (PaO₂=28 torr) in a glomectomized, vagotomized and paralyzed cat. Fig. 3B. The effect on phrenic nerve activity of hypoxia in a glomectomized cat that had previously been pretreated with theophylline.

In <u>Summary</u>, we showed that systemically administered theophylline causes a significant increase in phrenic nerve activity in paralyzed cats. The response is mediated at the level of the medulla and pons, and somehow involves the action of dopamine. We also showed that an exogenously administered analog of adenosine, L-2-phenylisophropyl-adenosine (PIA), causes a dose-related depression of respiration, involving both tidal activity and respiratory frequency. Moreover, the PIA mediated inhibition of respiration is reversed or prevented by theophylline. The most important finding however, was that brief exposure to hypoxia in peripheral chemoreceptor denervated animals causes inhibition of respiration that persists for more than an hour after return to hyperoxia. The post-hypoxia depression of respiration is prevented by pretreatment with theophylline. We conclude that the endogenous adenosine is released during hypoxia and causes long-lasting inhibition of respiration.

ACKNOWLEDGEMENTS

Dr. Millhorn is an Established Investigator of the American Heart Association. Dr. Kiley is a Parker B.Francis Fellow in Pulmonary Research. These studies were supported by a grant from the American Heart Association (82-640) and U.S.P.H.S. grants HL17689 and NS11132.

REFERENCES

1.   Richmond, F. H. (1949). Action of caffeine and aminophylline as respiratory stimulants in man. <u>J. Appl. Physiol.</u>, 2, 16-23.

2.   Gerhardt, T., McCarthy, J., and Bancalari (1977). Effects of aminophylline on the ventilation and metabolic rate in premature infants with apnea. <u>Pediatr. Res.</u>, 11, 533.

3.   Jezek, V., Ourendik, A., Stepanek and Boudik, F. (1970). The effect of aminophylline on the respiration and plumonary circulation. <u>Clin. Sci.</u>, 38, 549-554.

4.   Mueller, R. A., Lundberg, D. B. and Breese, G. R. (1981). Alternation of aminophylline-induced respiratory stimulation by pertubation of biogenic amine systems. <u>J. Pharmacol. Exc. Therap.</u>, 218, 593-599.

5.   Sollman, T. and Pilcher (1911). The actions of caffeine on the
     mammalian circulation. J. Pharmacol. Exp. Therap., 217,
     215-221.

6.   Lundberg, D. B., Breese, G. R. and Mueller, R. A. (1981).
     Aminophylline may stimulate respiration in rats by activation
     of dopaminergic receptors. J. Pharmacol. Exp. Therap., 217,
     215-221.

7.   Rall, T. W. (1980). The Xanthines. In: Gillman, A. G.,
     Goodman, L. S. and Gillman (eds.). The Pharmological Basis of
     Therapeutics. 6th edition. pp 592-607. (New York: Macmillan)

8.   Aubier, M.,DeTroyer, A., Sampson, M., Macklem, P. T. and
     Roussos, C. (1981). Aminophylline improves diaphragmatic
     contractility. N. Engl. J. Med., 365, 249-252.

9.   Aubier, M., Murciano, D., Viires, N., Lecoggic, Y., Palacios, S.
     and Pariente, R. (1963). Increased Ventilation caused by
     improved diaphragmatic efficiency during aminophylline
     infusion. Am. Respir. Dis., 127-154.

10. Eldridge, F. L., Millhorn, E. E., Waldrop, T. G. and Kiley, J.
     P. (1983). Mechanism of respiratory effects of methylxan-
     thines. Respir. Physiol., 53, 239-261.

11. Phyllis, J. W. and Wu, P. H. (1981). The role of adenosine and
     its nucleotides in central synaptic transmission. Progr.
     Neurobiol, 16, 187-239.

12. Eldridge, E. L., Millhorn, D. E. and Kiley, J. P. (1984).
     Respiratory effects of a long-acting analog of adenosine.
     Brain Research, 301, 273-280.

13. Snyder, S. H., (19781). Adenosine receptors and the actions of
     methylxanthines. Trends Neurosci., 4, 242-244.

14. Eldridge, F. L., Millhorn, D. E., and Waldrop, T. G. (1982).
     Adenosine receptors and the methylxanthines: Involvement in
     the neural control of respiration. Fed Proc., 41, 1960.

15. Millhorn, D. E., Eldridge, F. L., Kiley, J. P., Waldrop T.,G.
     (1984). Prolonged inhibition of respiration following acute
     hypoxia in glomectomized cats Respir. Physiol., in press.

16. Winn, H. R., Rubio, R., and Berne, R. M. (1981). Brain
     adenosine concentration during hypoxia in rats. Am. J.
     Physiol, 241, H235-H242.

69

Brainstem Unit Discharges During Slow and Fast Respiratory Rhythms Induced by Halothane

D. CAILLE, I. BRIOIS, C. POUJEOL, A. S. FOUTZ, J.-F. VIBERT
and A. HUGELIN

Several observations suggested that the fundamental organization of the respiratory pattern generator is state-dependent. Discharge pattern of units recorded in specific areas of the brainstem depend on physiological state, particularly natural sleep-wake patterns (1, 2), anesthesia (3), vagotomy (4), paralyzing agents (5), etc... A new concept is now emerging from these results: the respiratory rhythm could be generated by completely different mechanisms through a fundamental reorganization of the generator itself. Investigation of this reorganization requires experimental conditions providing distinct, stable and reproducible respiratory rhythms. In this study we first defined experimental conditions providing "slow" and "fast" rhythms, then we explored the brainstem with microelectrodes in these two states in order to reveal a possible reorganization of the pattern generator.

The first series was devoted to producing distinct and stable rhythms. Ten cat implanted for EEG recordings, head restraint and with a chronic tracheal fistula allowing pneumotachographic measurements, were studied for several hours through their natural sleep-wake cycles and when breathing increasing concentrations of halothane (0.5% to 3.5% in oxygen). A wide variety of respiratory frequencies, with rapid and unpredictable shifts, were observed in the undrugged state. These frequencies were correlated with the states of consciousness (waking, Slow Wave Sleep, REM sleep). Under light halothane anesthesia (0.5%), respiratory rhythm was slow, with a relatively high variability (albeit lower than in the waking state). When halothane concentration increased, respiratory frequency increased and variability progressively decreased, leading to a rapid and stable respiratory rhythm with halothane concentrations above 2.5% and up to

3.5%. Thus, in the chronic preparation at least two clearly distinct rhythms can be obtained with 0.5% and 2.5% halothane.

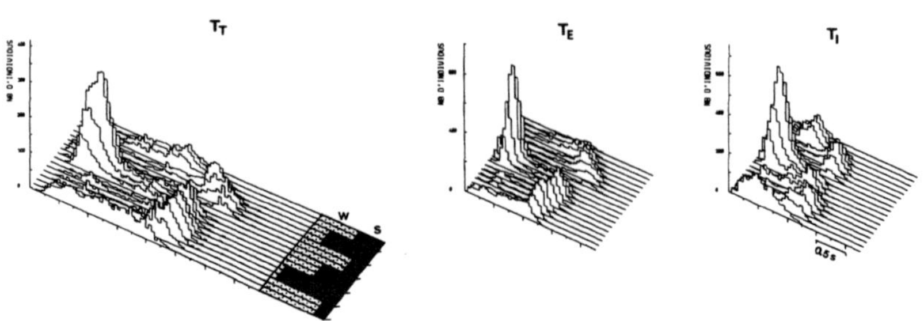

Figure 1

Spontaneous evolution of the respiratory period (Tt), the expiration (Te) and inspiration (Ti) durations in a chronic cat during a 6 hours recording. Each histogram represents 60 min. Black area indicates the sleep percentage (S) and dotted area the wakefulness percentage (W), computed each hour. Note the dominant respiratory periods correlated with the physiological state.

In a simplified preparation (cats vagotomized, decorticated, spinalised at Th1 level and paralyzed with gallamine), increasing halothane concentration also produced an increase in respiratory frequency with a highest frequency when 1.5% concentration was reached. Unlike chronic animals, this preparation had a very regular phrenic nerve discharge pattern in the control state as well as with all concentrations of halothane. Therefore it was possible to lock this preparation on two

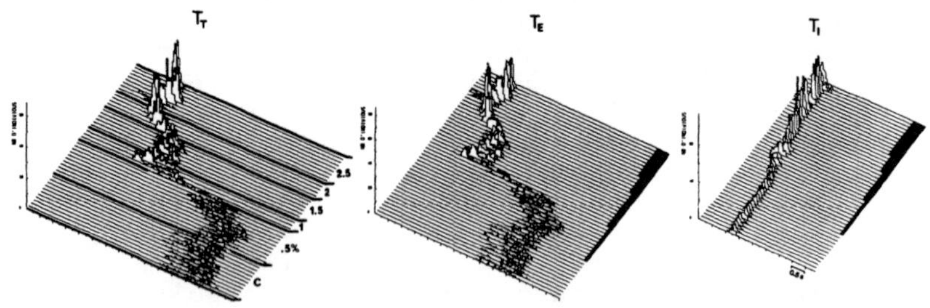

Figure 2

Effect of the concentration of halothane on the respiratory durations (Tt, Te and Ti) in the "simplified preparation". C: control; numbers indicate the halothane concentration in oxygen. Black strips in Te and Ti indicate increases in halothane concentration. Note that Tt variations are mainly due to Te.

stable frequencies: a "fast" one at 1.5% or above and a "slow" one at 0.5% halothane. These conditions were chosen for unit recordings. RRUs were recorded in structures currently considered as playing a role in the origin and/or control of respiratory activity: 1/the dorsal and ventral nuclei including nucleus ambigualis, retroambigualis, retrofacial and nucleus infra-solitarius; 2/the pneumotaxic complex, particularly the nucleus parabrachialis medialis and the brachium conjunctivum and 3/the bulbopontine reticular formation.

When the animals breathed with a slow rhythm, 41.8% of the 471 units recorded from the medulla to the pneumotaxic area were respiration related (RRUs); most units displayed a tonic discharge pattern. When the animals were locked on a fast rhythm, among 339 recorded units only 19.7% RRUs were found, mainly with a phasic pattern of discharge; most of these RRUs were recorded caudally to the facial nucleus. The RRU proportion was significantly lower during high frequency than during low frequency periods in the pneumotaxic complex and in the pontine reticular formation. Among all recorded units, 64 were recorded in both conditions. When shifting from low to high frequency, 24 RRUs lost their respiratory modulation, 19 RRUs retained their modulation, the other

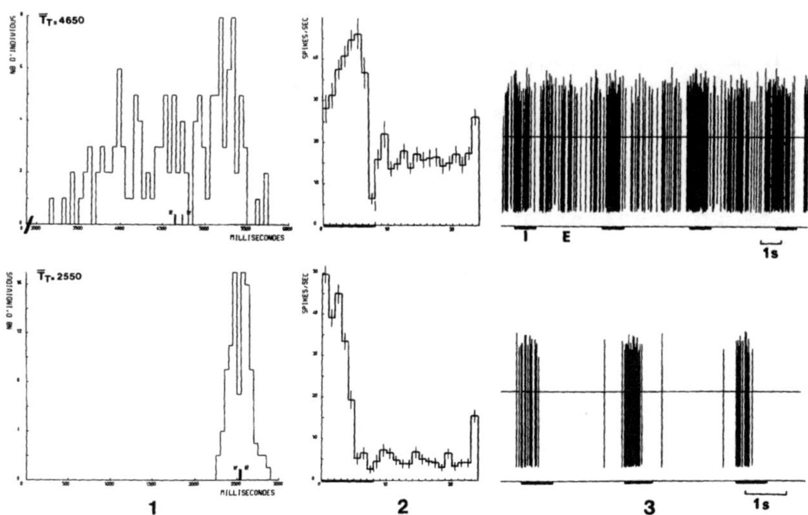

Figure 3

The same bulbar reticular unit recorded during long (Tt= 4650 ms with 0.5% halothane) -upper row- and short (Tt= 2550 ms with 2% halothane) -lower row- respiratory periods. Left: Histogram of Tt, Middle: Cycle triggered histogram of the unit shown on the right (black thick bars indicate the inspiration time). Note that the unit becomes more phasic and more early inspiratory.

21 non-modulated units remained non-modulated .   The 19 unchanged RRUs were only found in the bulbar respiratory nuclei and in the bulbar reticular formation. Their discharge   rate   decreased,   leading   to more phasic discharges, but the temporal discharge patterns were not or slightly modified.

It is suggested that low rhythm generation involves pontine mechanisms whereas fast respiratory rhythm induced by halothane can be locally generated in a bulbar network where RRU population is smaller, constituted with more phasic, and more respiratory modulated units. Functional connections in such a respiratory network where units fire with a high stability can lead to a better synchronization between units and could explain the low variability of the respiratory period under deep halothane anesthesia.

These results support the hypothesis of multiple brainstem sites of respiratory neurogenesis and of a reorganisation of the rhythm generator in different states.

References

1 - Sieck , G.C. and Harper , R.M.   (1980), Pneumotaxic area neuronal discharge during sleep-waking states in the cat. Exp. Neurol. , 67 79-102.

2 - Orem , J. , Montplaisir , J.   and Dement , W.C. (1974), Changes in the activity of respiratory neurons during sleep. Brain Res. , 82, 309-315.

3 - Caille , D. , Vibert , J.-F. , Bertrand , F. , Gromysz , H. and Hugelin , A. (1979) Pentobarbitone effects on respiration related units;selective depression of bulbopontine reticular neurones. Respir. Physiol. , 36 ,201-216.

4 - Feldman , J.L. , Cohen , M.I. and Wolotsky , P. (1976) Powerful inhibition of pontine respiratory neurons by pulmonary afferent activity. Brain Res. , 104, 341-346.

5 - Caille , D. , Foutz , A.S. , Vibert , J.-F. and Hugelin , A. (1984) Gallamine and vagotomy enhance respiratory modulation of reticular units. Brain Res. , 299 ,79-89.

70

Differential Sensitivity of Cranial Nerve Discharges to Diazepam and Protriptyline

M. BONORA, W. M. ST-JOHN and T. A. BLEDSOE

Phasic contraction of the muscles of the tongue and the larynx during inspiration participates in maintaining upper airway patency. A loss of genioglossal activity during sleep has been shown to be associated with obstructive apneas in some patients[1]. In this context, it has been recently reported that obstructive sleep apneas were exacerbated by alcohol or by the benzodiazepines flurazepam and diazepam[2,3]. In contrast, apneic episodes during sleep decreased in frequency and severity following treatment with the non-sedating tricyclic antidepressant protriptyline[4,5]. As demonstrated in previous studies[6,7], respiratory-related activity of nerves supplying upper airway muscles is selectively depressed by alcohol or by various anesthetics. We, therefore, hypothesized that a common action of all these drugs would be differential augmentations or reductions of upper airway respiratory motor activity relative to phrenic activity. Accordingly, we examined the influence of protriptyline and diazepam on the respiratory activity of hypoglossal and recurrent laryngeal nerves compared to phrenic nerve. A possible role of carotid chemoreceptors in the mechanism of action was also explored.

METHODS

Twenty-one adult cats were decerebrated, paralyzed and ventilated. The vagi were sectioned at the mid-cervical level on the left side and within the thorax on the right side to allow the recording of the recurrent laryngeal nerve (RLN). In addition, in eight of these cats, a carotid sinus nerve sectioning was performed bilaterally. Activities of phrenic, hypoglossal and recurrent laryngeal nerves were monitored. End-tidal fractional concentration of CO_2 and O_2 and arterial blood pressure were continuously monitored. Drugs were administered intravenously in boluses in various animals. The basic protocol was to record nerve activities under hypercapnic, hyperoxic conditions before and after administration of either protriptyline (0.8 mg/kg) (in 13 cats) or diazepam (0.05 mg/kg) (in 8 cats). Recordings were done at intervals for one hour after drug injections.

RESULTS

Protriptyline

In all eight cats studied with intact carotid sinus nerves, protriptyline induced a marked and sustained increase in hypoglossal and recurrent laryngeal nerve discharges, whereas the phrenic nerve discharge was not consistently altered. This differential effect was discernible during our first recording 5 minutes after injection and persisted for about 30 minutes. Mean values of data obtained at the 15 minute recording show that changes induced by protriptyline in the peak integrated activity of the hypoglossal nerve were significantly different from pre-drug values (Fig.1). Though a statistical analysis could not be done for the RLN, the peak integrated activity of this nerve had consistently increased after protriptyline in all cats studied. Interestingly, the effect of protriptyline was more pronounced on the expiratory than on the inspiratory discharge.

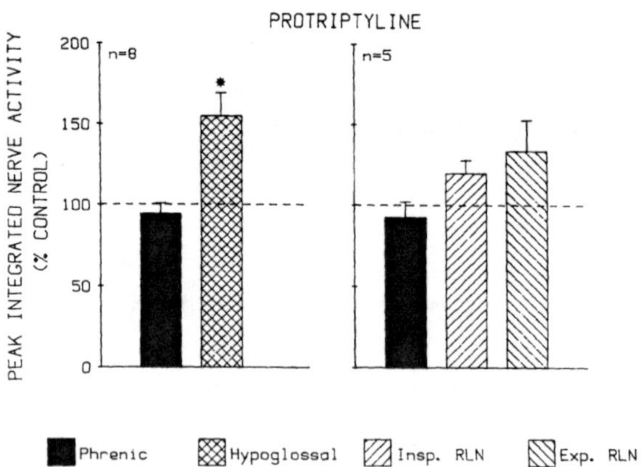

FIGURE 1. Mean values (+ SE) of peak integrated activity of phrenic, hypoglossal and recurrent laryngeal nerves, recorded 15 minutes after administration of protriptyline, expressed as percents of pre-drug values. (* = P < 0.05).

In the five cats with sectioned carotid sinus nerves, protriptyline induced also a large increase in the hypoglossal activity during inspiration. Though an elevation of phrenic discharge occured as well, hypoglossal activity increased to a much greater extent than that of the phrenic.

Diazepam

Administration of diazepam induced a dramatic reduction in hypoglossal and RLN activities in all cats with intact or sectioned carotid sinus nerves. This effect was observed 5 minutes after drug injection and was maintained for at least 45 minutes. Mean values of peak integrated activity of hypoglossal and RLN after diazepam were consistently reduced compared to control values (Fig.2). Phrenic nerve

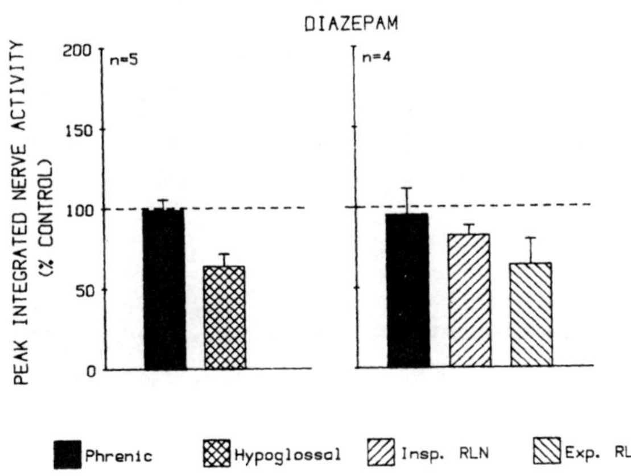

FIGURE 2. Mean values (+ SE) of peak integrated activity of phrenic, hypoglossal and recurrent laryngeal nerves, recorded 15 minutes after administration of diazepam, expressed as percents of pre-drug values.

activity was usually unchanged or decreased slightly after diazepam in cats with intact carotid sinus nerves ; a consistent diminution occured in carotid denervated cats. However, as for protriptyline, the effect was always greater for the hypoglossal than for the phrenic discharge.

CONCLUSIONS

These findings show that the respiratory activity of hypoglossal and recurrent laryngeal nerves is markedly enhanced by protriptyline and reduced by diazepam at doses which do not consistently affect phrenic nerve activity. It suggests that motoneurons innervating the upper airway muscles are much more sensitive to these drugs than are those of the bulbo-spinal phrenic system. This differential effect was still present in cats with sectioned carotid sinus nerves suggesting that carotid chemoreceptors do not play a major role in the mechanism of action involved. Also, suprapontile structures and vagal afferents are not required for these drug actions. Since other sedative drugs such as barbiturates and alcohol have been shown to depress as well cranial nerve discharges[6,7], a non-specific action on the brainstem reticular formation may be at the origin of these pharmacological influences. Whatever the precise mechanism, the exacerbation of obstructive sleep apnea by several sedative drugs could be associated with the differential depression of nerve activities supplying upper airway muscles. Conversely, the beneficial effect of protriptyline observed by several investigators could be due to a selective stimulation of these muscles.

REFERENCES

1. REMMERS, J.E., et al.- J. Appl. Physiol., 44 : 931-938, 1978.
2. DOLLY, F.E. and A.J. BLOCK.- Am. J. Med., 73 : 239-243, 1982.
3. SIMMONS, F.B., et al.- The Laryngoscope, 87 : 326-338, 1977.
4. CLARK, R.W., et al.- Neurology, 29 : 1287-1292, 1979.
5. CONWAY, W.A., et al.- Thorax, 37 : 49-53, 1982.
6. HWANG, J.C., et al.- J. Appl. Physiol., 55 : 785-792, 1983.
7. BONORA, M., et al.- Am. Rev. Respir. Dis., 130 : 156-161, 1984.

71

Effects of Almitrine on Hypoglossal and Phrenic Electroneurograms

D. E. WEESE-MAYER, R. T. BROUILLETTE, L. KLEMKA
and C. E. HUNT

Pharyngeal airway obstruction occurs in a wide variety of clinical situations including apnea of prematurity, general anesthesia, and obstructive sleep apnea. Remmers et al (1) have proposed that the pharyngeal airway obstructs when the dilating force of upper airway muscles is insufficient to counterbalance negative pharyngeal pressure during inspiration. A drug which would stimulate phasic inspiratory activity in upper airway-maintaining muscles might be useful in diseases involving pharyngeal airway obstruction. The recently synthesized respiratory stimulant almitrine has been found to increase ventilation in chronic obstructive lung disease and to alleviate narcotic-induced respiratory depression but little attention has been focused on its potential effect on upper airway dilating musculature.

METHODS

The experiments were performed in eight chloralose-urethane anesthetized, paralyzed, vagotomized, artificially ventilated cats. Records of P_ACO_2, PaO_2, blood pressure, rectal temperature, raw and integrated hypoglossal (HG) and phrenic (PHR) electroneurograms (ENGs) were made on a polygraph and tape recorder. The integrated ENGs were sampled at 25 Hz, displayed, stored and analyzed by a computer.

Three normoxic, eucapneic cats were studied before and after almitrine doses ranging from 0.1 to 4.0 mg/kg intravenously. To allow for cat-to-cat comparisons, PHR and HG ENGs were expressed as a percentage of maximal activity. The interactive effects of hypoxemia and almitrine were investigated in four cats. Holding P_ACO2 constant

405

(between 38-40 mmHg), PaO_2 was varied from 30 to 150 mmHg before and after 1 mg/kg of almitrine. The interactive effects of almitrine and hypercarbic stimulation were investigated in five cats. Maintaining PaO_2 above 150 mmHg, $PaCO_2$ was varied from 30-70 mmHg before and after 1 mg/kg of almitrine. The effects of almitrine on respiratory timing were investigated in six cats. Maintaining PaO_2 between 85-124 mmHg and $PaCO_2$ between 38 and 41 mmHg, T_i, T_e, F, V_T/T_i, and T_i/T_{tot} were recorded based upon the PHR ENG before and after 1 mg/kg of almitrine.

RESULTS

In all animals almitrine increased tidal inspiratory activity in both the HG and PHR nerves (Figure 1).

Dose response studies. The hypoglossal nerve demonstrated little or no phasic activity prior to almitrine infusion at normoxia, eucapnea, whereas the phrenic nerve demonstrated phasic activity during baseline recording. Phasic inspiratory HG ENG activity increased progressively with almitrine dosages between 0.1 and 1.0 mg/kg given intravenously, from 12 ± 12% at baseline to 65 ± 7% of maximal activity after 1.0 mg/kg of almitrine ($p < .01$). Tidal PHR ENG activity increased progressively with almitrine dosages between 0.1 and 1.0 mg/kg, from 51 ± 5% at baseline to 89 ± 6% of maximal activity after 1.0 mg/kg of almitrine ($p < .001$).

Oxygen response studies. In all four animals phasic inspiratory HG and PHR ENGs increased after 1 mg/kg almitrine given intravenously. A consistent increase in the ENGs was seen in the hyperoxic range (PaO_2 150-170 mmHg) with more variable increases in the hypoxic range (PaO_2 30-50 mmHg).

Carbon dioxide response studies. In all five animals phasic inspiratory HG and PHR ENGs increased after 1 mg/kg almitrine given intravenously. The increased activity was seen at all $PaCO_2$ values between 30 and 70 mmHg.

Ventilatory timing studies. In six normoxic, eucarbic cats, almitrine increased phasic PHR ENG activity ("tidal volume") from 47.2 ± 14.8% to 81.2 ± 14.7% of maximal activity ($p < .005$), whereas respiratory frequency remained unchanged. V_T/T_i approximately doubled after almitrine ($p < .05$), while T_i/T_{tot} decreased significantly ($p < .005$).

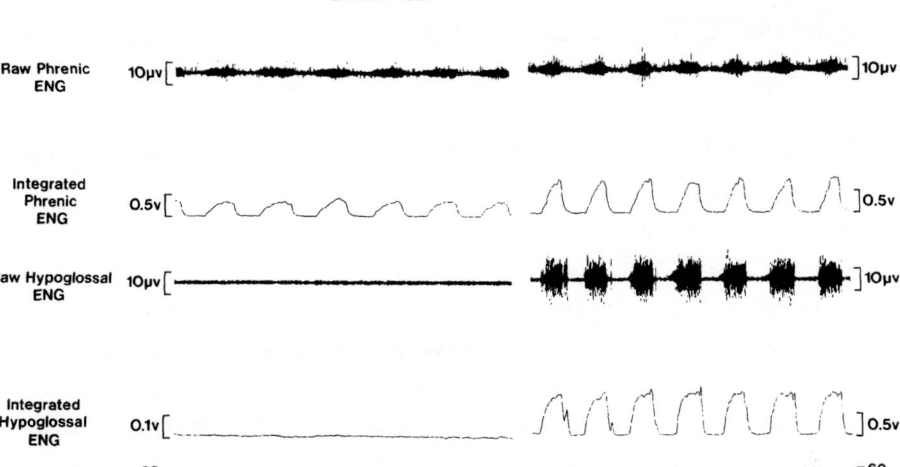

FIGURE 1   Polygraphic tracings before and after 1 mg/kg intravenous infusion of almitrine are shown. Note the marked increase in inspiratory PHR and HG activity after almitrine. P_ACO_2 and PaO_2 were similar before and after almitrine being 32–33 mmHg and 160–180 mmHg, respectively.

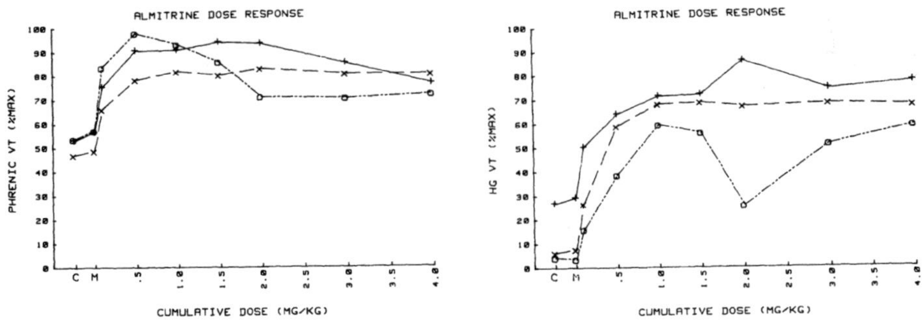

FIGURE 2   These figures show the changes in phasic PHR (left) and HG (right) ENG activity with increasing almitrine doses. Each symbol represents the mean of several (4.7 ± 2.3) minutes of data at a particular drug dosage. Different symbols and line types represent different animals. Abbreviations: C = control, pre-drug infusion; M = values after infusion of malic acid/saline.

DISCUSSION

We found that intravenous almitrine doses previously found to increase ventilation and peripheral chemoreceptor activity, increased inspiratory HG and PHR ENG activity.

Our data and previous studies suggest that almitrine and hypoxemia stimulate the peripheral chemoreceptors by partially different mechanisms. Bisgard (2), Roumy and Leitner (3), and Laubie and Schmitt (4) found that hyperoxia attenuates the carotid chemoreceptor response and/or the ventilatory response to almitrine. The additive interaction of almitrine and hypercarbia can presumably be explained by the differing sites of action, peripheral vs. central chemo-receptors, respectively. These results suggest that almitrine might be effective in preventing anesthetic-related upper airway obstruction. If similar results are found in unanesthetized animals, the drug might be useful in obstructive sleep apnea or apnea of prematurity.

REFERENCES

1. Remmers, J. E., deGroot, W. J., Sauerland, E. K. and Anch, A. M. (1978). Pathogenesis of upper airway occlusion during sleep. J. Appl. Physiol., 44, 931-39.

2. Bisgard, G. E. (1981). The response of few-fiber carotid chemoreceptor preparations to almitrine in the dog. Can. J. Physiol. Pharmacol., 59(4), 396-401.

3. Roumy, M. and Leitner, L. M. (1981). Stimulant effect of almitrine (S2620) on the rabbit carotid chemoreceptor afferent activity. Bull. Eur. Physiopathol. Respir., 17(2), 255-59.

4. Laubie, M. and Schmitt, H. (1980). Long-lasting hyperventilation induced by almitrine: Evidence for a specific effect on carotid and thoracic chemoreceptors. Eur. J. Pharmacol., 61, 125-36.

Section 6
Clinical and Physio-pathological Correlates

72

Systems Analysis Applied to Studies of Respiratory Control

D. C. SHANNON and D. CARLEY

We have developed a new, quantitative, computer-based model of the respiratory control system. Our purposes are 1) to explore interactions between system parameters and their effects on respiratory pattern 2) to identify the sources of stability or instability 3) to determine whether pathologic oscillations in respiratory output might represent extremes of normally occurring periodicities and 4) the extent to which the model might explain the occurrence of pathologic oscillations seen in various clinical settings. We describe herein the rationale and structure of the model and some preliminary evidence that oscillations in breathing are a normal phenomenon in which the period or cycle time is determined primarily by the phase delay and the amplitude as seen in healthy or in pathophysiologic states, is the result of measurable alterations in one of several system parameters.

Previous models (1-5) have assumed that ventilation is stable in the steady state and have found that oscillatory behavior develops when the gain between ΔP_{CO2} and $\Delta \dot{V}_A$ is increased or the time delay between $\Delta \dot{V}_A$ and ΔP_{CO2} sensed at the chemoreceptor is increased. In each case, many fold increase in gain was needed to produce oscillations.

However, these models and their predictions are not consistent with empiric observations. We and others have found that spontaneous fluctuations typify the output of cardiovascular and respiratory systems in healthy subjects (6-10). For example, adult man in standing posture exhibits short-term oscillations in heart rate that occur at two frequencies, one at breathing frequency and the other centering at 0.05 Hz ie., every 20 sec. The heart rate of human infants oscillates at similar but somewhat higher frequencies (Fig. 1)(11). Respiratory activity collected from chest wall movement transducers and selected for apparent stability also oscillates at about once in 15 sec (Fig. 2).

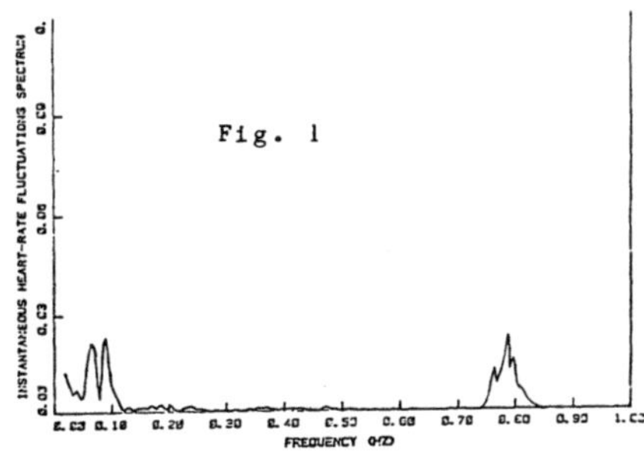

The instantaneous heart rate power spectrum of a sleeping 1 month old infant breathing at 48/min showing respiratory sinus arrhythmia at that frequency and a two lobed low frequency peak.

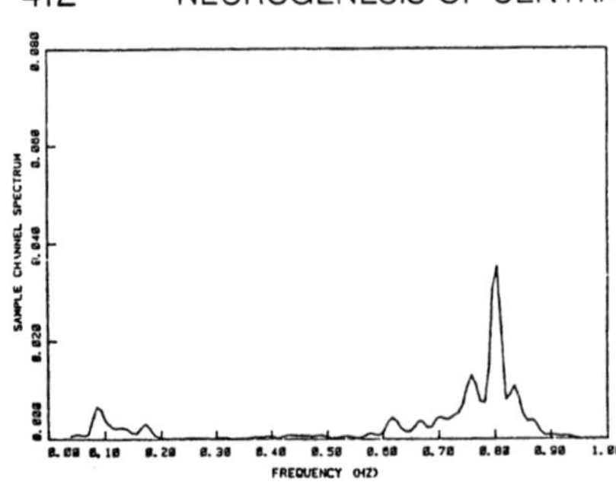

Fig. 2

The respiratory power spectrum showing peaks at frequencies similar to those in the heart rate spectrum.

Thus, it appears, that both ventilation and heart rate oscillate every 15-20 sec. in healthy infants and adults. Such fluctuations may then become exaggerated or diminished under conditions of both health and disease. In adult volunteers, for example, we have found that low frequency heart rate oscillations grow an order of magnitude on standing (6), fall an order of magnitude with age from 20-65 yrs (unpublished observation) and can be modified by autonomic agonists and antagonists (6). In infants, we have found that excessive periodicity of both ventilation and heart rate characterize some infants who will die of Sudden Infant Death syndrome, SIDS (11). In a prospective study Southall et al also found excess periodic breathing in some babies that died of SIDS (12). We have noted that the period of both the heart rate and breathing cycles in these babies is 15 $\pm$ 1 sec. Spectral analysis of heart rate in one such baby yielded a peak at .07 Hz (4/min) that was 4000% greater than in a healthy baby.

We now turn to a description of the model. The
controller is a switch that phasically varies "neural"
output to respiratory muscles in response to input.
This output is translated into a change in lung volume
as a function of time $V(t)$ depending on the dynamic
properties of lung and thorax. $V(t)$ then affects the
partial pressures of CO_2 and O_2 $P_{CO2}(t)$ and $P_{O2}(t)$,
according to the dynamics of alveolar-capillary gas
exchange.

The inputs which govern the controller output are
dependent on the respiratory phase. Inspiratory drive
is initiated when the controller input crosses a fixed
threshold. During inspiration, the input is composed
of a ramp-like term proportional to inspiratory
duration. This term is analogous to the central neural
input suggested by Von Euler (13). A second input is
directly proportional to the instantaneous lung volume
and represents the effect of vagal feedback. When the
combination of these two terms reaches a fixed
threshold, passive expiration begins.

During the expiratory phase, the controller input is
derived solely from chemical feedback. This feedback
is the integral of a term which is directly
proportional to arterial P_{CO2} and inversely
proportional to arterial PO_2. When this ramp-like term
crosses a threshold inspiration is begun and the cycle
repeats. The switching of the controller occurs
instantaneously, while the time course of alveolar

volume (V(t) is determined by the mechanical RC time constants of inspiration and expiration Ti and Te.

Chemical feedback plays one other important role. The magnitude of the step increase in inspiratory neural drive is scaled by chemical feedback.

Studies have shown that acute hypoxic feedback interacts with hypercapnic feedback in a multiplicative fashion (14). The inspiratory neural step-size is scaled by a term which is inversely proportional to PO_2 and directly proportional to (PaCO2 - Pset). Thus both the rate and depth of breathing are governed by chemical feedback.

This model does not include a cortical or thermal influence on respiration but is congruent with most observations of the behavior of the controlled system.

We will now examine briefly the general model behavior. In order to facilitate discussion, we will simplify the above model by linearizing the controller. We define the controller for ventilation; $\dot{V}_A = (P_{CO2} - P_{set})|A(f)|$ where A = Controller gain; $\dot{V}_A = 0$ when $P_{CO2} = P_{set}$. Increasing A to achieve a smaller steady-state error $(P_{CO2} - P_{set})$ increases the tendency of the system to fluctuate in response to transient or maintained external perturbations. Using published values for $\dot{V}_A$ and $|A(f)|$ in human adults and infants during a DC or non-time-varying state we find that $(P_{CO2} - P_{set}) \cong 2$ mmHg in both cases. Conservation of mass requires that PACO2 = $\dot{V}_{CO2}/\dot{V}_A$ (PB

- 47) and implies a linear relation between ($\dot{V}_{CO_2}$ and $\dot{V}_A$ if P_{ACO_2} is controlled. Therefore, the infant who typically has $\dot{V}_{CO_2} = 2(\dot{V}_{CO_2}$ adult) must also have 2 $\dot{V}_A$ adult to keep P_{CO_2} constant. This implies that the infant $|A (f)|$ must also be twice the adult $|A (f)|$. Empirically $|A (f)|$ normalized by body weight is 28.5 ml min^{-1}/torr kg in adults, and 63 ml min^{-1}/torr kg in infants, values that agree well with the model prediction of a 2 fold difference.

While we define $|A (f)|$ as the transfer characteristic between the error signal $(P_{CO_2} - P_{set})$ and the controller output $\dot{V}_A$, we define $|B (f)|$ as the transfer characteristic of the plant between $\dot{V}_A$ and Pa_{CO_2}. The phase delay which might generate spontaneous oscillations in the system is the sum of the delays in these two transfer functions.

A system can sustain oscillations if the phase delay (ϕ) through the entire feedback loop is 360°; oscillations at other values of ϕ will cancel rather than sustain. The frequency (f) of oscillations can be predicted if one can characterize the phase shifting elements. The amplitude of oscillations will be stable if the magnitude of $|A(f)|$ $|B(f)|$ is unity and will grow to saturation and sustain if the magnitude is greater than unity. It is clear that respiratory control is mediated by a "negative" feedback system; changes in PCO_2 lead to compensatory rather than reinforcing adjustments in $\dot{V}_A$. This sign inversion represents a -180° phase shift; thus we must account for another -180° in order for sustained oscillations to occur. The physiologic control system is a "distributed" one and information transfer does not occur instantaneously between its elements. This can be modelled as a circulatory "pure delay" between the plant and the controller; an element whose phase is a linear function of frequency. If this delay were the only significant phase shifting element in the system, we would predict oscillations to occur at a frequency of $1/2Tc$, where Tc is the circulatory delay. The experimental observation that oscillations typically occur at a frequency of approximately 4 times the normal delay of 6 to 7 seconds implies that -90° of phase shift is contributed by system elements other than the pure circulatory delay. These would include damping by FRC, diffusion of CO_2 to receptor sites and implementation of a change in neural output.

We have mathematically modelled ventilatory control and have simulated breathing in adults compared to infants using empirically observed values for system parameters (Table). The ventilatory outputs of the model in a typical adult and infant are consistent with typical values for VT, f, Pa_{CO2}, and Pa_{O2}. Note that VT is only the alveolar volume change. A reduction in FRC elicits periodic breathing in the infant at a cycle time of about 10 sec. Fig. 3. However, the same change in FRC in the adult only increases breath-by-breath changes in PCO_2 and PO_2 (Fig. 4). The relatively higher FRC in the adult tends to stabilize ventilation.

Fig. 3

Fig. 4

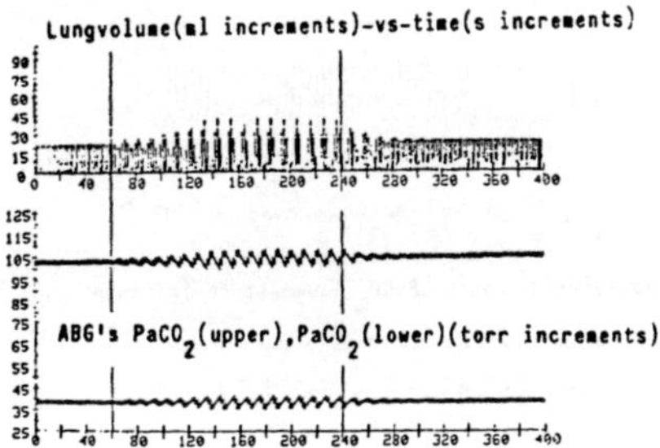

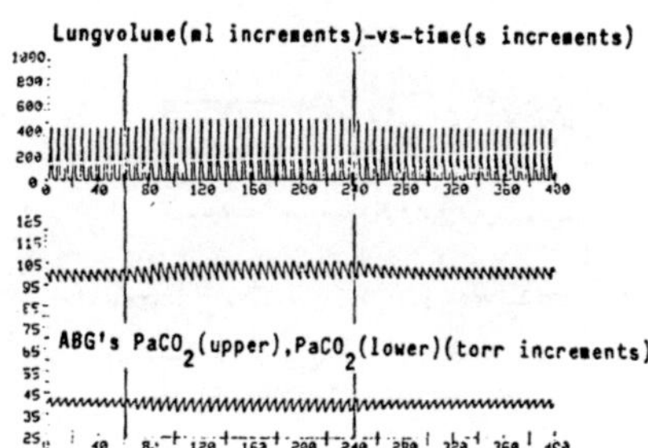

If A (f) is doubled, pronounced periodic breathing occurs in the infant at a cycle time of about 12 sec. (Fig. 5) and in the adult at 22 sec. (Fig. 6).

Fig. 5

Fig. 6

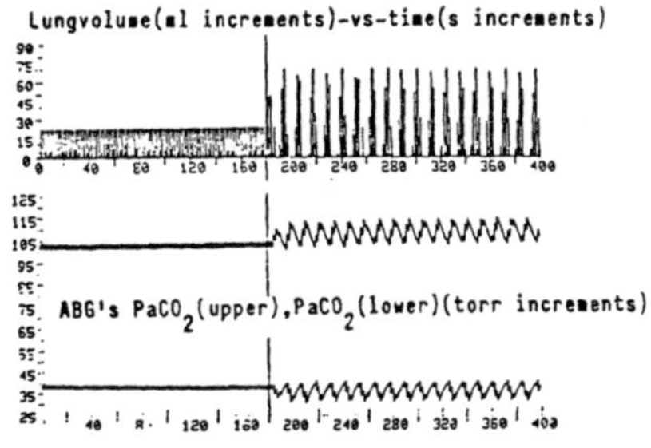

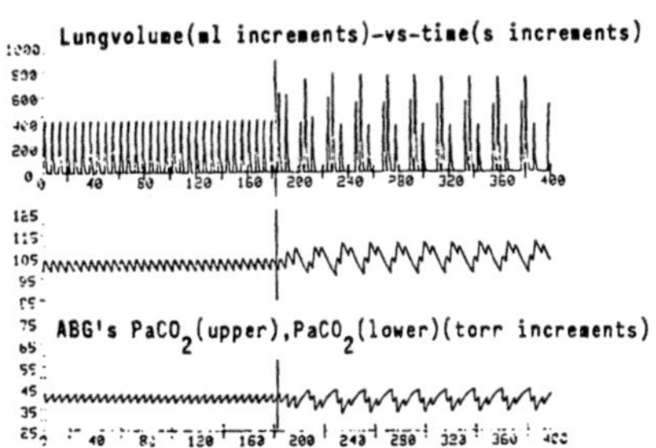

On the other hand, a 75% increase in Tc elicits periodic breathing in the infant (Fig. 7) but not in

the adult (Fig. 8). The period in the infant is 17s,
an increase in cycle time that is expected from the
increased phase delay.

Fig. 7

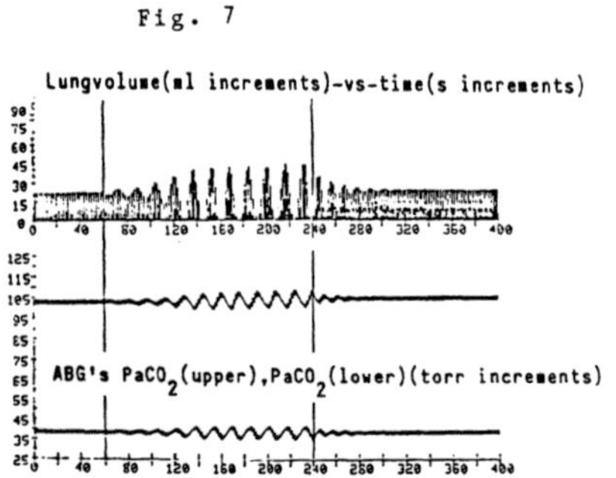

Fig. 8

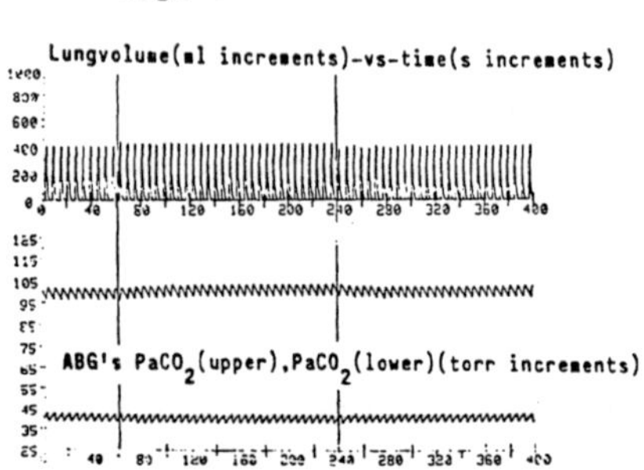

If in addition to the 75% increase in Tc we now
increase | A (f) | by 60% in the adult, we see striking
periodicity at a cycle time of about 33s again
reflecting the increased phase delay (Fig. 9).

Fig. 9

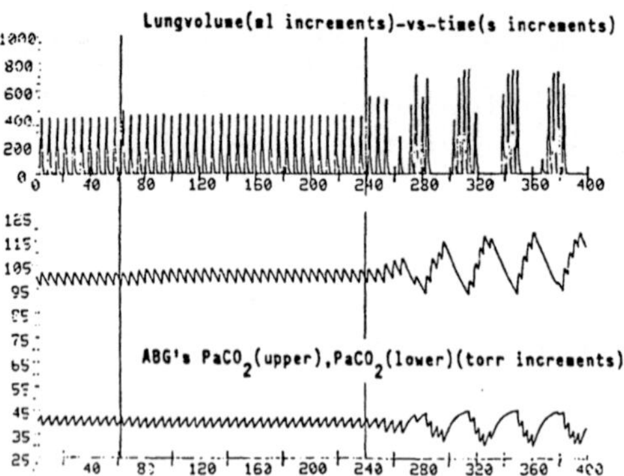

TABLE

	Adult	Infant		Adult	Infant
V_{CO_2}	35 L	.3L	$\dot{V}_{CO_2}$	240 ml Min^{-1}	35 ml Min^{-1}
V_{O_2}	35 L	.3L	$\dot{V}_{O_2}$	300 ml Min^{-1}	44 ml Min^{-1}
P_{50}	25 torr	19 torr	$P_{97.5}$	100 torr	94 torr
A (f)	1.8 L Min^{-1} $torr^{-1}$ (Adult)			.3 L Min^{-1} $torr^{-1}$ (Infant)	
Tc	7 sec	3 sec			
Tau Ti	.5 sec	.25 sec.	Tau Te	.5 sec	.25 sec.
FRC	3000 ml	250 ml			

Volume dependent vagal feedback

The period of these oscillations and the thresholds of
instability in infants and adults are similar to
empiric observations; infants typically exhibit
periodicity of 15 sec. and adults of 30 sec. The period
increases with Cheyne Stokes breathing during
congestive heart failure; this is also consistent with
the model. The model will be used to investigate
abnormalities in breathing pattern and the transition
from stable to oscillatory behavior.

REFERENCES

1. Grodins, F.S. Gray, J.S. Schroeder, K.R., Norins, A.I. and Jones, R.W. (1954). Respiratory responses to CO_2 inhalation; A theoretical study of a nonlinear biological regulator. J Appl Physiol., 7:283-306.

2. Horgan, J.D.and Lange, D.L. (1962). Analog computer studies of periodic breathing. IRE Trans. on Biomed Elec. October:221-228.

3. Milhorn, H. and Guyton, A.C. (1965). An analog computer analysis of Cheyne-Stokes breathing. J Appl Physiol. 20:328-333.

4. Longobardo, G.S. Cherniack, N.S. and Fishman, A.P. (1966). Cheyne-Stokes breathing produced by a model of the human respiratory system. J Appl Physiol. 21:1839-1846.

5. Grodins, F.S. Buell, J. and Bart, A. (1967). A mathematical analysis and digital simulation of the respiratory control system. J Appl Physiol. 22:260-276.

6. Pomeranz, B. Macaulay, R.J.B. Caudill, M.A., et al (In Press). Assessment of autonomic function in man by heart rate spectral analysis. Am J of Physiol.

7. Akselrod, S. Gordon, D. Ubel, F.A. Shannon, D.C. Barger, A.C. and Cohen, R.J. (1981). Power spectrum analysis of heart rate fluctuation: a quantitative probe of beat-to-beat cardiovascular control. Science. 213:220-222.

8. Eckberg, D.L. Abboud, F.M. and Marx, A.L. (1976).
Modulation of carotid baroreflex responsiveness in man:
effects of posture and propranolol. J Appl Physiol.
41: 282-287.

9. Hirsch, J.A. and Bishop, B. (1981). Respiratory
sinus arrhythmia in humans: how breathing pattern
modulates heart rate. Am J Physiol. 241:H620-H529.

10. Kitney, R.I. (1975). An analysis of the nonlinear
behavior of the human thermal vasomotor control system.
J Theor Biol. 52:231-248.

11. Gordon, D. Cohen, R.J. Kelly, D. Akselrod, S. and
Shannon, D.C. (In Press). Sudden infant death
syndrome: Abnormalities in short term fluctuations in
heart rate and respiratory activity. Pediatric
Research.

12. Southall, D.P. Richards, J.M. and Shinebourne,
E.A. (1983). Prospective population-based studies into
heart rate and breathing patterns in newborn infants:
prediction of infants at risk of SIDS? Eds. Tildon,
J.T. Roeder, L.M. and Steinschneider, A. Sudden
Infant Death Syndrome. Academic Pres, New York. pp.
621-652.

13. Euler, C. von (1977). The functional organization
of the respiratory phase-switching mechanisms. Fed
Proc. 36:2375-2380.

14. Edelman, N.H. Epstein, P.E. Lahiri, S. Cherniack,
N.S. (1975). Ventilatory responses to transient
hypoxia and hypercapnea in man. Resp Physiol.
17:302-314.

73

Hypoxia and Endogenous Neurotransmitters in Newborns

E. E. LAWSON, W. A. LONG, J. GINGRAS-LEATHERMAN
and M. C. McNAMARA

INTRODUCTION

Control of breathing in newborns has many similarities to that of
adults. However, the pronounced susceptibility of newborns to develop
hypoventilation and apnea suggest that there are fundamental respiratory control differences from that of adults. We speculated that this
susceptibility may be due to imbalance of excitatory and inhibitory
neurotransmitters during early postnatal life. To test this
hypothesis we have performed two types of experiments. In the first
experiments we determined the effect of naloxone or naltrexone which
block opiate receptors, and aminophylline, an adenosine receptor
blocker, on the biphasic respiratory response to hypoxia. In the
second series of experiments we determined the developmental patterns
and the effects of hypoxia on the concentrations of met-enkephalin
(ME) and biogenic amines in discrete brainstem nuclear regions.
RESULTS

Piglet Studies

Exogenous opiates have long been recognized as central respiratory
depressants and recent studies indicate a similar function for the
endogenous opiate-ligand, methionine enkephalin (ME) (1). Endorphins
and enkephalins are present in the brainstem of fetal and newborn rats
in relatively high concentration (2,3) indicating that they may have
physiologic importance during early life. Additional evidence
suggests that the respiratory depressant action of the endogenous
opiate neurotransmitters is more pronounced in early development (4).

To further study the effects of transmitter antagonists on the
respiratory drive during development and the responses to hypoxia we

utilized anesthetized (chloralose and urethane), paralyzed, vagotomized, mechanically ventilated piglets. The mechanical ventilator rate was continously adjusted by an electronic circuit to maintain the end-tidal PCO_2 constant. Respiratory activity was measured by integration of phrenic nerve activity. The respiratory response to 15% O_2 was assessed before and after administration of naloxone, naltrexone, or aminophylline.

Intravenous naloxone causes phrenic activity to increase in piglets who are less than a week of age and in piglets over three weeks of age (5). This finding confirmed our earlier finding of a similar increase in acutely prepared cats who were either anesthetized or decerebrate. The difference in the per cent increase of respiratory output between young and old piglets was statistically significant and was consistent with the earlier findings of Hazinski et al (3).

<u>Respiratory responses to Hypoxia</u>: We hypothesized that the biphasic respiratory response of newborns to hypoxia may be due to central release of inhibitory neurotransmitters such as enkephalins or adenosine. Using the above piglet model we demonstrated that the biphasic breathing pattern during hypoxia is due to decreased central respiratory output (6). Figure 1 shows the changes in respiratory output which occur on changing the inspiratory gas from 100% O_2 to 15% O_2. The pretreatment pattern is characterized by increasing respiratory output for 1-3 minutes followed by a steady decline towards or below the initial control value. Subsequent exposures to 15% O_2 without any pretreatment did not significantly alter this pattern (7).

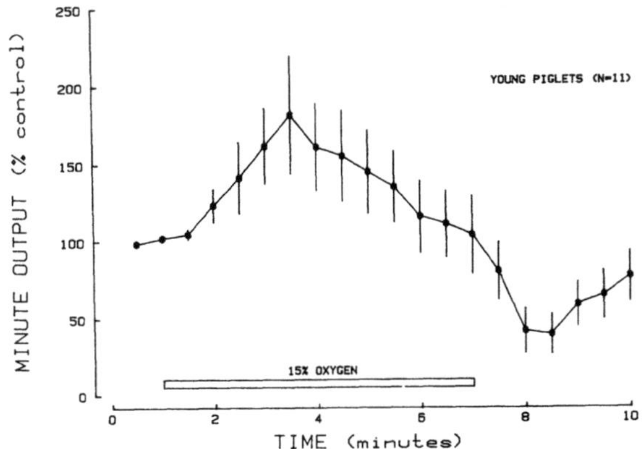

Figure 1. Shown is the average respiratory response of 11 young piglets before, during, and after mechanical ventilation with 15% O_2.

Administration of naloxone (N=3) or naltrexone (N=3) caused respiratory output to increase as shown above. In another group of six piglets, aminophylline administration also caused respiratory output to increase. However, whether pretreated with naloxone or aminophylline the biphasic pattern of respiratory output was unchanged by subsequent ventilation with 15% O_2 (figs. 2 and 3).

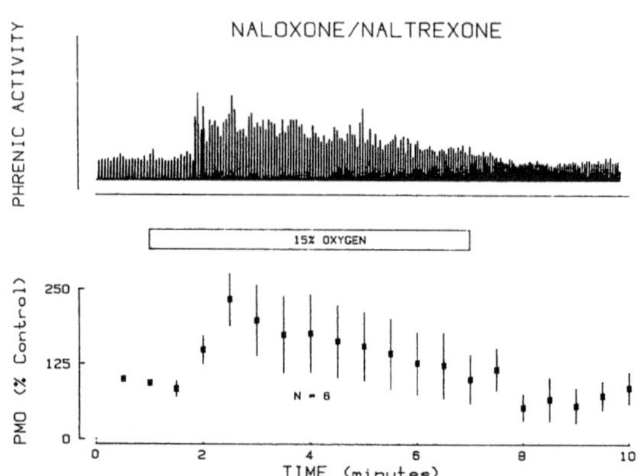

Figure 2. The top graph represents integrated phrenic activity in a naloxone pretreated piglet before during and after exposure to 15% O_2. The bottom graph represents the average response of six piglets after pretreatment with naloxone or naltrexone.

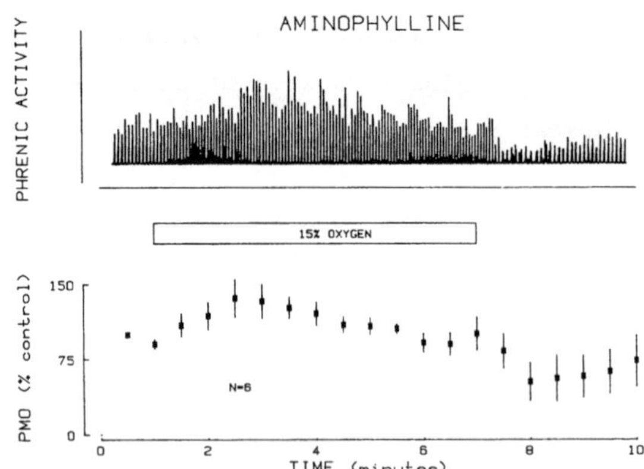

Figure 3. The top graph represents integrated phrenic activity in an aminophylline pretreated piglet before, during, and after ventilation with 15% O_2. The bottom graph represents the average response of six piglets after treatment with aminophylline.

Rabbit Studies

<u>Development of Brainstem Neurotransmitters</u>: To further assess the
hypothesis that imbalances of excitatory and inhibitory neurotransmitters during the postnatal period may account for central neuronal
instability we investigated the development of brainstem ME and
biogenic amines in rabbits. Further, we assessed the effect of
inhalation of 10% O_2 for varying periods of time on the brainstem
concentrations of ME and the biogenic amines in 3 day-old and 21
day-old rabbits.

In the following studies rabbits of varying ages are rapidly
killed by cervical dislocation. Their brains are then removed,
blocked, and rapidly frozen in isopentane and dry ice. Chosen
brainstem nuclear regions were then removed under direct visualization
using the microdissection techniques described by Palkovits et al (8).
Met-enkephalin (ME) was measured using a commercially available
radioimmunoassay kit (Immuno Nuclear, Stillwater, MN - USA). The
biogenic amines were measured using sensitive radioenzymatic assays as
previously described (9). Both ME and the biogenic amines are
expressed as ng transmitter per mg protein.

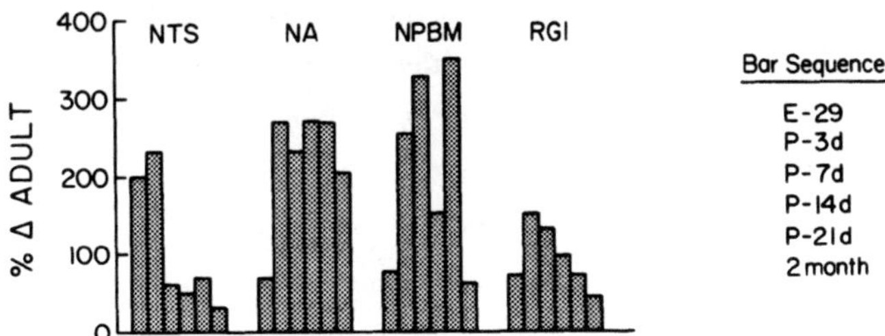

Figure 4. Developmental profiles for ME in the n. tractus solitarius
(NTS), n. ambiguus (NA), n. parabrachialis medialis (NPBM), and the n.
reticulogigantocellularis (RGI). The values are expressed as the per
cent difference from the adult value for each nucleus.

ME was measurable in all nuclear groups analyzed and at all ages examined. All the forebrain nuclear regions demonstrated a statistically significant increase in [ME] throughout development whereas the major ontogenic pattern observed within the brainstem nuclei was one of decreasing [ME] throughout development. Two exceptions to these patterns were noted, one in the forebrain region of the lateral septum, the other in the locus coeruleus. Two developmental patterns emerged when the brain nuclei were grouped according to those having or those not having major respiratory significance. In the respiratory related nuclei ME was higher in prenatal and early postnatal life than in adults. In contrast, the prenatal and early postnatal ME levels in non-respiratory nuclei were lower or equal to the adult values. The developmental profiles for the brainstem respiratory related nuclei are demonstrated in figure 4.

<u>Hypoxia and Brainstem Enkephalin levels</u>: To assess the effect of hypoxia on endogenous opiates in developing rabbits we exposed 3 day-old and 21 day-old rabbits to 10% O_2 for varying periods. All rabbits were placed in water jacketed environmental chambers which were continuously flooded with warmed humidified 21% O_2 or 10% O_2. Four groups of rabbits were studied at each age. Group I remained in a chamber for six hours and always breathed 21% O_2 (balance nitrogen). Group II breathed 10% O_2 for 10 minutes followed by 21% O_2. This was repeated 12 times over 6 hours. Group III breathed 10% O_2 for 2 hours followed by 4 hours breathing 21% O_2. Group IV breathed 21% O_2 for 4 hours then 10% O_2 for the remaining 2 hours. At the end of the six hour period the pups were killed and prepared as described above.

The effect of hypoxia on ME in the n. parabrachialis, n. ambiguus, and the n. tractus solitarius is demonstrated in figure 5. There was no significant effect of hypoxia on [ME] in any nucleus in the three day-old pups. Variable effects of hypoxia were observed in the nuclei of the 21 day-old pups. The findings in the three day-old pups reinforce the conclusion of the naloxone studies above that endogenous opiate-like transmitters are not importantly involved in physiologic responses to moderate hypoxia.

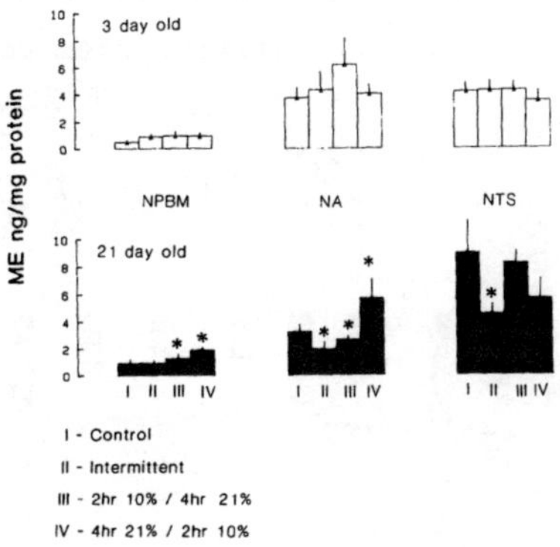

Figure 5: Demonstrated are the effects of six hours exposure on
met-enkephalin concentrations of the n. parabrachialis medialis
(NPBM), the n. ambiguus (NA), and the n. tractus solitarius (NTS) to
different gases in environmental chambers.

Hypoxia and Biogenic Amines: Serotonin has been reported as both
stimulating (10) and depressing (11,12,13) of respiration. It is
highly likely that these differences are due to co-responses of
different neurotransmitters (12) and pre- and postsynaptic effects of
serotonin. To study the effect of hypoxia on brainstem biogenic amine
levels during development we measured dopamine, norepinephrine, and
serotonin (5HT) levels in the dorsal raphe (DR), locus coeruleus (LC),
the substantia nigra (SN), and the n. reticularis pontis oralis (rpo)
in rabbits.

Three day-old and 21 day-old rabbits were exposed to different
inspiratory gases as described for the above ME studies. Compared
with control levels no changes in DA concentration were observed in
any of the nuclei under any of the test hypoxia conditions (e.g.
groups II, III, and IV). Occasional, but inconsistent, changes were
noted in norepinephrine values. However, under all of the hypoxia
conditions serotonin was consistently diminished in the 3 day-old
rabbits and unchanged or elevated in the 21 day-old rabbits (fig. 6).
The 3 day-old results are supported by similar findings of 5-HTP

levels in young rats (14) but the 21 day-old results are distinctly different from results of earlier experiments (15) and suggest a developmental sensitivity of this transmitter system to hypoxia.

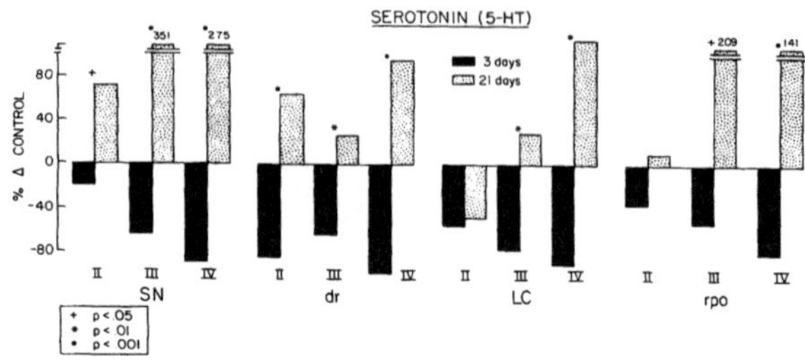

Figure 6. Demonstrated are the per cent change from control in 3 day-old and 21 day-old rabbits for serotonin in four brainstem and medullary nuclei and three hypoxia conditions.

Our findings suggest that serotonin utilization in these brainstem nuclei, and their fibers of passage, exceeds the synthesis rate under various hypoxia conditions. The 5HT results in the 3 day-old rabbits suggest increased 5HT utilization or impaired synthesis during moderate hypoxia and inability to replenish 5HT levels quickly in the recovery period (e.g. Group III). It appears that the central metabolism of serotonin is sufficiently mature by 21 days of age to accomodate the increased utilization. Under basal conditions, we demonstrated that 5HT levels and 5HT turnover are reduced in young rabbits, but we did not demonstrate any difference in basal metabolic turnover between 3 and 21 day old rabbits (16).

CONCLUSION

Results of others showing no respiratory effect of naloxone administration on completely unoperated and unanesthetized animals (17) indicates that opiate-like transmitters do not provide tonic respiratory inhibition under basal conditions. We have shown that naloxone administration in anesthetized, paralyzed, mechanically ventilated and acutely prepared piglets stimulates respiratory output. Our naloxone and developmental ME data indicate that this effect is more prominent in young piglets suggesting an increased potential effect of endogenous opiate-like transmitters in younger animals.

This finding is in harmony with the finding of increased [ME] levels in the brainstem nuclei specifically related to respiratory control of young animals. However, that naloxone does not affect the biphasic respiratory response to short periods of hypoxia and that in young animals brainstem nuclear [ME] levels are not affected by hypoxia suggest that ME does not have a major role in the respiratory depression during acute hypoxia in newborns. In addition the aminophylline results indicate that adenosine, a putative inhibitory neurotransmitter, does not determine the respiratory patterns of newborns during moderate hypoxia.

On the other hand, we have shown that brainstem nuclear levels of 5HT are differentially affected in young rabbits by moderate hypoxia compared with 21 day-old rabbits. This finding suggests, but does not prove, a functional role for disturbances in 5HT release or metabolism as an explanation of aberrant behavior patterns seen in newborns, including the biphasic respiratory response to hypoxia. Finally, it should be noted that the neurotransmitter systems may only have long term modulatory effects on respiratory control. Acute effects of hypoxia, such as respiratory depression seen several minutes after onset of hypoxia, may be mediated by other as yet undetermined mechanisms.

REFERENCES

1. Denavit-Saubie, M., Champagnat, J. and Zieglgansberger, W. (1978). Effects of opiates and methionine-enkephalin on pontine and bulbar respiratory neurones in the cat. Brain Res., 155,55-67.

2. Palmer, M.R., Miller, R.J., Olson, L. and Seiger, A. (1982). Prenatal ontogeny of neurons with enkephalin-like immuno reactivity in the rat central nervous system: An immunohistochemical mapping investigation. Medical Biol. 60, 61-88.

3. Bayon, A., Shoemaker, W.J., Bloom, F.E., Mauss, A. and Guillemin, R. (1979). Perinatal development of the endorphin and enkephalin containing systems in the rat brain. Brain Res. 179, 93-101.

4. Hazinski, J.A., Grunstein, M.M., Schlueter, M.A. and Tooley, W.H. (1981). Effect of naloxone on ventilation in newborn rabbits. J. Appl. Physiol. 50, 713-717.

5. Long, W.A. and Lawson, E.E. (1983). Developmental aspects of the

effect of naloxone on control of breathing in piglets. Resp. Physiol. 51, 119-129.

6. Lawson, E.E. and Long, W.A. (1983). Central origin of biphasic breathing pattern during hypoxia in newborns. J. Appl. Physiol:Respirat, Environ. Exercise Physiol. 55, 483-488.

7. Long, W.A. and Lawson, E.E. (1984). Neurotransmitters and biphasic respiratory response to hypoxia. J. Appl. Physiol.:Respirat. Environm. Exercise Physiol. 57, 213-222.

8. Palkovits, M., Brownstein, M., Saavedra, J.M. and Axelrod, J. (1974). Norepinephrine and dopamine content of hypothalamic nuclei of the rat. Brain Res. 97, 137-149.

9. McNamara, M.C. and Lawson, E.E. (1983). Ontogeny of biogenic amines in respiratory nuclei of the rabbit brainstem. Developmental Brain Res. 7, 181-185.

10. Millhorn, D.E., Eldridge, F.L. and Waldrop, T.G. (1980). Prolonged stimulation of respiration by endogenous central serotonin. Resp. Physiol. 42, 171-176.

11. Lundberg, D., Mueller, R.A. and Breese, G.R. (1980). An evaluation of the mechanism by which serotonergic activation depresses respiration. J. Pharmacol. Exper. Theraputics, 212, 397-404.

12. Meldrum, M.J. and Isom, G.E. (1981). Role of monoaminergic systems in morphine-induced respiratory depression. Neuropharmacology, 20, 169-175.

13. Armigo, J.A. and Florez, J. (1974). The influence of increased brain 5-hydroxytryptamine upon the respiratory activities of the cat. Neuropharmacology, 13, 971-986.

14. Hedner, T. and Lundborg, P. (1980). Serotonin metabolism in neonatal rat brain during asphyxia and recovery. Acta Physiol. Scandanavia, 109, 163-168.

15. Hedner, T. and Lundborg, P. (1979). Regional changes in monoamine synthesis in the developing rat brain during hypoxia. Acta Physiol. Scandanavia, 106, 139-143.

16. McNamara, M.C. and Lawson, E.E. (1984). Turnover and synthesis of biogenic amines in discrete brainstem nuclei of the rabbit. Brain Res. 299, 259-264.

17.Wheeler, M.E. and Farber, J.P. (1983). Naloxone administration and ventilation in adult cats. Brain Res. 258, 343-346.

74

Three-Dimensional Tissue Reconstruction Reveals Integrative Structural Features among Neurons within Central Respiratory Centers of the Brainstem

J. J. QUATTROCHI and J.H. RHO

INTRODUCTION

We propose that several neuronal groups in the brainstem respiratory network have dendritic interrelationships that are crucial to the central control of respiratory function. Physiological properties of <u>individual neurons</u> in different brainstem nuclei have been well characterized in relation to respiratory activity; however, a serious deficit of most studies to date has been failure to define a more macroscopic view of how synaptic integration among these neurons may coordinate synchronous and oscillatory respiratory activity. We have focused on spatial relationships among apparently disparate neurons in order to investigate if there might be a multicellular assembly modeled to generate a rhythmic synchronization beyond the scope of any individual neuron. While the brainstem reticular formation appears to be involved in central respiratory control[1], synaptic relationships between neurons of this medullary region and other nuclear respiratory groups have not been identified. We now postulate an integrative role based on dendritic interactions between the reticular formation and dorsal and ventral respiratory neuronal groups within the lower brainstem in coordinating synchronous respiratory activity.

Neuronal integration in the brainstem has been proposed by Scheibel[2] to be mediated through a system of dendritic bundles which may guide rhythmic repetitive output sequences. Rather than focus on how each respiratory group sustains its own rhythmicity, we have extended Scheibel's concept to the system level on the basis of initial data on the orientation in a given 3-dimensional space of dendrites from neurons in different brainstem groups. Our working hypothesis is that these

431

anatomically-defined respiratory neuronal groups, the dorsal
respiratory component of the nucleus tractus solitarius (NTS)
the ventral respiratory neurons of the nucleus ambiguus (NA)
and the reticular formation, are integrated through a 3-dimen-
sional chain of elongated dendritic processes within the matrix
of the brainstem. To test this hypothesis we are performing
rapid Golgi, combined Golgi-EM, and computer-assisted serial
contour analyses, and are developing a new technique which
allows specific brainstem neurons proved to innervate final
common pathway phrenic motor neurons to be visualized in structural relationship with other brainstem neurons. This level of
organizational precision has been unattainable previously.

DENDRITIC BUNDLES

The dendritic bundle, defined as a length-wise closely apposed
grouping of dendrites, has been demonstrated in various regions
of the CNS[3], but has not been described in the human brainstem. Dendritic bundling within the brainstem of the kitten is
thought to serve a synchronizing integrative function for
control of respiration and cardiovascular rhythm[4]. By means
of rapid Golgi impregnation, we now identify dendritic bundling
within the reticular formation of the human infant brainstem
(Fig. 1). Bundle segments included 2 to 5 dendrites forming
units approximately 60-150 um in length. After several neurons
were digitized in each transverse 150 um section of the medulla
in proper spatial relationship with each other and simultaneously computer-reconstructed into a registered series, dendritic bundling was found to occur in the transverse plane of
the brainstem perpendicular to the rostral-caudal axis. Real
time computer-image rotation of reconstructed 3-dimensional
neuronal matrices simultaneously with global contour perimeters
of the infant medulla revealed that dendrites radiate within
restricted transverse planes so as to demonstrate spatial interrelationships among widely separate neurons of the NTS, NA, and
reticular formation (Fig. 2). This non-random dendritic distribution may well participate in a synchronous coordination
between these different respiratory-related areas.
    No previous ultrastructural analysis has been reported in
the human reticular formation. We have examined a single case,
a six-month-old infant (7 hours postmortem), by the Golgi-EM
method[5] and have found dendrodendritic gap junctions within
the medullary reticular formation (Fig. 1). The presence of gap
junctions, the ultrastructural correlate of synchronizing electrotonic synapses[6], within a dendritic bundle provides a possible mechanism by which components of the reticular formation
might participate in central system integration of respiration.

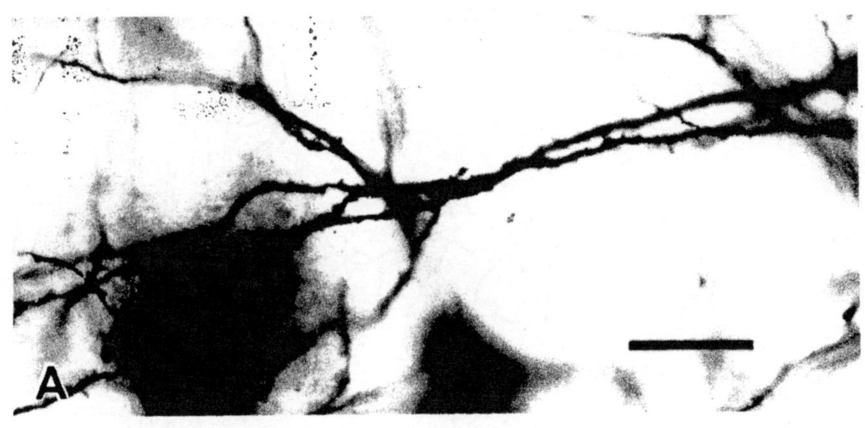

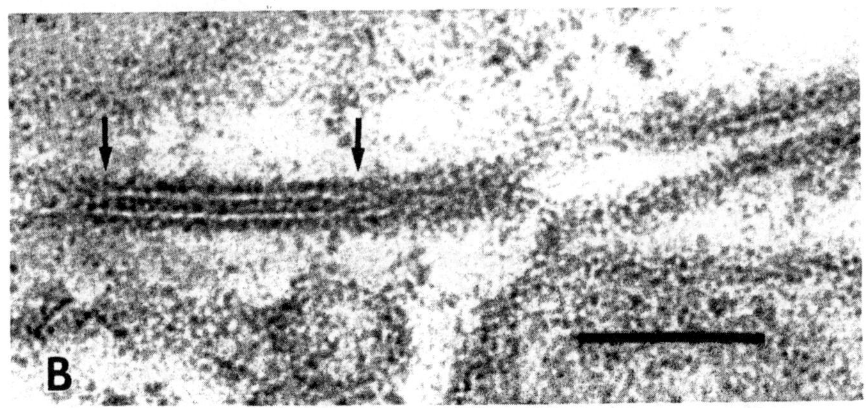

FIGURE 1   Dendrodendritic apposition in the medulla of a six-
month-old infant. (A) Bundle segment of two Golgi-
impregnated dendrites (2 um diameter) in close linear
apposition (<1 um) with each other extends through
the transverse plane of the magnocellular reticular
formation. Scale bar = 40 um.
(B) Electrotonic synapse between two reticular formation dendrites is indicated by a gap junction (arrows)
measuring 1.6 nm in width. Scale bar = 50nm.

FIG. 2

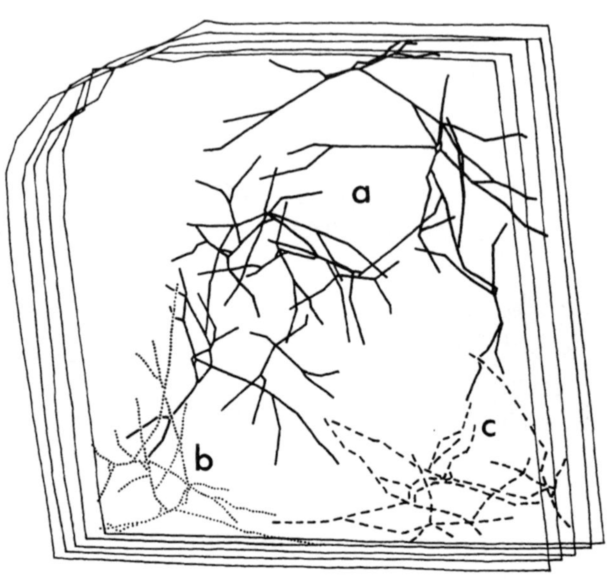

FIG. 3

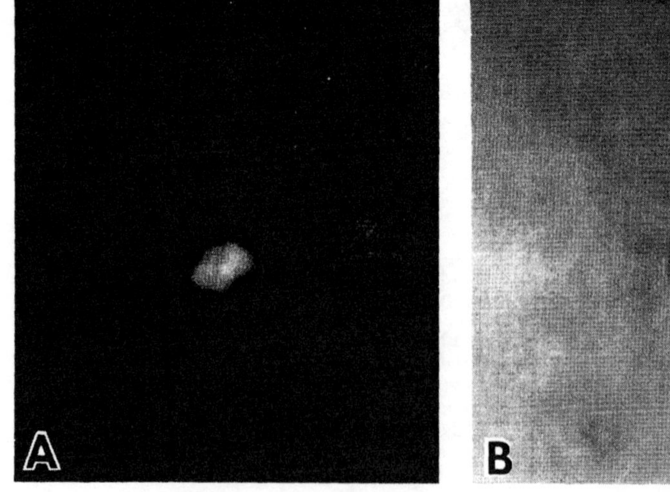

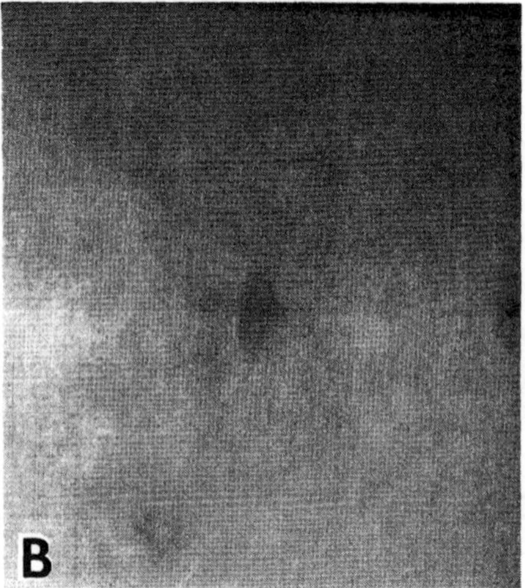

INTRACELLULAR LUCIFER YELLOW FILLING OF IDENTIFIED NEURONS

To pursue system integration further, we are developing a new
methodology that will allow us to study in _fixed_ _tissue_ slices
dendrodendritic ultrastructural relationships between HRP-
labeled neurons which are identified within a specific respiratory group and neurons of the reticular formation. Using methodology adapted from electrophysiological _in_ _vitro_ brain slice
recording techniques, we have found that the use of fixed tissue
slices in place of _in_ _vitro_ slices has several advantages in
neuroanatomical studies. Gelfoam is saturated with 30% HRP and
implanted within the phrenic motor nucleus between cervical
segments 3 and 5 in 60-day-old mice. After 48 hours, animals are
anesthetized with 0.02 cc Avertin (1.2% 2,2,2-tribromo-ethanol
in amyl alcohol)/gm body weight and killed by cardiac perfusion
with 30 ml 0.9% buffered saline followed by 4% paraformaldehyde
and 0.5% glutaraldehyde in 0.1M phosphate buffer, pH 7.3, at a
rate of 10 ml/min for 20 minutes, and then with 100 ml 0.9%
buffered saline. All fluids are at room temperature. Brainstems
are sectioned on a Vibratome into a series of transverse 200 um
slices which then are reacted with 0.1% diaminobenzidine (DAB)
as an HRP stain so that retrogradely labeled cell bodies in the
dorsal respiratory group of the NTS can be identified. Slices
are mounted in physiological saline in a tissue chamber so that
HRP-labeled cells and dendritic processes can be visualized.
Such cells are filled iontophoretically with 5% Lucifer Yellow
(LY). In addition, unlabeled cell bodies within the reticular
formation also are penetrated with a microelectrode and filled
with LY. Intracellular filling is visualized directly with a

FIGURE 2   Three-dimensional computer reconstruction of trans-
            verse hemisections of the human infant brainstem
            medulla. Several Golgi-impregnated neurons and
            boundary contours were digitized simultaneously in
            each of 5 serial 150 um sections. Angular rotation
            of sections in a registered series demonstrates a
            radiating dendritic pattern through the transverse
            plane among neurons of the reticular formation (a),
            n. tractus solitarius (b), and n. ambiguus (c).
            Reconstructed at 80X.

FIGURE 3   LY filling of an HRP-labeled neuron in the NTS.
            (A) Fixed tissue slice preparation demonstrates an
            LY-filled neuron which first was HRP-labeled from
            the phrenic motor nucleus. Four LY-filled dendrites
            are seen to originate from a multipolar cell body
            (20 um in diameter). 250X. (B) DAB electron dense
            reaction product is illustrated within the same
            neuron after photo-oxidation of LY with the addition
            of 0.1% DAB to the preparation. 250X.

compound epi-fluorescence microscope (Fig. 3). Then, under blue illumination, all LY-filled cells are photo-oxidized in the presence of 0.1% DAB in the bathing fluid so as to convert LY to a DAB electron dense reaction product (Fig. 3). This method permits us to visualize the detailed dendritic morphology of a specific NTS neuron whose axon has been proven by retrograde HRP labeling to innervate the phrenic motor nucleus monosynaptically and to determine its spatial relationships with other LY-filled cells within the reticular formation. After this final reaction with DAB, slices are osmicated, embedded in plastic, and sectioned for electron microscopy (EM). EM analysis of such labeled cells will be a significant component of our ongoing endeavor to elucidate detailed synaptic relationships and EM contours between neurons of the dorsal respiratory group and the reticular formation which otherwise could not be recognized by conventional morphological methods.

SUMMARY AND FUTURE DIRECTIONS

We have presented a view of the central respiratory network of the human and murine brainstem that emphasizes hitherto unrecognized structural relationships among neurons of different neuronal groups. The data suggest a system model of central respiratory control based upon the unusual spatial organization of neuronal processes and specific types of synaptic interrelationships between respiratory nuclear groups and the reticular formation. Specifically, dendritic bundling within the neuropil of the reticular formation may provide the apparatus for synchronous oscillatory functional integration between these apparently separate neuronal groups. We intend to ask how this spatial arrangement of dendrites among respiratory groups comes to develop during the stages of functional maturation of the brainstem. This has relevance to general developmental neuroscience and to the study of human clinical conditions such as the sudden infant death syndrome[7].

This work was supported in part by N.I.H. Grants NS06965 to J. H. Rho; NS20820 and NS07264 to Professor R.L. Sidman; and a Mental Retardation Core Grant HD18655 to Children's Hospital.

REFERENCES

1. Hugelin, A. (1980). Does the respiratory rhythm originate from a reticular oscillator in the waking state? In: Hobson, J.A. and Brazier, M.A.B. (eds). The Reticular Formation Revisited. pp. 261-74. (New York: Raven Press)

2. Scheibel, M.E. and Scheibel, A.B. (1975). Dendrites as neuronal couplers: the dendrite bundle. In: Santini, M. (ed). Golgi Centennial Symposium. pp. 347-54. (New York: Raven Press)

3. Roney, K.J., Scheibel, A.B., and Shaw, G.L. (1979). Dendritic bundles: survey of anatomical experiments and physiological theories. Brain Res. Rev., 1, 225-71.

4. Scheibel, M.E., Davies, T.L., and Scheibel, A.B. (1973). Maturation of reticular dendrites: loss of dendritic spines and development of bundles. Exp. Neurol., 38, 301-10.

5. Fairen, A., Peters, A., and Saldanha, J. (1977). A new procedure for examining golgi impregnated neurons by light and electron microscopy. J. Neurocytol., 6, 311-37.

6. Sotelo, C. (1975). Morphological correlates of electrotonic coupling between neurons in mammalian nervous system. In: Santini, M. (ed). Golgi Centennial Symposium. pp. 355-65. (New York: Raven Press)

7. Quattrochi, J., McBride, P.T., and Yates, A. (1984). Brainstem immaturity in SIDS: a quantitative rapid golgi study of dendritic spines in 95 infants. Brain Res., In Press.

75

Control of Breathing in Patients with Chronic Obstructive Pulmonary Disease with Acute Ventilatory Failure

D. MURCIANO, M. AUBIER, R. PARIENTE and J. MILICI-EMILI

Acute ventilatory failure in patients with chronic obstructive pulmonary disease (COPD) is characterized by increased arterial PCO_2 ($PaCO_2$) and severe hypoxemia. Although the literature is replete with measurements of blood gases in these patients, until recently little quantitative information was available concerning their breathing pattern, respiratory drive and respiratory mechanics. In 1980, Aubier et al.(1) were the first to demonstrate that, contrary to previous belief, the minute ventilation ($\dot{V}_E$) and the inspiratory drive (measured in terms of mouth occlusion pressure, PO.1) are not depressed in COPD patients in acute ventilatory failure. In fact, as shown in Table 1, $\dot{V}_E$ is about twice as large as in normal resting subjects and PO.1 is increased about 8-fold (the value of PO.1 for normals at rest amounting to $\tilde{}$ 1 cmH_2O). Since COPD is invariably associated with pulmonary hyperinflation (12),it is axiomatic that in COPD patients the inspiratory muscles operate at a marked mechanical disadvantage as pressure generators

(8). Accordingly, the high values of PO.1 found in the acutely ill COPD patients imply that the activity of their inspiratory muscles must be extremely high. Clearly, such high levels of inspiratory muscle activity cannot be sustained for a prolonged period without the development of muscle fatigue (1Ø). As a result, mechanical ventilation is required.

TABLE 1.   Mean Values ($\pm$ SE) in 2Ø COPD Patients in Acute Respiratory Failure Breathing Room Air and O_2-Enriched Air for 3Ø Minutes ($FI_{O2} \simeq$ Ø.4) (1).

	Room Air	O_2-enriched air (3Ø min)
Pa_{CO2} (mmHg)	61 $\pm$ 2	68 $\pm$ 3
Pa_{O2} (mmHg)	38 $\pm$ 2	12Ø $\pm$ 13
$\dot{V}_E$ (L/min)	11.4 $\pm$ Ø.9	9.8 $\pm$ Ø.7
PO.1 (cmH_2O)	8.3 $\pm$ Ø.8	4.9 $\pm$ Ø.9

Since $\dot{V}_E$ in acutely ill COPD patients is relatively high, their increased arterial PCO_2 (Table 1) cannot be attributed to a "depressed minute ventilation", but rather mainly reflects an increase of wasted (dead space) ventilation due both to increased physiological dead space and reduced tidal volume (V_T) (1,2). In fact, the breathing pattern of acutely ill COPD patients is characteristically very rapid and shallow (1,2). Apart from decreasing the effective alveolar ventilation (and hence, increasing $PaCO_2$), the increased respiratory frequency (f) exhibited by these patients leads to a further "dynamic" increase in end-expiratory lung volume. This is due to the combined effect of decreased duration of expiration (Te) and of expiratory flow limitation due to dynamic airways compression (1,5). The latter is often present even in stable COPD patients (6), and is further exacerbated by the airway infection which commonly triggers the acute ventilatory failure in COPD patients (4,5). The combination of (a) decreased Te which is a characteristic feature of airway infection in COPD patients (1,2) and (b) expiratory flow limitation during tidal breathing (5) necessarily results in a "dynamic" increase of lung volume, i.e. with the shortened Te there is not enough time during expiration for the respiratory system to return to its resting (elastic equilibrium) volume. As a result, acutely ill COPD patients invariably exhibit a positive end-expiratory alveolar pressure which has been termed intrinsic or auto-PEEP (5,9). Intrinsic PEEP has been shown to amount to up to 13 cmH_2O in acutely ill COPD patients (7). The presence of intrinsic PEEP implies that in addition to the pressure required to produce the actual breathing movements (i.e. V_T), the inspiratory muscles of the COPD patients must also develop a substantial pressure to overcome the intrinsic PEEP. In fact, the latter can be regarded as an inspiratory threshold load . In addition, abnormally high inspiratory pressures are also required to produce the tidal breaths because of increased resistance and decreased lung compliance. In fact, the inspiratory pulmonary flow resistance in acutely ill COPD patients averages about 17 $cmH_2O.1^{-1}.s$ (4), as compared to about 3 $cmH_2O.1^{-1}.s$ in normal individuals.

Similarly, the dynamic lung compliance in COPD patients in acute ventilatory failure averages about 34 ml cmH_2O^{-1}, as compared to about 200 $ml.cmH_2O^{-1}$ in normal individuals (4). It should also be stressed that the "dynamic" increase of lung volume caused by airway infection further reduces the effectiveness of the inspiratory muscles as pressure generators, hence promoting the failure of the ventilatory pump due to fatigue of the inspiratory muscles.

Based on the above premises a tentative schematic representation of the factors leading to acute ventilatory failure (respiratory muscle fatigue) in COPD patients is presented in the following diagram.

Acute Ventilatory Failure in COPD

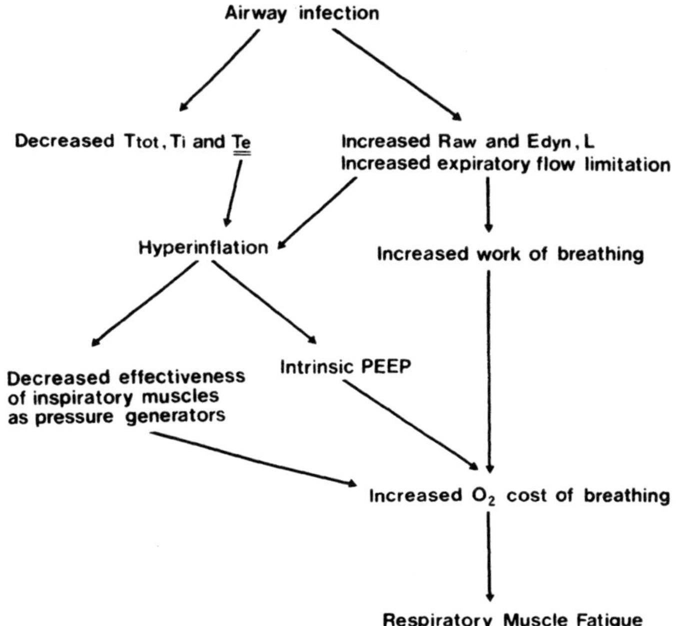

As shown in the above diagram, an excessive (i.e. fatiguing) O_2 cost of breathing in acutely ill COPD patients can be brought about <u>directly</u> by increased mechanical work of breathing due to the high airway resistance (Raw) and dynamic pulmonary elastance (Edyn,L - the reciprocal of dynamic lung compliance), and <u>indirectly</u> by the increased respiratory frequency (shorter total respiratory cycle duration, T_{tot}) which is associated with a shortened duration of inspiration and expiration. The shorter Te leads to "dynamic" pulmonary hyperinflation which, by virtue of intrinsic PEEP and decreased efficiency of the inspiratory muscles, leads to a further increase of the O_2 cost of breathing. In this scheme, the rapid and shallow breathing pattern plays a paramount role in promoting acute ventilatory failure, and indeed during the recovery period of these patients, the respiratory frequency decreases and Te increases (1).

The nature of the increased f in acutely ill COPD patients is not fully understood. Conceivably, airway infection stimulates the pulmonnary irritant receptors, and this is known to cause rapid and shallow breathing (1,11). It is also possible that stimulation of other pulmonary receptors (e.g. stretch and J-receptors) is involved (3). Whatever the nature of the increased f (and decreased Te) in acutely ill COPD patients, this appears to be an important mechanism leading to ventilatory failure. Accordingly, further studies designed to explain the cause(s) of the rapid and shallow breathing in acutely ill COPD patients are needed.

REFERENCES

1.   Aubier, M, Murciano, D.,Fournier, M., Milic-Emili, J., Pariente, R. and Derenne, J.Ph.(1980). Central respiratory drive in acute respiratory failure of patients with chronic obstructive lung disease. Am.Rev.Respir.Dis., 122, 191-99

2.   Aubier, M., Murciano, D, Milic-Emili, J., Touaty, E., Daghfous, J., Pariente, R. and Derenne, J.Ph. (1980). Effects of O_2 administration on ventilation and blood gases in acute respiratory failure of patients with chronic obstructive lung disease. Am.Rev.Respir.Dis.,122, 747-54

3.   Coleridge, J.C.G. and Coleridge, H.M. (1984). Afferent vagal C fibre innervation of the lungs and airways and its functional significance. Rev.Physiol.Biochem.Pharmacol., 99, 2-110

4.   Fleury, B., Murciano, D., Talamo, C., Aubier, M., Pariente, R. and Milic-Emili, J. (1985). Work of breathing in patients with chronic obstructive pulmonary disease in acute respiratory failure. Am.Rev.Respir.Dis., (Accepted for publication)

5.   Gottfried, S.B., Rossi, A., Higgs, B.D., Calverley, P.M.A., Zocchi, L., Bozic, C. and Milic-Emili, J. (1985). Non-invasive determination of respiratory system mechanics during mechanical ventilation for acute respiratory failure. Amer.Rev.Respir. Dis., (In press)

6.   Leaver, D.G. and Pride, N.B. (1971). Flow-volume curves and expiratory pressures during exercise in patients with chronic airways obstruction. Scand.J.resp.Dis., 52, (suppl.77) 23:7

7.   Murciano, D., Aubier, M., Bussi, S., Derenne, J.Ph., Pariente, R. and Milic-Emili, J. (1981).   Comparison of esophageal, tracheal, and mouth occlusion pressure in patients with chronic obstructive pulmonary disease during acute respiratory failure. Am.Rev.Respir.Dis., 126, 837-41

8.   Pengelly, L.D., Alderson, A.M. and Milic-Emili, J. (1971). Mechanics of the diaphragm. J.Appl.Physiol.30, 797-805

9.   Pepe, P.E. and Marini, J.J. (1982). Occult positive end-expiratory pressure in mechanically ventilated patients with airflow obstruction. Am.Rev.Respir.Dis., 126,166-70

10. Roussos, Ch. (1982). The failing ventilatory pump. Lung, 160, 59-84

11. Sant'Ambrogio, G. (1982). Information arising from the tracheobronchial tree of mammals. Physiol.Rev., 62, 531-69

12.   Sharp, J.T., van Lith, P., Nuchprayoon, C.V., Briney, R., and Johnson, F.N. (1968). The thorax in chronic obstructive lung disease. Amer.J.Med., 44, 39-46

76

The Role of Ambient Temperature, Airway Mechanosensory and Chemical Stimuli to Breathing During Sleep in Developing Lambs

P. JOHNSON, D. C. ANDREWS, L. FEDORKO and J. C. WOLLNER

Introduction

The respiratory system is usually considered to be primarily under chemical control with other factors such as metabolism, thermo regulation and airway mechanosensory reflexes acting as modulaters. However in late fetal life in a precocial species such as the sheep, breathing activity is a feature of the low voltage electrocortical rapid eyemovement (REM) state and is probably inhibited during the high voltage slow wave electrocortical (SWS) state. The usual chemical factors are unable to initiate breathing during SWS whereas the infusion of indomethicin(1) (even after denervation of the peripheral chemoreceptors) or the cooling of amniotic and thus fetal temperature by skin surface cooling (2)induced breathing activity to occur continuously during SWS and REM. Fetal breathing is energy consuming,increasing the resting oxygen consumption of the fetus by 30% while causing a fall in pa02 and a rise in paCO2 during REM (3). Thus fetal breathing ,which is essential for lung development (4) but unnecessary for respiration, is heavily modulated by behavioural state and environmental temperature.
Convincing clinical observations have clearly shown the immense importance of mechano-sensory reflexes in the establishment of effective lung volume (5) and breathing in the immediate postnatal period. The active expiratory

445

component was well-described, both in the healthy term baby and particularly in the preterm infant grunting with the respiratory distress syndrome (6), although its significance in normal respiratory control mechanisms was not fully appreciated. Nonetheless it was clearly shown that effective gas exchange depended on the expiratory grunt which was enhanced by hypoxemia. Glottic constriction in response to hypoxia, as opposed to the dilation observed in deeply anaesthetized animals (7) has only recently been recognized (8).

Despite the recognition that breathing activity, albeit intermittent, was a normal feature of fetal life (9), the activation of chemoreceptors at, or shortly after, birth has been considered critical in the onset of independent respiration after birth. However in tracheostomised lambs it was found that carotid body denervation or dopamine infusion caused a fall in breathing frequency by lengthening of expiratory time in the second week of life but not in the immediate newborn period (10). This suggestion that carotid body function was of greater importance in the older lamb was supported by the observation that 4 of 7 carotid body denervated lambs died ´unexpectedly´ at four to six weeks of postnatal age (11). Recently it has been shown that the arterial chemoreceptors in the newborn lamb at birth were silenced when the arterial oxygen tension rose abruptly with the onset of air breathing and remained so for a day or so before ´resetting´ to adult oxygen values had occurred (12). So what is the stimulus that sustains breathing in the newborn period?

In earlier studies in which chronically monitored telemetred lambs reared with their ewes had been observed for sleep state, breathing frequency and heart rate, from 1 until 32 days of age, it had been noted that both breathing frequency and heart rate rose after birth before declining between the second and third week of postnatal life (13). It has been tacitly assumed that the higher respiratory rates of the newborn are due to the high metabolic demands of the neonatal period. However, the rise in the heart rate, not commented on in earlier studies, has now been well described in a number of species. The time course of the rise and fall in the heart rate and respiratory frequencies in our chronically monitored lambs were similar to those in a group of unanaesthetised lambs in which heart rate, cardiac output, oxygen consumption, amongst other variables were measured serially (14) during postnatal life. It was assumed in the latter studies that oxygen consumption was close to the minimal level as they were studied when quiet at 22-26 °C.In a large study of free-ranging lambs the progressive rise in minimal oxygen consumption over the first 36 hours of life was clearly shown; however these studies included observations on breathing frequencies as well (15). They showed the marked increase in breathing frequency that occurred during the first 36 hours of postnatal life as minimal oxygen consumption rose while thermal efficiency, as judged by a marked fall in the lower critical temperature, was increasing. Notably breathing frequencies were lower when the lambs were below the lower critical temperature suggesting that increased metabolic demands were probably met by an increase in tidal volume and a reduction in frequency. Joining these two sets of observations with our own, in which oxygen consumption was not measured, and those in which oxygen consumption, cardiac output and heart rate were (16), but minute ventilation and breathing frequency were not, it is not difficult to speculate at the close relationship between metabolic demands and breathing frequency.

In our earlier studies, we had noted that as the breathing frequency fell during SWS in the second, third and fourth weeks of postnatal life, so the appearance of prominent laryngeal expiratory constrictor activity appeared. The fall in frequency was almost entirely accounted for by an increased expiratory time and in many instances anything from 60-100% of breaths in SWS were associated with marked expiratory glottic constriction (16). Breathing frequency in rapid eye movement (REM) sleep at this stage was signficantly higher than that observed in SWS whereas, remarkably perhaps, in the first 10 days of life breathing frequencies had been higher in SWS than in REM sleep. At a month of age a tracheal window was implanted under halothane anaesthesia. When this 'window' was opened during SWS there was a marked drop in breathing frequency, often an arrhythmia, with prolonged and enhanced laryngeal constrictor activity. This only occurred during SWS and was entirely consistent with the absence of this phasic laryngeal expiratory activity, as with the loss of intercostal muscle activity, that occurs during REM sleep. The ventilatory response to hypoxia, after the acute arousal response, was not sustained if the tracheal window was open(17). It was evident that laryngeal constrictor activity was enhanced by hypoxemia. Applying positive expiratory pressure, reconnecting the upper airway, stimulating to arousal, adding carbon dioxide, sometimes pulsing air through the redundant upper airway, and the chance increase in ambient temperature(>20 $^{\circ}$C) would restore effective breathing. Although we had defined an age between 2 and 6 weeks in which respiratory failure could be readily induced by hypoxemia, the variability both within an animal and between animals was considerable suggesting that other factors modulated the response. It was evident that the vagally mediated 'braking' mechanism, elegantly demonstrated in unanaesthetised cats (18) as a response to opening a tracheal window, was in fact an essential component of spontaneous breathing in the intact post-neonatal lamb, especially during SWS when breathing frequencies were lowest. Rhythmogenesis and sometimes effective breathing was apparently entirely dependent on

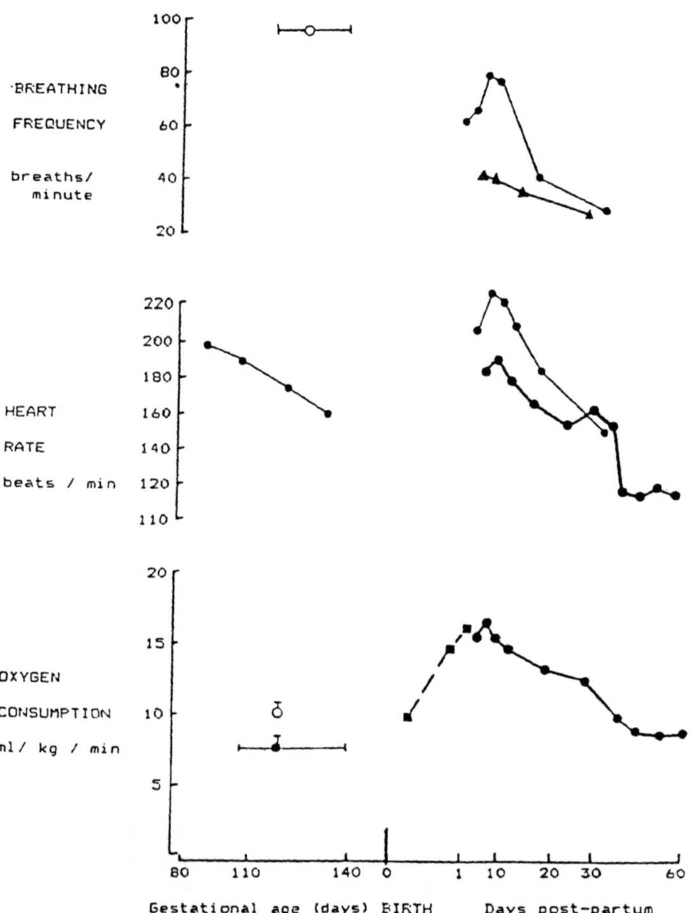

FIGURE 1 This is a composite illustration with data from
several studies replotted on the same axes.   Oxygen
consumption. Fetal values in SWS   (●) and in REM (o)
are from Rurak (3).Neonatal values.The broken line (■) is
that of Mercer (15). The solid line that of Lister (14).
Heart rate. Fetal data from Johnson (13). Postnatal upper
solid line -Johnson (13).Lower solid line - Lister (14).
Breathing frequency. Fetal values in REM (3). Postnatal
values from (13) and this study [▲]. Note the postnatal
rise and fall in heart and breathing rates related to
oxygen consumption.

the presence of this mechanism. Older lambs demonstrated
the same mechanism and a qualitatively similar response,
although the marked arrhythmia, lung collapse and
asphyxia did not occur. We reasoned then that either
chemical control was still developing during this period
or that lung volume was not so readily compromised in the
older lambs with a more rigid thoracic cage.
    It was decided to study in closer detail the
developmental effects of ambient temperature on these

mechanisms after birth:in particular the effects on the
high breathing frequencies after birth, and secondly on
the lower breathing rate observed in later postnatal
life. Thus a group of lambs (chronically instrumented
under fluothane anaesthesia after separation from their
ewes and establishment on an ad-lib artificial milk
regime) were then studied serially. Nine lambs were then
studied while in SWS at 10, 15 and 20 $^{\circ}$C (in some 5,25
and 30 $^{\circ}$C) at 3, 6, 12 and 28 daysof age. The initial
breathing frequencies in the first 12 days of life in the
artifically-reared lambs were markedly lower than those
observed in the ewe-reared lambs(fig 1). Although studied
in the same ambient temperature (Ta) range we were able
to see from closed circuit video that the younger lamb,
raised with its ewe, slept against the maternal flank
which is known to be a potent heat source. We are unable
to say whether the marked difference in breathing
frequencies were due to this factor or some other factor
related to natural feeding. Breathing frequencies fell
significantly with age (39.7 + 9 SD breaths / minute at
3 days to 27.5 + 6 SD at 28 days, p > 0.001) regardless
of a Ta between 10 and 20 $^{\circ}$ C. No consistent change in f
occurred between 20 and 10 C at 3 or 6 days of age, but
a significant fall in f occurred between 20 and 10 $^{\circ}$C at
12 days (p <0.01) and 28 days (p <0.o5).Shivering was
observed in lambs < 6 days at 10 $^{\circ}$ C and f increased
markedly between 25 and 30 $^{\circ}$ C in two lambs tested.Thus we.

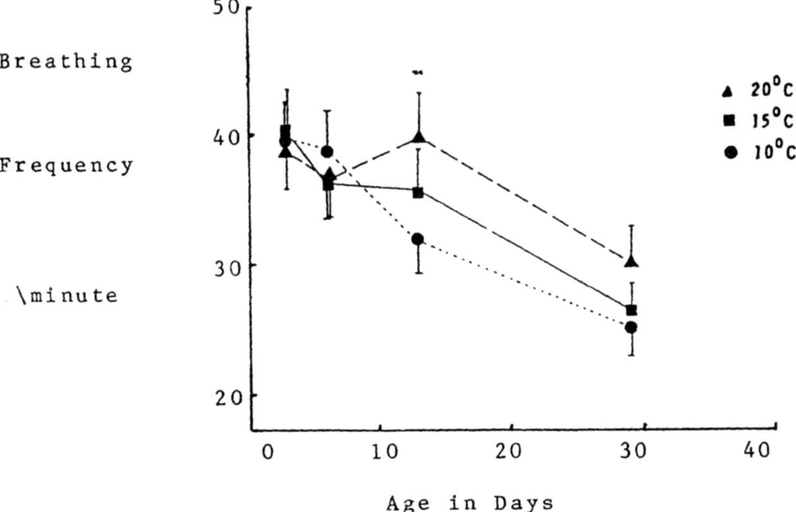

Age in Days

FIGURE 2
     Breathing frequencies during SWS in 9 lambs in 3
ambient temperatures at 3, 6, 12, and 28 days of age.
Note that breathing frequencies only changed after 6
days of age within this thermal range; becoming lower
at the lower Ta. There was an overall fall with age.

were able to influence breathing frequency in the very age group in which we had demonstrated a specific vulnerability to breathing irregularity and respiratory failure when the regulation of the upper airway had been removed. Since we had observed shivering in lambs at 3 and 6 days of age at 10oC Ta, it seemed reasonable to conclude that these animals were subject to a cold stimulus to breathing and that any Ta between this and 30oC when breathing frequencies were noted to abruptly increase, represented a cold stimulus to thermogenic metabolism and with it breathing. Only at 12 days of age was it observed that breathing frequency was influenced by Ta such that comparatively low breathing frequencies were observed at Ta´s within the thermo-neutral range.

Thus we had an explanation for the possible variability in the degree of dysrhythmia that followed the bypassing of the upper airway in lambs aged between 2 and 6 weeks. In a number of these animals it had been noted that the fall in breathing frequency and irregularity was sufficient to lead to hypoxemia and hypercapnia, clearly demonstrating that the chemical drive to breathing was insufficient to restore effective breathing. Either a cold stimulus or a hot stimulus were adequate to stimulate breathing and restore effective gas exchange. Our previous results had suggested that lambs older than about 50 days of age were relatively immune from this form of failure although the dysrhythymia was still apparent. This finding accords with the report that carotid denervated fetal lambs will establish effective breathing after birth (19) and that they, as well as lambs carotid denervated in the neonatal period, while having a measurably lower oxygen tension and higher CO2 tension than normals, gradually improve these variables over the first weeks of life. However, 4 of 7 chemo-denervated animals were found unexpectedly dead at about 40 days of age (11). This finding if substantiated might be interpreted as indicating that chemoregulation of breathing becomes particularly important at this age when the metabolic and thermal stimuli to respiration decrease in their importance except ,of course, at greater extremes.

It is concluded that, in the development of control of breathing after birth, there are at least three phases during which different factors are prominent in sustaining rhythmic and effective breathing. First a major metabolic demand, largely to achieve effective thermoregulation, with feeding and ingestion playing a part, followed by a period when mechano-sensory pathways are prominent as thermal efficiency improves and ambient thermal factors assume an importance before the chemical regulation of breathing achieves primacy. It should be appreciated that even in this final phase there are other factors which play a role in the maintenance of effective lung volume such as the increasing elastic recoil as the chest wall develops rigidity. It follows from this

SEQUENCE OF POSTNATAL CONTROL OF BREATHING IN THE LAMB

Temperature— — — — —
     — — Metabolism— — —
               — — Airway mechanosensory system— —
                              Chemoreception — — —

Birth | 1-2   4-7 weeks

proposal that deficiencies in any of these three (or four) factors will only become apparent as the relevant developmental periods are reached. For example a depression of metabolism such as with starvation (15) or hypothyroidism, will decrease the metabolic demands and increase dependence on mechano-sensory reflexes for sustaining breathing before the chemical drive is adequately developed. Thus although the regulation of breathing is clearly multi-factorial, there is a changing hierarchy of these factors with increasing postnatal age. The fact that these effects were prominent during SWS suggest that, as in fetal life, SWS, or an associated factor, can operate as a potential inhibition to breathing if an adequate afferent sensory input is not sustained.

What is also quite clear from these studies is that rhythmogenesis is also sustained by lung afferents, presumptively pressure, particularly in the expiratory phase of breathing in older lambs breathing at lower frequencies. Prior to that high breathing frequencies stimulated by a high metabolic rate achieve an adequate lung volume by virtue of a short expiratory time. Since the metabolic requirements of temperature regulation are a major stimulus to breathing in the newborn the much discussed biphasic ventilatory response to hypoxia is very likely to simply reflect a reduction in metabolic rate and is certainly not a depression of breathing even in the unanaesthetized newborn (20).

References

1. Kitterman, J.A., Liggins, G.C., Clements, J.A. & Tooley, W.H. (1975) J. Devel. Physiol. 1, 453-466.
2. Gluckman, P.D., Gunn, T.R. & Johnston, B.M. (1983) J. Physiol. 343, 495-506.
3. Rurak D.W. and Gruber N.C (1983). J.Appl.Physiol.:Respirat. Environ. Exercise Physiol. 54(3): 701-707.
4. Fewell J.E. Hislop A.A. Kitterman J.A. Johnson

P.(1983) J.Appl.Physiol.: Respirat. Environ. Exercise Physiol. 55(4): 1103-1108.

5. Karlberg, P., Cherry, R.B., Escardo, F.E. & Koch, G. (1962) Acta Paediatrica 51, 121-136.

6. Harrison, V.C., de Heese, H. & Klein, M. (1968) Pediatrics 41, 549.

7. Dixon, M., Szereda-Przestazewska, M., Widdicombe, J.G. & Wise, J.C.M. (1974) J. Physiol. 239, 347-363.

8. Fewell, J.E. & Johnson, P. (1981) J. Physiol. 320, 57.

9. Boddy, K., Dawes, G.S., Fisher, R., Pinter, S. & Robinson, J.S. (1974) J. Physiol. 243, 599.

10. Maycock, D.E., Standaert, T.A., Guthrie, R.D. & Woodrun, D.E. (1983) J. Appl. Physiol. 54(3), 814-820.

11. Bureau, M.A. (1983). Am. Rev. Resp. Dis. 127, 215.

12. Blanco, C.E., Dawes, G.S., Hanson, M.A. & McCooke, H.B. (1984). J. Physiol. 351, 25-38.

13. Johnson, P. (1978) In: Central Nervous Control Mechanisms in Breathing; Eds. C. von Euler and H. Lagercrantz, pp. 337-351.

14. Lister, G., Walker, T.K., Versmold, H.T., Dallman, P.R. & Rudolph, A.M. (1979) Am. J. Physiol. 237, H668-H675.

15. Mercer, J.B. (1974) Ph. D. Dublin University.

16. Harding, R., Johnson, P. & McClelland, M.E. (1980) Respir. Physiol. 40, 165-180.

17. Fewell, J.E. & Johnson, P. (1983) J. Physiol. 339, 495-504.

18. Remmers, J.E. & Bartlett Jr. D. (1977). J. Appl. Physiol. 42, 80-87.

19. Jansen, A.H., Ioffe, S., Russell, B.J. & Chernick, V. (1981). J. Appl. Physiol. 51(3), 630-633.

20. Blanco C.E. Hanson M.A. Johnson P. Rigatto H. (1984) J.Appl.Physiol.: Respirat.Environ.Exercise Physiol.56 (1): 12-17.

77

Endorphins and Respiration in the Neonatal Period

I. R. MOSS

The perinatal period and early infancy are accompanied by perturbations of respiratory rhythm and responsivity, the causes of which are still unknown. Of major concern and interest are factors that establish sustained regular breathing in the neonate following transition from the fetal state, in which breathing movements alternate naturally with periods of prolonged apnea. Sometime after birth, particularly in the immature neonate, prolonged episodic apneas may develop which can be fatal. In addition, other, more subtle and relatively benign disturbances are frequently encountered, including decreased fetal breathing movements during labor, periodic breathing with short respiratory pauses in preterm infants, and immature respiratory drive and responsivity apparent in both preterm and term infants (reviewed in 1).

Suppression of normal respiratory control mechanisms by neuromodulators could explain some of these pnenomena. In an effort to evaluate the role of endorphins as natural inhibitory respiratory neurotransmitters/neuromodulators, we had established that B-endorphin, when instilled in minute quantities (14 n moles) into the cisterna magna of adult dogs, produced marked depression of ventilation by decrease of both tidal volume and breathing frequency (2). Detailed analysis of this response led us to conclude that endorphins acted at central chemosensory mechanisms and at central afferent vagal pathways (3). We had also shown that endorphin antagonism with naloxone enhanced respiratory drive in adult animals (2,3) initiated and enhanced the onset of fetal breathing movements (FBM) in the apneic fetus, and enhanced the rate of rise (respiratory drive), amplitude, frequency and ventilation analogue of FBM in response to a "fetal CO_2 test" (4-6).

Because endorphin levels are elevated postnatally (7), when respiratory irregularities are common, we and others have recently directed our effort to the exploration of the relationship between endorphins and respiratory control in the newborn period.

We studied the maturation of respiration and the effect of endorphin displacement (ED) with naltrexone (3 mg/kg i.v.) during breathing at rest, progressive hyperoxic hypercapnia and progressive isocapnic hypoxia in 1-42 day old (do) piglets. The piglets breathed spontaneously under anesthesia with age-adjusted doses of pentobarbital. Measured functions included arterial pressure, pHa, PaO_2 and $PaCO_2$, gaseous PCO_2, tracheal pressure and integrated air flow; the rate of rise of intratracheal inspiratory occlusion pressure, breathing frequency, tidal volume and ventilation were calculated at rest and from regression of response curves to hypoxia and hypercapnia at 40 Torr PaO_2 and 55 Torr $PaCO_2$, respectively, before and after treatment with naltrexone. ED enhanced several respiratory functions at rest as well as during hypercapnia and hypoxia within the youngest

(YG = 1-4 do) and the intermediate group (IG = 10-18 do), but not in the oldest piglet group (OG = 25-42 do). Breathing pattern indicated maturation with age at rest as well as during hypercapnia and hypoxia. Whereas this maturation pattern was sustained after ED at rest and during hypercapnia, during hypoxia, the differences in respiratory pattern among the age groups were abolished after ED, and the pattern became that of the OG as shown in the following figure:

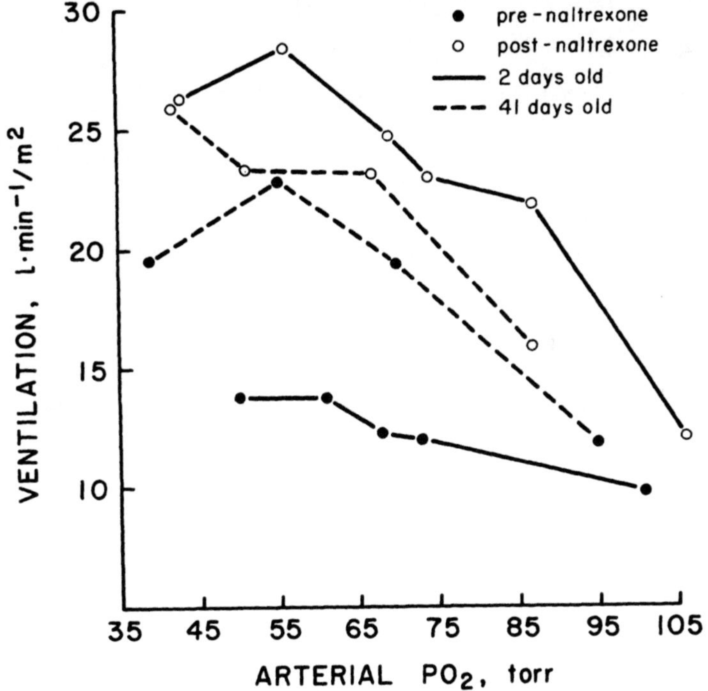

FIGURE: Response to isocapnic progressive hypoxia with and without naltrexone (3 mg/kg) in two piglets.

We concluded that (1) endorphins modulate breathing at rest as well as during hypercapnia and hypoxia in young animals; (2) maturation with age of breathing at rest and during hypercapnia is at least in part independent of endorphin effect; (3) maturation of breathing response to hypoxia is dependent on endorphins; and (4) endogenous opiod activity may be enhanced during hypoxia in young animals. We plan to expand these studies using chronically instrumented, unanesthetized newborn piglets.

References

1. Scarpelli EM, and Moss IR. (1980). Control of fetal and neonatal breathing and its disturbances, Clinics in Chest Med. 1:145-159.

2. Moss, I.R. and Friedman, E. (1978). Effects on respiratory regulation. Life Sci. 23:1271-1276.

3. Moss, I.R. and Scarpelli, E.M. (1981). B-endorphin central depression of respiration and circulation. J. Appl. Physiol. 50: 1011-1016.

4. Moss, I.R., Condorelli, S. and Scarpelli, E.M. (1981). The progressive onset of spontaneous and induced fetal breathing. Resp. Physiol. 45:299-308.

5. Moss, I.R. and Scarpelli, E.M. (1979). Generation and regulation of breathing in utero: Fetal CO_2 response test. J. Appl. Physiol. 47:527-531.

6. Moss, I.R. and Scarpelli, E.M. (1984). CO_2 and naloxone modify sleep/wake state and activate breathing in the acute fetal lamb preparation, Resp. Physiol. 55:325-340.

7. Moss, I.R., Conner, H., Yee, W.F.H., Iorio, P. and Scarpelli,E.M. (1982). Human B-endorphin-like immunoreactivity in the perinatal/ neonatal period. J. Pediat. 101:443-446.

78

Brainstem Neural Development and Apnoea in the Newborn

D. J. HENDERSON-SMART, A. G. PETTIGREW, D. A. EDWARDS
and M. BUTCHER-PEUCH

Instability of the breathing rhythm and apnoea is observed frequently in preterm infants. With the increase in survival rate of very low gestation infants, recurrent prolonged episodes (>20 sec) have posed an increasing clinical problem. In a retrospective analysis [1] of the case histories of infants born at King George V Hospital over a 6 year period, 249 (1%) of infants had this problem. It recurred in over 80% of those born at <30 weeks gestation, 54% of those at 30-31 weeks, 14% at 32-33 weeks and 7% at 34-35 weeks and in only 0.06% at term. In those with apnoea, the duration of the problem (age at last apnoea) also decreased with advancing gestational age at delivery. These results suggest that immaturity plays a major part in its causation.

Polygraphic analysis of 1522 apnoeas of >10 sec in 28 preterm infants revealed that the majority of apnoeas are of a central type (cessation of breathing efforts) or mixed apnoea (predominant central component together with obstructed breaths at the resumption of breathing efforts) [2]. This result, together with the clinical observations above, suggest that immaturity of the brainstem rhythm generator is a likely underlying factor. To assess brainstem maturity, the objective auditory evoked response (AER) technique was used to evaluate a neural pathway lying close to the respiratory neurones. [3]

Brainstem conduction times (BCT) of the AER (Wave I-Wave V and IIn to V) were measured in 58 infants at between 30 and 37 weeks postmenstrual age. The mean BCT decreased at increasing postmenstrual

ages, with a time course similar to that for decreasing apnoea. At
gestations between 32 and 35 weeks there were sufficient data to show
that the mean conduction times were shorter in infants without
clinical prolonged apnoea. Longitudinal studies of individuals
indicated that apnoea disappeared when the BCT decreased to levels
obtained in non apnoeic infants.

Preterm infants who had undergone intrauterine growth retardation
(IUGR) were found to have shorter BCTs for their age[4] and an
associated decrease in the incidence of apnoea. Hypertension in
pregnancy, usually treated with α-methyl dopa or clonidine Hcl, was
the most common factor associated with IUGR. Although these drugs are
known to alter brainstem monoamine metabolism, they could not account
for the shorter BCTs; since results in 19 normally grown preterm
infants born to hypertensive mothers were similar to those for 68
normally grown controls. Growth retardation therefore appeared to be
the major factor associated with altered brainstem development in
utero.

These findings suggest that immaturity is the major factor
associated with apnoeic episodes in preterm infants. Immaturity of
brainstem neural function, demonstrated in the AER studies, seems the
most likely underlying factor. However, associated immaturity of
central and peripheral neural inputs to the respiratory neurones or of
the respiratory pump may also contribute to the instability of
breathing rhythm.

References:

1. Henderson-Smart DJ. The effect of gestational age on the
incidence of recurrent apnoea in newborn babies. Aust. Paediatric.J.,
1981;17:273-276.

2. Butcher-Peuch M., Holley D. and Henderson-Smart DJ. Apnoea type,
bradycardias and neurological outcome in preterm infants. Aust.
Paediatr. J., 1983;19:260.

3. Henderson-Smart DJ, Pettigrew AG and Campbell DJ. Clinical apnea
and brainstem neural function in preterm infants. New Eng.J.Med.,
1983;308:353-357.

4. Edwards DA, Pettigrew AG and Henderson-Smart DJ. The effect of intrauterine growth retardation (IUGR) on brainstem development of preterm infants. *Exerpta Medica, Asia Pacific Cong. Series*, 1983; 18:59.

4. Edwards DA, Pettigrew AG and Henderson-Smart DJ. The effect of intrauterine growth retardation (IUGR) on brainstem development of preterm infants. *Exerpta Medica, Asia Pacific Cong. Series*, 1983; 18:59.

79

Respiratory Arrests in "Normal" Premature and Full-Term Newborn

M. F. RADVANYI-BOUVET, M. MONSET-COUCHARD
and F. MOREL-KAHN

Using a pneumotachograph and polygraphic recordings in "normal premature" and full-term newborns, in a previous work (1) we described sudden short arrests of airflow (SSAA $\leq$ 2 sec.) occuring mainly during expiratory phase, in addition to peculiar variations of expiratory flow (retarded expiratory flow = R.E.F.)

The present work used the same techniques to look into various types of apneas, central apneas (without respiratory movements), obstructive apneas (with respiratory movements), mixed apneas (partly central, partly obstructive). The study was aimed at assessing the number of SSAA and of those various types of apnea according to sleep states and gestational age (GA), and their relationship to other respiratory events (sigh, percentage of R.E.F.).

MATERIAL

50 healthy newborns (APGAR score $\geq$ 7 à 1') without any respiratory or other problem have been recorded during the first week of life. They were subdivided in 3 groups according to their G.A.

		M + SD
Group I. (n = 10) G.A. = 33 - 36 weeks		(35.2 + 1.1)
Group II. (n = 19) G.A. = 37 - 39 weeks		(38.2 + 0.8)
Group III. (n = 21) G.A. = 40 - 43 weeks		(40.8 + 0.7)

METHOD

Polygraphic recordings were performed in neonates lying supine in their bed or incubator. Respiratory data were recorded with 4 channels : Respiratory movements at thoracic and abdominal level (with graphite rubber belt), pulmonary flow : $\overset{\circ}{V}$ (from a Fleisch model 00 Flowmeter set in a small naso-oral mask), Tidal volume = TV (obtained by integration of flow through a pulmostar pneumotachograph) Intraoesophageal pressure was recorded in some cases (through an open saline filled catheter). Sleep states were monitored using EEG and eye movements. Active sleep (A.S.) was identified on presence of rapid eye movements (REM) and continuous EEG, quiet sleep (Q.S.) on absence of REM, discontinuous EEG in premature newborn, tracé alternant in full-term newborn.These criteria had to be present for at least 2 minutes. Transitional and intermediate sleep were discarded (the same duration of A.S. and Q.S. was recorded for groups of different G.A.).

RESULT

The number of SSAA showed large individual variations (from 0 to 2.7/min. but was significantly higher in A.S. in group II and III. (P $<$ .02 and P $<$ .001_ T test), without relation to sleep state in group I.

Frequency of SSAA increased significantly with G.A., the more so during A.S. than during Q.S.

Apnea : The mean total duration of obstructive or mixed apnea/100 min. was greater during A.S. than in Q.S. in the 3 groups (gr. I : P $<$ .001 ; gr. II P $<$ .05 ; gr. III P $<$ .02). The same trend was observed for central apnea but without statistical significance. The number of babies who did not have apnea in the 3 groups was as follows :

gr. I	1 out of 10 during A.S.	5 during Q.S.
gr. II	5 out of 19 during A.S.	8 during Q.S.
gr. III	6 out of 21 during A.S.	12 during Q.S.

Central apneas decreased markedly while G.A. increased ; obstructive or mixed apneas decreased less. Thus, the percentage of these obstructive apneas relative to the total duration of apneas varied from 40 % in group I to 73 % in group III.

The number of SSAA and apneas was not related to the number of sighs nor to the percentage of R.E.F.

DISCUSSION

Short glottis closure increasing during the first days of life has bee
described in lambs (2 - 3). We suggest that SSAA which increased with GA
might be related to the same phenomena. SSAA did not seem related to swal
lowing : there was no important simultaneous variation of intraoesophageal pressure during the same breath, as described during swallowing (4).
The frequency of obstructive apneas showed some discrepancies with other
results (5), this might be related to our method using a naso-oral mask
and to supine position of the babies.

CONCLUSION

Both SSAA and apneas prevailed in A.S., but the number of SSAA increased and apneas decreased with G.A. (obstructive apneas less than central
apneas). We suggest therefore, that different mechanisms account for the
occurence of SSAA and obstructive apnea.

REFERENCES

1. Radvanyi-Bouvet, M.F., Monset-Couchard, M., Morel-Kahn, F., Vicente, C
Dreyfus-Brisac, C. (1982). Expiratory patterns during sleep in normal
full-term and premature neonates. Biol. neonate, 41, 74-84.
2. Harding, R., Johnson, P., McClelland, M.E. (1980). Respiratory functic
of larynx in developing sheep and the influence of sleep state. Resp.
Physiol, 40, 165-169.
3. Johnson, P., (1979). Comparative aspects of the control of breathing
during development. In : Von Euler and Lagercrantz (eds), Central nervous
control mechanism breathing pp. 336-352 (Oxford : Pergamon Press).
4. Wilson, S.L., Thach, B.T., Brouillette, R.T. and Abu-Osba, Y.K. (1981)
Coordination of breathing and swallowing in human infants. J. Appl. Physiol., 50, 851-858.
5. Flores-Guevara, R. Plouin, P., Curzi-Dascalova, L., Radvanyi, M.F.,
Guidasci, S., Pajot N., and Monod, N. (1982). Sleep apneas in normal neonates and infants during the first 3 months of life. Neuropediatrics, 13
Suppl., 21-28.

80

Lung Mechanics and Breathing Pattern During Wakefulness and Sleep in Children with Enlarged Tonsils

J.-P. PRAUD, C. GAULTIER, A. BUVRY, M. BOULE and F. GIRARD

Thirteen children (mean age : 45 mth) with nocturnal symptoms of upper airway obstruction due to enlarged tonsils were tested during wakefulness (W) and sleep (S) induced by chloral hydrate ($<$ 50 mg/kg).

During W, lung mechanics, blood gases, breathing pattern and airflows during tidal breathing were in the normal range.

During S, total lung resistance increased significantly, and dynamic lung compliance and transcutaneous PO_2 decreased significantly. Moreover, during S, the tidal volume (V_T), and the mean inspiratory flow normalized for BW decreased while the ratio of the inspiratory time (T_I) over the total duration of the respiratory cycle (T_{TOT}) rose indicating an increase in the time of contraction of respiratory muscles. In seven children the time to reach peak inspiratory flow as a percentage of T_1 (dT_I/T_I) increased, with no change in the ratio of the expiratory flow over the inspiratory flow, both measured at 50 % of V_T (EF_{50}/IF_{50}). In three other cases, dT_I/T_I decreased with an increase in EF_{50}/IF_{50}. We conclude that in children with enlarged tonsils, S modified lung mechanics, gas exchange and the inspiratory components of the breathing pattern.

The present data will be published in a next issue of Sleep (Raven Press).

81
Shape of Inspiratory Neuromuscular Drive and its Implications

J. MILIC-EMILI and W. A. ZIN

Mechanical analysis of the breathing movements consists of three steps: (a) description of the forces applied to and developed by the respiratory system and its parts; (b) description of the motion of the respiratory system and its parts; and (c) relation of the motion and forces in accordance with physical lows. Implicit in such analysis is the need to precisely describe the pressure wave which drives respiration. Yet, until recently, little attention has been paid to this fundamental aspect of the control of the breathing movements.

<u>Inspiratory Driving Pressure.</u>

The deeply anesthetized animals and humans, the pressure developed at the airway opening during an inspiratory effort against occluded airway at functional residual capacity, or airway occlusion pressure ($P°ao$), respresents the <u>net</u> potential pressure available for inspiration (3,8,12).

In both deeply anesthetized cats (4,5,7,8,12) and dogs (6), the rising part of the occlusion pressure wave can be approximated by a simple power function:

$$P°ao = at^b \qquad (1)$$

where $\underline{t}$ is time (s) after the onset of inspiration, $\underline{a}$ is constant representing the value of $P°ao$ (cm H_2O) at 1 s, and $\underline{b}$ is a dimensionless constant describing the shape of the occlusion pressure wave. When the occlusion pressure-time profile is linear, $\underline{b}$ = 1,; when the shape is concave upward, b $>$ 1; when the shape is convex upward, $\underline{b} <$ 1.

Table 1 provides values of the constants $\underline{a}$ and $\underline{b}$ of Eq.1 found in air breathing cats anesthetized with different agents. It can be seen that in cats the shape of the occlusion pressure waveform (constant $\underline{b}$) changes markedly with different anesthetics. Furthermore, with any given anesthetic, there are marked interanimal differences in shape.

Table 1.   Constants $\underline{a}$ and $\underline{b}$ of Eq.1 in cats anesthetized with different agents.

Ref.	No. of Cats	Anesthetic	Dose	a, cmH_2O	b
5	4	Enflurane*	3.6%	8.1 (5.2-12.0)	0.70 (0.47-1.20)
4	10	Ketamine	40 mg/kg im	9.5 (4.2-14.0)	0.50 (0.30-0.65)
8	8	Pentobarbital Sodium	35 mg/kg ip	8.1 (4.6-16.0)	1.02 (0.80-1.41)
12	6	Pentobarbital Sodium	35 mg/kg ip	8.6 (6.1-11.9)	0.94 (0.73-1.14)

Values are means with ranges in parentheses. $\underline{a}$, Negative pressure developed 1 s after onset of inspiration; $\underline{b}$, dimensionless index of shape of curve. * Inspired gas composition: 3.6% enflurane, 50% O_2, balance N_2.

In dogs deeply anesthetized with halothane and enflurane, the value of $\underline{b}$ averages ($\pm$ SD) 0.63 $\pm$ 0.03 and 0.67 $\pm$ 0.06, respectively (6).

In anesthetized humans, the time-course of the occlusion pressure wave does not fit Eq.1. With methoxyflurane the human occlusion pressure wave exhibits an upward convexity (3), while with both enflurane (S. Shore, unpublished results) and halothane (2,11) it is sigmoidal in shape.

In both anesthetized animals and humans, the effect of CO_2 and hypoxia on the airway occlusion pressure wave is an increase in amplitude without any appreciable change in shape (1,3,4,8). The shape, however, can be affected by changes in body temperature (10).

The nature of the differences in occlusion pressure waveform within and between species is not understood. It is clear, however, that these different driving pressure waveforms profoundly affect the time-course of volume and flow during inspiration (4,7,9,11,12), and consequently also the ventilatory output (7,9). Accordingly, there is a great need to elucidate the nature of the neuromuscular factors determining of the occlusion pressure waveform.

REFERENCES

1.   ALTOSE, M.D., S.G. KELSEN, N.N. STANLEY, N.S. CHERNIACK, AND A.P. FISHMAN. Effects of hypercapnia and flow-resistive loading on tracheal pressure during airway occlusion. J. Appl. Physiol. 40: 345-351, 1976.

2.   BEHRAKIS, P.K., B.D. HIGGS, A. BAYDUR, W.Z. ZIN, AND J. MILIC-EMILI. Active inspiratory impedance in halothane-anesthetized humans. J. Appl. Physiol.: Respirat. Environ. Exercise Physiol. 54: 109-121, 1983.

3.   DERENNE, J.-P., J. COUTURE, S. ISCOE, W.A. WHITELAW, AND J. MILIC-EMILI. Occlusion pressures in men rebreathing CO_2 under methoxyflurane anesthesia. J. Appl. Physiol. 40: 805-814, 1976.

4.   JASPER, N., M. MAZZARELLI, C. TESSIER, AND J. MILIC-EMILI. Effect of ketamine on control of breathing in cats. J. Appl. Physiol.: Respirat. Environ. Exercise Physiol. 55: 851-859, 1983.

5.   MAZZARELLI, M. Effect of various anesthetics and drugs on control of breathing in cats. Ph.D. Thesis (Physiology), McGill University, Montreal, Canada, 1978.

6. RICH, C.R., K. REHDER, T.J. KNOPP, AND R.E. HYATT. Halothane and enflurane anesthesia and respiratory mechanics in prone dogs. J. Appl. Physiol.: Respirat. Environ. Exercise Physiol. 46: 646-653, 1979.

7. SIAFAKAS, N.M., M. BONORA, B. DURON, H. GAUTIER, AND J. MILIC-EMILI. Dose effect of pentobarbital sodium on control of breathing in cats. J. Appl. Physiol.: Respirat. Environ. Exercise Physiol. 55: 1582-1592, 1983.

8. SIAFAKAS, N.M., H.K. CHANG, M. BONORA, H. GAUTIER, J. MILIC-EMILI, AND B. DURON. Time course of phrenic activity and respiratory pressures during airway occlusion in cats. J. Appl. Physiol.: Respirat. Environ. Exercise Physiol. 51: 99-108, 1981.

9. SIAFAKAS, N.M., R. PESLIN, M. BONORA, H. GAUTIER, B. DURON, AND J. MILIC-EMILI. Phrenic activity, respiratory pressures, and volume changes in cats. J. Appl. Physiol.: Respirat. Environ. Exercise Physiol. 51: 109-121, 1981.

10. TRIPPENBACH, T., AND J. MILIC-EMILI. Temperature and CO_2 effect on phrenic activity and tracheal occlusion pressure. J. Appl. Physiol.: Respirat. Environ. Exercise Physiol. 43: 449-454, 1977.

11. ZIN, W.A., P.K. BEHRAKIS, S.C.M. LUIJENDIJK, B.D. HIGGS, A. BAYDUR, A. BÖDDENER, AND J. MILIC-EMILI. Immediate response to resistive loading in anesthetized humans. J. Appl. Physiol.: Respirat. Environ. Exercise Physiol. (Submitted for Publication).

12. ZIN, W.A., L.D. PENGELLY, AND J. MILIC-EMILI. Active impedance of respiratory system in anesthetized cats. J. Appl. Physiol.: Respirat. Environ. Exercise Physiol. 53: 149-157, 1982.

82
Postinspiratory Activity of Inspiratory Muscles

C. D. SHEE, Y. PLOYSONGSANG and J. MILIC-EMILI

In humans, postinspiratory activity has been shown to occur in the diaphragm and the inspiratory intercostal muscles (1,2). The time-course of pressure developed by the inspiratory muscles during expiration ($Pmus_I$) has only recently been quantified. Agostoni and Citterio (3) determined $Pmus_I$ decay during the period of zero flow after the end of inspiration under a discontinuous elastic load, but their method only measures $Pmus_I$ decay early in expiration. Zin et al. (4) have reported a non-invasive method to compute $Pmus_I$ based on the measurement of (a) total respiratory system elastance (Ers) and resistance (Rrs), and (b) changes in lung volume (V) and airflow ($\dot{V}$) during spontaneous expiration. This "$V-\dot{V}$ method" has been used in anesthetized humans (5), and we have now extended this approach to compute $Pmus_I$ in spontaneously breathing conscious humans.

METHODS

We studied 8 healthy male non-smokers (age range 25-35 years). Subjects were seated, and breathed through a mouthpiece attached to a pneumotachograph to enable measurement of $\dot{V}$, and by electrical integration, V. Pressure at the airway opening was measured with a differential pressure transducer. A pneumatically operated valve was used for rapid occlusion of the airway opening. Equipment dead space was 70 ml and equipment flow resistance was linear (0.6 cm H_2O $1^{-1}.s$). Variables were recorded on a chart recorder. Ers was measured by the relaxation method, and Rrs by the interrupter method. For each subject, 10 breaths were selected at random from periods of quiet breathing, and V and $\dot{V}$ were measured at 0.2s intervals throughout spontaneous expiration. $Pmus_I$ decay in these breaths was determined from the formula

$$Pmus_I(t) = Ers. V(t) - Rrs. \dot{V}(t) \qquad (1)$$

where V is volume and $\dot{V}$ flow at any instant t, and Rrs includes equipment resistance (4).

RESULTS

The mean ($\pm$SD) value of Ers was 8.15 $\pm$ 1.35 cm H_2O. 1^{-1}, and of Rrs (including equipment) was 2.78 $\pm$ 0.34 cm H_2O. 1^{-1}.s.   McIlroy et al. (6) pointed out that, for constant Ers and Rrs, the slope of the V-$\dot{V}$ relationship during relaxed expiration corresponds to the passive time constant of the respiratory system ($\mathcal{T}$rs)

$$V/\dot{V} = -\mathcal{T}rs = - Rrs/Ers \qquad (2)$$

Fig.1(A) shows a representative V-$\dot{V}$ loop during spontaneous expiration, along with the relaxed V-$\dot{V}$ relationship predicted for that subject according to Eq. 2. In all subjects, as exemplified in Fig. 1 (A), $\dot{V}$ for the major part of expiration falls to the left of the $\mathcal{T}$rs line, indicating substantial braking by $Pmus_I$.

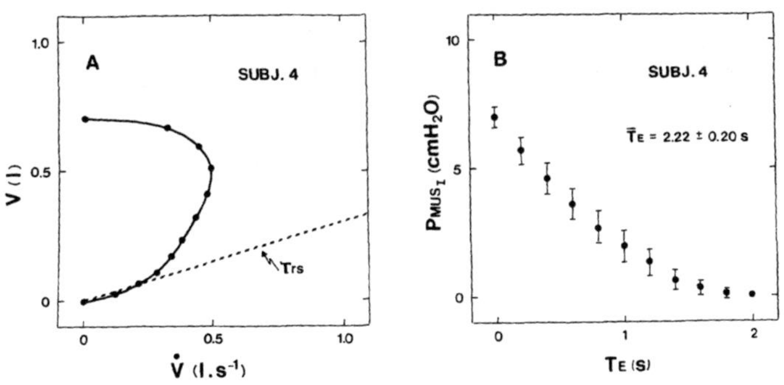

Figure 1. (A) Expiratory V-$\dot{V}$ curve during quiet breathing in a representative subject. Filled circles are mean values of V and $\dot{V}$ measured at 0.2s intervals in 10 breaths.   Broken line represents the passive expiratory V-$\dot{V}$ relationship, the slope corresponding to the passive time constant ($\mathcal{T}$rs) for that subject. (B) Time-course of $Pmus_I$ during spontaneous expiration. Each filled circle is the average of 10 determinations; bars indicate $\pm$1 SD. Average ($\pm$SD) total duration of expiration (T$_E$) is indicated.

The time course of $Pmus_I$ decay, calculated from Eq. 1, is shown in Fig.1 (B) for a representative subject. For the 8 subjects, the mean time for $Pmus_I$ to reduce to 50% (T_{50}) and to zero (T_Z) amounted, respectively, to 23% and 79% of expiratory time, T_E (Table 1).

Table 1. Expiratory decay of inspiratory muscle pressure for awake and anesthetized adults, and sleeping infants.

Authors	Condition	n	T_E,s	T_{50}/T_E,%	T_Z/T_E,%
Present study	Awake adults	8	2.54	23	79
Ref. 5	Anesthetized adults	6	2.10	13	70
Ref. 7	Sleeping infants	12	0.50	22	83

n, number of subjects; T_E, expiratory time; T_{50} and T_Z, time required for inspiratory muscle pressure to decay to 50 and 0% of initial values, respectively.

DISCUSSION

There was an initial fairly rapid $Pmus_I$ decay followed by a later slower decay. This late slow decay has now been documented in infants (7), anesthetized adults (5) and conscious adults (8). Although T_E differs between infants and adults, when T_{50} and T_Z are expressed as percent of T_E, the findings are rather similar (Table 1).

The inspiratory muscle activity during expiration involves energy expenditure. Indeed, in our subjects, $60 \pm 12\%$ ($\pm SD$) of the elastic energy stored during inspiration was wasted as negative inspiratory muscle work.

The results shown in Table 1 might suggest that the duration of $Pmus_I$ actually modulates T_E. Agostoni and Citterio (3) found that the relative decay rate of $Pmus_I$ was inversely related to T_E, and similarly Martin et al. (8) found a correlation between T_{50} and T_E. In our study, if T_E is measured up to zero $\dot{V}$ (i.e., excluding the end-expiratory pauses exhibited by 2 subjects), the group results show that both T_{50} and T_Z are correlated with T_E ($p < 0.01$). In view of its functional significance, $Pmus_I$ should not be regarded simply as an incomplete off-switch of the inspiratory activity (9).

REFERENCES

1.   PETIT, J.M., G. MILIC-EMILI, AND L. DELHEZ. Role of the diaphragm in breathing in conscious normal man: an electromyographic study. J. Appl. Physiol. 15: 1101-1106, 1960.

2.   GREEN, J.H., AND J.B.L. HOWELL. Correlation of intercostal muscle activity with respiratory airflow in conscious human subjects. J. Physiol. London 149: 471-476, 1959.

3.   AGOSTONI, E., AND G. CITTERIO. Relative decay rate of inspiratory muscle pressure during expiration in man. Respir. Physiol. 38: 335-346, 1979.

4.   ZIN, W.A., L.D. PENGELLY, AND J. MILIC-EMILI. Single-breath method for measurement of respiratory mechanics in anesthetized animals. J. Appl. Physiol.: Respirat. Environ. Exercise Physiol. 52: 1266-1271, 1982.

5.   BEHRAKIS, P.K., B.D. HIGGS, A. BAYDUR, W.A. ZIN, AND J. MILIC-EMILI. Respiratory mechanics during halothane anesthesia and anesthesia-paralysis in humans. J. Appl. Physiol.: Respirat. Environ. Exercise Physiol. 55: 1085-1092, 1983.

6.   McILROY, M.B., D.F. TIERNEY, AND J.A. NADEL. A new method for measurement of compliance and resistance of lungs and thorax. J. Appl. Physiol. 17: 424-427, 1963.

7.   MORTOLA, J.P., J. MILIC-EMILI, A. NOWORAJ, B. SMITH, G. FOX, AND S. WEEKS. Muscle pressure and flow during expiration in infants. Am. Rev. Respir. Dis. 129: 44-53, 1984.

8.   MARTIN, J., M. AUBIER, AND L.A. ENGEL. Effects of inspiratory loading on respiratory muscle activity during expiration. Am. Rev. Respir. Dis. 125: 352-358, 1982.

9.   EULER, C. VON. On the central pattern generator for the basic breathing rhythmicity. J. Appl. Physiol.: Respirat. Environ. Exercise Physiol. 55: 1647-1659, 1983.

Index

Lightning Source UK Ltd.
Milton Keynes UK
17 November 2009

146387UK00001B/68/A